Advanced Research on Animal Cell Technology

NATO ASI Series

Advanced Science Institutes Series

A Series presenting the results of activities sponsored by the NATO Science Committee, which aims at the dissemination of advanced scientific and technological knowledge, with a view to strengthening links between scientific communities.

The Series is published by an international board of publishers in conjunction with the NATO Scientific Affairs Division

A Life Sciences	Plenum Publishing Corporation
B Physics	London and New York
C Mathematical	Kluwer Academic Publishers
and Physical Sciences	Dordrecht, Boston and London
D Behavioural and Social Sciences	
E Applied Sciences	
F Computer and Systems Sciences	Springer-Verlag
G Ecological Sciences	Berlin, Heidelberg, New York, London,
H Cell Biology	Paris and Tokyo

Series E: Applied Sciences - Vol. 156

Advanced Research on
Animal Cell Technology

edited by

Alain O. A. Miller

Université de l'Etat à Mons, Faculté de Médecine,
Service de Biochimie Moléculaire et
Unité de Biotechnologie Appliquée, Mons, Belgium

Kluwer Academic Publishers

Dordrecht / Boston / London

Published in cooperation with NATO Scientific Affairs Division

Proceedings of the NATO Advanced Research Workshop on
Advances in Animal Cell Technology
Brussels, Belgium
September 21–24, 1987

Library of Congress Cataloging in Publication Data

```
NATO Advanced Research Workshop on "Advances in Animal Cell
   Technology" (1987 : Brussels, Belgium)
     Advanced research on animal cell technology / editor, Alain O.A.
Miller.
       p.   cm. -- (NATO ASI series.  Series E, Applied sciences ; no.
   156)
     ISBN 0-7923-0031-9
     1. Animal cell biotechnology--Congresses.  2. Cytology-
-Congresses.   I. Miller, Alain O. A., 1936-   . II. Title.
III. Series.
TP248.27.A53N37 1987
660'.6--dc19                                                88-13841
                                                              CIP
```

ISBN 0–7923–0031–9

Published by Kluwer Academic Publishers,
P.O. Box 17, 3300 AA Dordrecht, The Netherlands.

Kluwer Academic Publishers incorporates the publishing programmes of
D. Reidel, Martinus Nijhoff, Dr W. Junk, and MTP Press.

Sold and distributed in the U.S.A. and Canada
by Kluwer Academic Publishers,
101 Philip Drive, Norwell, MA 02061, U.S.A.

In all other countries, sold and distributed
by Kluwer Academic Publishers Group,
P.O. Box 322, 3300 AH Dordrecht, The Netherlands.

TABLE OF CONTENTS

vi

A C K N O W L E D G E M E N T S

The Director of the Advanced Research Workshop wishes to
express his appreciation for the sponsorship of the

North Atlantic Treaty Organization

and the financial support of the following organizations:

Athena Programme	Organon International BV
Becton Dickinson	Pharmacia Belga
Crédit Communal de Belgique	Setric Génie Industriel
Diosynth BV	Smith Kline-RIT
Fonds National de la Recherche Scientifique	Solvay et Cie
Générale de Banque	Ville de Bruxelles

Part I.

Cell-Cell Communication and Cell Extracellular Matrix Interactions

EXTRACELLULAR MATRIX AND CELL DIFFERENTIATION

J.M. FOIDART, A. NOEL, H. EMONARD, B. NUSGENS and Ch.M. LAPIERE

Laboratory of Pathophysiology of Pregnancy and Experimental Dermatology
University of Liege - C.H.U., Sart-Tilman (B 23), B 4000 Liege (Belgium).

Epithelial-mesenchymal interactions are believed to be an important factor determining tissue form and function (1-3) with each adjacent tissue influencing the others differentiation and maintenance. By placing epithelial and stromal cells on opposite sides of a filter, Grobstein (1955)(4) showed that factors secreted by one type of cell were active in changing the synthetic activity of the other cell type. Many studies indicate that these components may include matrix molecules (5-8).

Epithelial and stromal tissues contain distinct extracellular matrices composed of specific collagens, glycoproteins and proteoglycans. Epithelial cells abut on basement membranes composed of type IV collagen, laminin, entactin and a heparan sulfate proteoglycan (9). Laminin is a large basement membrane glycoprotein that binds to type IV collagen, entactin, heparan sulfate proteoglycan and to the cell surface *via* at least two types of specific membrane receptors (10-14). It is a multifunctional protein with diverse biological activities. Like fibronectin, it influences cell adhesion, growth, morphology, differentiation, migration and agglutination as well as the deposition and assembly of the extracellular matrix (15). Laminin primarily affects cells of epithelial origin while fibronectin acts preferentially on mesenchymal cell activities (15, 16).

Fibroblasts abut on matrices containing types I and III collagen, fibronectin, and small chondroitin sulfate proteoglycans (8, 17). Fibronectin binds to all collagen types as well as to glycosaminoglycans and to receptors on the surface of various cells including fibroblasts, myoblasts, hepatocytes, and glial cells (1, 18, 19). Fibronectin can influence the attachment and shape of the cells and cell differentiation (8, 17). It has also been shown to promote cell growth, motility, and matrix production (6, 20-23).

Interactions between cells and these matrix molecules have been characterized in culture by exposing primary cultures or established cell lines to defined components. Monolayer cell cultures on plastic impose an abnormal environment and thus induce "nonphysiologic", morphologic, and biosynthetic properties of cells. Studies *in vitro* of cell-matrix interactions require culture conditions reproducing at best environments found *in vivo*. Classic monolayer cell cultures on plastic do not allow such studies.

Different cell properties such as adhesion, survival, and proliferation have been characterized on defined substrates by plating freshly trypsinized cells in the presence of appropriate components either in soluble forms or after coating on plastic dishes (7, 24, 25). More complex substrates, extracellular matrices deposited by cells such as endothelial cells, have also been utilized as a support for subsequent cultures of fibroblasts or cancer cells (26, 27). Noncultivable cells in usual conditions of culture on plastic can be cultured on matrices prepared

3

A. O. A. Miller (ed.), Advanced Research on Animal Cell Technology, 3–13.
© *1989 by Kluwer Academic Publishers.*

using the tissue from which they are derived. For instance, matrices prepared from liver allow long-term cultures of hepatocytes (28). Matrix derived from whole mammary glands promotes mammary epithelial cell growth and differentiation (29).

I. INTERACTIONS BETWEEN CELLS AND INTERSTITIAL CONNECTIVE TISSUE

Three-dimensional cultures of fibroblasts in a solid type I collagen lattice provide a physiologic extracellular environment for the study of interactions between these cells and the interstitial connective tissue (30-33). In such cultures dermal fibroblasts are mixed with type I collagen, serum and medium. In the appropriate proportions a gel forms at 37°C and is contracted by the fibroblasts. The mixture can be poured into a mold whose shape determines the geometry of the contracted tissue. If poured into a petri dish, the tissue forms as a disk, of diameter much smaller than that of the plate. A fivefold decrease in diameter is usual. The contracted tissue equivalent floats just below the surface of the medium expressed from it. The rate of contraction is proportional to cell number and inversely proportional to lattice collagen content. The cells collect and condense the collagen fibrils as they extend and withdraw cytoplasmic podia that attach to the fibrils. They are drawn towards the cell body in the course of podial contractions.

Fibroblasts, after 4 to 7 days in this matrix are growth-regulated. They cease to incorporate DNA precursors (34). This regulation is not due to contact inhibition because cell density is low and cells are well-separated from one another, as they are *in vivo*.

A number of enzyme systems i.e. perinuclear peroxidase, prolyl-hydroxylase, collagenolytic activity, procollagen peptidases and synthesis of collagen, behave differently in this tissue-equivalent matrix *in vitro* than in monolayer cell cultures. These findings suggest that cells in the tissue matrix *in vitro* are led to adopt a state of differentiation at variance from that of conventionally cultured fibroblasts. In general, cellular activities in the tissue-equivalent model more closely resemble those of these same cells *in vivo* (35). They exhibit during the early and active phase of gel compaction, a contractile phenotype characterized by an important intracellular microfilamentous material and a scarce rough endoplasmic reticulum and rare Golgi membranes (31). In the same culture conditions, this contractile phenotype has been correlated with a large decrease in collagen biosynthesis and an increase in collagenolytic activity (37). Given that there is both polymerization and deposition of newly made collagen in the tissue-equivalent, as well as collagen degradation, the system provides an opportunity to study tissue remodeling *in vitro*. One of its major applications is the reconstitution of "living skin" by plating human keratinocytes on top of this contracted dermal equivalent (30, 35, 36). They adhere rapidly, multiply and spread to form a continuous sheet. This model provides an ideal opportunity to study keratinocytes-fibroblasts interactions as well as their biosynthesis, matrix modeling, differentiation, morphogenesis and cell contractile activities. It also could be useful as a skin replacement by grafting this reconstituted matrix for example in burned patients.

II. INTERACTIONS BETWEEN CELLS AND BASEMENT MEMBRANES

Recently, a new support for cell cultures, termed matrigel, has been described (38). It is composed of several major membrane macromolecules : laminin, entactin, type IV collagen, and heparan sulfate proteoglycan, which polymerize at 37°C to form a porous gel (39, 40). When cultured on matrigel,

melanocytes differentiate and accumulate melanin (38) while Sertoli cells form three-dimensional tubular structures resembling seminiferous tubes (41).

1. Mesenchymal cells

Fibroblasts are not usually in contact with basement membranes. However, during wound healing, in granulation tissue and various pathological conditions such as liver cirrhosis, they become closely apposed to basement membranes. Their interactions with basement membrane macromolecules has been studied in monolayer cultures and on matrigel. In monolayer, human embryonic fibroblasts possess specific laminin receptors and attach to laminin-coated dishes (14, 42). Certain fibroblasts do not recognize, however, laminin and will not survive well in culture if excess laminin is present (11). Only a small proportion of the fibroblasts can bind to laminin. In cultures on plastic, addition of fibronectin promotes fibroblast proliferation while laminin has no stimulatory effect on DNA synthesis (16).

When plated on matrigel, fibroblasts form a cellular network after one day of culture on top of the gel. They penetrate into the matrix, deposit fibronectin and collagen and retract the matrigel (43). They display a large cytoskeleton and a poor synthetic apparatus (rough endoplasmic reticulum and Golgi vesicles). Their cytoplasm contains pinocytotic vesicles in which laminin can be demonstrated by immuno-electron-microscopy (43).

The proliferation rate of normal fibroblasts on matrigel is stimulated (3 X) as compared with monolayered cells. Protein and collagen biosynthesis are considerably reduced (6 to 10 fold). They incorporate, however, more type III collagen in the matrigel (44) and exhibit a higher collagenolytic activity than cells on plastic. Thus basement membrane proteins may participate in the regulation of cell populations *in vivo* on both side of this matrix and may be involved in regulation of epithelial-mesenchymal regulations.

2. Malignant cells

There is much similarity in the basic processes involved in morphogenesis, cancer growth and metastasis : parenchymal cells dissociate to form individual cells which migrate to another location. In their new site, these cells attach to structural elements and again reassociate or multiply to form new clusters. These processes are directed by cell-cell and cell-matrix interactions. They can be monitored in cultures. Two different organization patterns of malignant cells cultures (human breast adenocarcinoma MCF7, T47D, SA52, human choriocarcinoma cells BeWo) can be induced by different extracellular matrix proteins (45, 46)(Figure 1).

When plated on plastic or polymeric type I collagen gel (used as a model of interstitial matrix), MCF7 cells spread and grow in monolayer. When cultured on matrigel (used as a model of basement membrane), cells form clusters attached to the matrix (Figure 1). Matrix proteins regulate these two types of cells organization by preferentially promoting cell-cell or cell-support interactions (45). On plastic in the presence of soluble laminin or on laminin-coated dishes, cells also form clusters. Addition of soluble fibronectin induces spreading of the cells suggesting that laminin and fibronectin have competitive antagonistic effects on MCF7 cells morphology. Anti-laminin antibodies inhibit clusters formation and attachment emphasizing the important role of this glycoprotein not only in promoting clusters attachment but also in cell-cell contacts formation. Such effects of extracellular matrix proteins could play significant roles in tumor progression and metastasis.

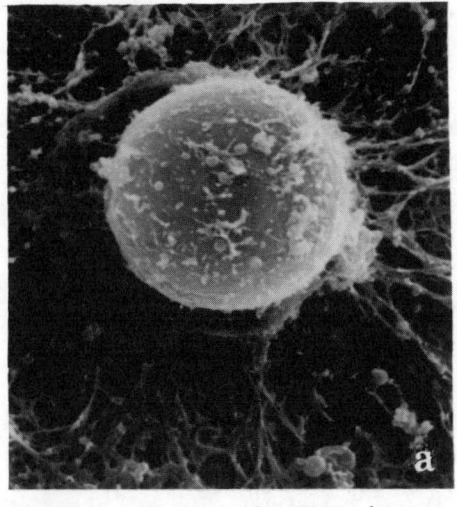 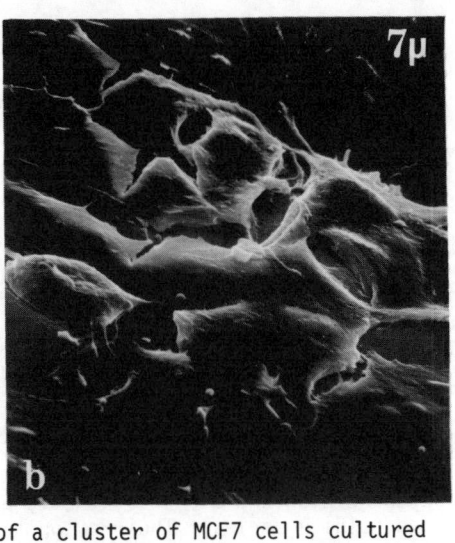

FIGURE 1. Scanning electron micrographs of a cluster of MCF7 cells cultured on matrigel (a) and flattened MCF7 cells growing as a monolayer on plastic (b) (Bars 7 μm).

CONCLUSIONS
    The interactions between extracellular matrix and cells play crucial roles in maintaining tissue architecture as well as during morphogenesis wound healing and metastasis. Additional studies are in progress to answer a series of questions : How do cells bind to the matrix ? What are the molecular links between the extracellular matrix and the cytoskeleton ? Several fibronectin and laminin receptors are being described (47). Several cell surface proteins with an affinity for fibronectin each consisting of at least two subunits were recently purified. They belong to the "Integrins" family. Peptides containing the recognition sequences of the fibronectin or laminin cell binding domains are synthesized and their biological activities are being tested. It is thus plausible that we will soon better understand the molecular events involved in such cell-matrix interactions and that we will learn how to manipulate them to the benefit of our patients.

ACKNOWLEDGMENT
    This work was supported by grants n° 3.4514.85 and 1.5067.87F of the FNRS in Belgium and by a grant of the CGER.

REFERENCES

1. Kleinman HK, Klebe RJ, Martin GR: Role of collagenous matrices in the adhesion and growth of cells. J Cell Biol 88, 473-485, 1981.
2. Alpin JD, Hughes RC: Complex carbohydrates of the extracellular matrix. Structures, interactions and biological roles. Biochem Biophys Acta, 694, 375-418, 1982.
3. Hay ED: Cell Biology of the Extracellular Matrix: Plenum Press, New York, 1982.
4. Grobstein C: Tissue interaction in morphogenesis of mouse embryonic rudiments in vitro. In Aspects of Synthesis and Order in Growth. Rudnick D(ed). Princeton: Princeton Univ. Press, pp 233-256, 1955.

5. Klebe RJ: Isolation of a collagen-dependent cell attachment factor. Nature 250, 248-251, 1974.
6. Ali IU, Hynes RO: Effects of LETS glycoprotein on cell motility. Cell 14, 439-446, 1978.
7. Vlodavsky I, Gospodarowicz D: Respective roles of laminin and fibronectin in adhesion of human carcinoma and sarcoma cells. Nature 289, 304-306, 1981.
8. Yamada KM: Fibronectin and other structural proteins. In Cell Biology of the Extracellular Matrix. Hay ED(ed). New York: Plenum Press, pp 95-114, 1982.
9. Timpl R, Martin GR: Components of basement membranes. In Immunochemistry of Collagen. Furthmayr H(ed). Cleveland, USA: CRC Press, vol. 2, pp 119-150, 1982.
10. Terranova VP, Rohrbach DH, Martin GR: Role of laminin in the attachment of PAM 212 (epithelial) cells to basement membrane collagen. Cell 22, 719-726, 1980.
11. Terranova VP, Rao CN, Kalebic T, Margulies TM, Liotta LA: Laminin receptor on human breast carcinoma cells. Proc Natl Acad Sci (USA) 80, 444-448, 1983.
12. Del Rosso M, Cappelletti R, Vita M, Vannuchi S, Chiarugi V: Binding of the basement-membrane glycoprotein laminin to glycosaminoglycans. An affinity-chromatography study. Biochem J, 199, 699-704, 1981.
13. Lesot H, Kühl A, von der Mark K: Isolation of a laminin binding protein from muscle cell membranes. EMBO J 2, 861-865, 1983.
14. Aumailley M, Nurcombe V, Edgar D, Paulsson M, Timpl R: The cellular interactions of laminin fragments. J Biol Chem 262, in press, 1987.
15. Kleinman HK, Cannon FB, Laurie GW, Hassel JR, Aumailley M, Terranova VP, Martin GR, Dubois-Dalcq M: Biological activities of laminin. J Cell Biochem 27, 317-325, 1985.
16. Terranova VP, Aumailley M, Sultan LH, Martin GR, Kleinman HK: Regulation of cell attachment and cell number by fibronectin and laminin. J Cell Physiol 127, 473-479, 1986.
17. Mosher DF: Fibronectin. Prog Homostasis Thromb 5, 111-151, 1980.
18. Pytela R, Pierschbacher MD, Ruoslahti E: Identification of a 140Kd cell surface glycoprotein with properties expected of a fibronectin receptor. Cell 40, 191-198, 1985.
19. Grancotti FG, Tarone G, Knudsen K, Damsky C, Comogleo PR: Cleavage of a 135 KD cell surface glycoprotein correlates with loss of fibroblast adhesion to fibronectin. Exp Cell Res 156, 182-190, 1985.
20. Gauss-Müller V, Kleinman HK, Martin GR, Schiffmann E: Role of attachment factors in fibroblast chemotaxis. J Lab Clin Med 96, 1071-1080, 1980.
21. Akers RM, Mosher DF, Lilien J: Promotion of retinal neurite outgrowth by substrate bound fibronectin. Dev Biol 86, 179-188, 1981.
22. Foidart JM, Berman JJ, Paglia L, Rennard S, Abe S, Perantoni A, Martin GR: Synthesis of fibronectin, laminin, and several collagens by a liver-derived epithelial line. Lab Invest 42, 525-532, 1980.
23. McDonald JM, Kelley DG, Brockelmann TJ: Role of fibronectin in collagen deposition: Fab' to the gelatin-binding domain of fibronectin inhibits both fibronectin and collagen organization in fibroblast extracellular matrix. J Cell Biol 92, 485-492, 1982.
24. Murray JC, Stingl G, Kleinman HK, Martin GR, Katz SI: Epidermal cells adhere preferentially to type IV (basement membrane)collagen. J Cell Biol 80, 197-202, 1979.

25. Wicha MS, Liotta LA, Garbisa S, Kidwell WR: Basement membrane collagen requirements for attachment and growth of mammary epithelium. Exp Cell Res 124, 181-190, 1979.

26. Vlodavsky I, Lui GM, Gospodarowicz D: Morphological appearance, growth behavior and migratory activity of human tumor cells maintained on extracellular matrix versus plastic. Cell 19, 607-616, 1980.

27. Kao RT, Hall J, Engel L, Stern R: The matrix of human breast tumor cells is mitogenic for fibroblasts. Am J Pathol 115, 109-116, 1984.

28. Rojkind M, Gatmaitan Z, MacKensen S, Giambrone MA, Ponce P, Reid LM: Connective tissue biomatrix : its isolation and utilization for long-term cultures of normal rat hepatocytes. J Cell Biol 87, 255-263, 1980.

29. Wicha MS, Lowrie G, Kohn E, Bagavandoss P, Mahn T: Extracellular matrix promotes mammary epithelial growth and differentiation in vitro. Proc Natl Acad Sci (USA) 79, 3213-3217, 1982.

30. Bell E, Ivarsson B, Merrill CH: Production of a tissue-like structure by contraction of collagen-lattices by human fibroblasts of different proliferative potential in vitro. Proc Natl Acad Sci (USA) 76, 1274-1278, 1979.

31. Bellows CG, Melcher AH, Bhargava U, Aubin JE: Fibroblasts contracting three-dimensional collagen gels exhibit ultrastructure consistent with either contraction or protein secretion. J Ultrastruct Res 78, 178-192, 1982.

32. Allen TD, Schor SL: The contraction of collagen matrices by dermal fibroblasts. J Ultrastruct Res 83, 205-219, 1983.

33. Farsi JMA, Aubin JE: Microfilament rearrangements during fibroblast-induced contraction of three-dimensional hydrated collagen gels. Cell Motility 4, 29-40, 1984.

34. Sarber R, Hull B, Merrill C, Soranno T, Bell E: Regulation of proliferation of fibroblasts of low and high population doubling levels grown in collagen lattices. Mech Ageing Dev 17, 107-117, 1981.

35. Bell E, Ehrlich MP, Buttle DJ, Nakatsuji T: Living tissue formed in vitro and accepted as skin-equivalent tissue of full-thickness. Science 211, 1052-1054, 1981.

36. Bell E, Sher S, Hull B, Merrill C, Rosen S, Chamson A, Asselineau D, Dubertret L, Coulomb B, Lapière C.M., Nusgens B, Neveux Y: The reconstitution of living skin. J Invest Dermatol 81, 2S-10S, 1983.

37. Nusgens B, Merrill C, Lapière CM, Bell E: Collagen biosynthesis by cells in a tissue equivalent matrix in vitro. Collagen Rel Res 4, 351-364, 1984.

38. Kleinman HK, McGarvey ML, Hassel JR, Star VL, Cannon FB, Laurie GW, Martin GR: Basement membrane complexes with biological activity. Biochemistry 25, 312-318, 1986.

39. Martin GR, Kleinman HK, Terranova VP, Ledbetter S, Hassel JR: The regulation of basement membrane formation and cell-matrix interactions by defined supramolecular complexes. In Basement Membranes and Cell Movement, Ciba Foundation Symposium 108. Porter R(ed). London: Pitman, pp 197-212, 1984.

40. Grant DS, Kleinman HK, Leblond CP, Inoue S, Chung AE, Martin GR: The basement membrane-like matrix of the mouse EHS tumor.II: Immuno-histochemical quantitation of six of its components. Am J Anat 174, 387-398, 1985.

41. Hadley MA, Byers SW, Suarez-Quian CA, Kleinman HK, Dym M: Extracellular matrix regulates Sertoli cell differentiation, testicular cord formation and germ cell development in vitro. J Cell Biol 101, 1511-1522, 1985.

42. Couchman JR, Höök M, Rees DA, Timpl R: Adhesion, growth, and matrix production by fibroblasts on laminin substrates. J Cell Biol 96, 177-183, 1982.
43. Emonard H, Callé A, Grimaud JA, Peyrol S, Castronovo V, Noël A, Lapière Ch.M., Kleinman H.K., Foidart J.M.: Interactions between fibroblasts and a reconstituted basement membrane matrix. J Invest Dermatol 89, 156-164, 1987a.
44. Emonard H, Grimaud JA, Nusgens B, Lapière ChM, Foidart JM: Reconstituted basement membrane matrix modulates fibroblast activities in vitro. J Cell Physiol, in press, 1987b.
45. Noël A, Callé A, Emonard H, Nusgens B, Foidart JM, Lapière ChM: Antagonistic effects of laminin and fibronectin in cell-cell and cell-matrix interactions in MCF7 cultures. In vitro, in press, 1987.
46. Callé A, Noël A, Nusgens B, Lapière ChM: Invasion of a reconstituted basement membrane (Matrigel) by malignant tumor cells. Submitted, 1987.
47. Leptin M: The fibronectin receptor family. Nature, 1986.

Foidart, J.M. : Functions of the extracellular matrix

Miller :   I doubt that your system can be used for skin grafts because first of the retractation and second because of the human antigen complex which is present on fibroblasts and absent on keratinocytes. You are I presume probably aware of Thivolet's work grafting keratinocytes. If you graft keratinocytes they will not be rejected, they will not retract and underneath, the fibroblasts, after a while, will reconstitute the complete skin.
So, why is it necessary to perform these complicated manipulations you describe ?

Foidart :   This study is a collaborative study that has been performed and is in progress with Gene BELL in the United States as well as with THIVOLET in Lyon and DUBERTRET in Paris. You only have to have keratinocytes. You are restricted by the amount of cells that you can plate on glass surfaces. Now, once you use such a collagen gel, you first pour the gel, you then take the fibroblasts of the patient himself, you let them proliferate in two dimensions and you put them in the three dimensional gel. They first start to retract the matrix but you do not let that happen as you immediately apply keratinocytes, not too many so as to make them subconfluent. They are just scattered throughout the surface of the gel and will produce a diffusible peptide which then subsequently inhibits the retraction by the fibroblasts. As soon as the fibroblasts obtain the information that keratinocytes are some way above them, then the lattice stops retracting allowing the generation of square decimeters of artificial skin.
You know that you can prepare artificial skin large enough to be grafted to the patient. There is no immune tolerance problem since you are using the fibroblasts and the

keratinocytes of the patient himself. So, the tolerance is rather good and there is no antigenicity due to the collagen itself. As a consequence, it is a useful technique which has been used exprimentally in the United States and in France. We grafted patients which were severely burned with this artificial skin. The skin that you obtain in this way will however give you some problems. It will never develop sweat glands nor sebaceous glands and the extensively burned patient will have problems since significant areas of the skin will have sweating problems as well as deshydrotic ones.

Miller :   I have been told that people are growing sebaceous glands, hair, adventitious organs and so on separately. Is that true ?

Foidart :   That is exactly what we are working on, trying to supplement the skin with appendages such as hair, sebaceous gland and sweat glands but it takes a lot of effort to artificially reconstruct a skin. This is already quite promising and useful enough to be tried not only in culture but also on human beings.

David :   Jean-Michel, have you looked on the reconstruction of the basal membrane at the interface ?

Foidart :   Yes.

David :   And what are the results ?

Foidart :   There is a basement membrane, it appears further that the laminins are being deposited first. It is exactly like in epiboly. I do not know if you are familiar with this model. The epiboly is what you obtain when you take pieces of skin biopsy and you put them in culture. The

epidermis will just go all over the piece and completely surround. the dermis. In the model that we have here, first laminin is being deposited and a few microns after, the electron-dense and amorphous basement membrane appears.

Baserga :   If I remember correctly from my studies when I was young, the hair has to have an amino acid called citrulline. Do you have to add it to grow hair in culture ?

Foidart :   No we don't.

Baserga :   Where is the citrulline ?

Foidart :   We do not add citrulline : we do not need citrulline.

Chowdhury : How many rounds of cell division do you allow the fibroblasts to go through and what is the chance of developing fibrosarcoma in the long run in this patient ?

Foidart :   First, once the cells are taken out from the patient, we do not know what the cell division numbers are up to the time when we get them. We try to grow them at a very high cell density by adding not only fetal calf serum but growth factors. Once they have divided, we place them into the matrigel where they almost stop dividing completely. But how many divisions occur in culture, I cannot tell you. They are not subcultured according to the one-to-five schedule before being transplanted. It is a very quick phenomenon. You first take the fibroblasts coming from the patient by cellular outgrowth and then you add PDGF and a series of growth factors in addition to calf serum and once the cells have reached confluency, you add them to the culture.

Chowdhury : Is there any long time experience with fibrosarcoma ?

Foidart :   The answer is no. The first patients were treated in the
            United States by G. BELL in 1981. There is a large
            background of accumulated data obtained with experimental
            animals such as rats and until now there has been no
            fibrosarcomas.

STRUCTURE AND FUNCTION OF THE HEPARAN SULFATE PROTEOGLYCANS

G. DAVID
Onderzoeksleider NFWO
Centrum voor Menselijke Erfelijkheid
Campus Gasthuisberg O&N6, Herestraat
B-3000   Leuven

Heparan sulfate proteoglycans are a complex and heterogeneous family of macromolecules. They are composed of linear sulfated polysaccharide chains, the heparan sulfate moieties, that are covalently attached to protein. Their distribution is nearly ubiquitous. They occur on plasma membranes, in the extracellular matrix, and as constituents of secretory vesicles. Their biological effects appear to depend on the selective reactivity of the proteoglycans with constituents of cell membranes and with soluble and insoluble ligands in the cellular micro-environment. They appear to modulate or to participate in processes that regulate cellular adhesion, cell shape and cell growth; that control matrix permeability, assembly and function; and that may determine the antithrombotic and protease-inhibitory properties of vessel walls and pericellular environments. These properties may define them as strategically positioned catalysts or regulators of cell-cell and cell-matrix interactions. Yet, structure-activity relationships remain poorly defined, both at the level of the heparan sulfate chains and at the level of the core protein. This information will be required, however, to understand these functions and also to evaluate the relevance of the evidence which implicates heparan sulfate proteoglycans in the pathogenesis of a variety of diseases. These include atheromatosis, thrombo-embolic disorders, nefrotic syndromes, diabetic basement membrane thickening, amyloidosis, fibrosis etc... and processes that lead to the genesis of malignant cell behavior, which implies loss of growth control (tumorigenicity) and loss of positional control (invasion and metastasis).

Excellent reviews on the progress in heparan sulfate proteoglycan research have appeared recently (Höök et al., 1984; Gallagher et al., 1986; Hassell et al., 1986). By themselves they are indicators of the rapidly growing interest in these components. The following sections summarize current knowledge on the structure, biosynthesis and possible functions of heparan sulfate proteoglycans and attempt to provide a rationale for our own investigations on the core protein structure of these proteoglycans. These studies have provided evidence for extensive structural heterogeneity, reminiscent of the multiple functions that have been ascribed to these components. The development of uniquely specific monoclonal antibody reagents has been instrumental to this progress, and holds promises for the exploration of the significance of this heterogeneity at the genomic and transcriptional level, for the determination of structure-function relationship, and for the evaluation of the involvement of heparan sulfate proteoglycan in the pathogenesis of disease.

A. O. A. Miller (ed.), Advanced Research on Animal Cell Technology, 15–24.
© 1989 by Kluwer Academic Publishers.

1. Heparan sulfate

Heparan sulfate and heparin are structurally related glycosaminoglycans, and form probably the most complex of all mammalian carbohydrates. They are initially synthesized as a repeating sequence of α, β 1→4-linked N-acetyl glucosamine and glucuronic acid residues. The nascent chains, however, undergo complex modifications (Lindahl et al. 1977, 1986). Part of the amino sugars are deacetylated and converted to N-sulfated glucosamines, a structure that is unique to this family of carbohydrates. In addition, some of the glucuronic acids are converted to their C5 epimers. Substitution with O-sulfate, mostly at C2 of some of the iduronic acids and at C6 of the glucosamines, further augments the anionic charge and complexity of the polymers. Some of these modifications, such as the formation of 3-O-sulfated GlcNSO$_3^-$, occur only at very low frequency, yet form essential components of specific sugar sequences that endow heparin and some heparan sulfate chains with high binding affinity to antithrombin III (Thunberg et al., 1982) and others with cell growth inhibitory properties (Castellot et al., 1986) There appear to be considerable differences in the O-sulfate concentration of heparan sulfates from different cell types, despite relative similarities in N-sulfate content (Gallagher and Walker, 1985). This cellular polymorfism may be genetically or developmentally determined. Cellular transformation, in general, reduces the extent to which sulfation and epimerisation occur (see below), which may have severe implications for the proposed functions of these polymers.

2. Heparan sulfate proteoglycan biosynthesis

The biosynthesis starts with the formation of a primed core protein, followed by the assembly of a non-sulfated polysaccharide, which is then extensively modified by sulfation and epimerisation (Rodén L. 1980). Very little is known on the structure of the protein cores of the heparan sulfate proteoglycans and there is no direct information on the number of genes that encode potential core proteins. Priming of the protein involves the (cotranslational?) attachment of a xylose-gal-gal sequence to some serine (-gly) residues in the core protein. Formation of the linkage and of the repeating disaccharide sequences involves the sequential transfer of sugar units from nucleotide-sugar precursors, catalysed by specific glycosyl transferases. Transfer is thought to occur directly, without the intermediate of carriers comparable with dolichol phosphate in the assembly of N-linked sugars. Interestingly, chondroitin sulfate chains, which are structurally different glycosaminoglycans, are linked to serine (-gly) residues in their respective core proteins, through a similar xylose-gal-gal-sequence. It is not known what features direct the further substitution of the core proteins with either heparan sulfate or chondroitin sulfate chains.

Modification at the level of the polymer seems to occur in a step-wise fashion and appears to be localized to particular regions of the glycosaminoglycan chains. This implies that a particular set of modifications must be completed before the next can occur, and that these may be realized by functionally integrated multi-enzyme complexes (Lindahl et al., 1986). Likely, these occur in some distinct Golgi-compartment.

3. Proteoglycan localisation

Indirect evidence suggests the occurrence of membrane-intercalated heparan sulfate proteoglycans in most, if not all cell types (Kraemer, 1971; Kjellén et al., 1981; Oldberg et al., 1979; Rapraeger and Bernfield, 1983; Lories et al., 1986, 1987). This evidence includes the requirement for detergent for complete solubilization of the proteoglycans, the occurrence of the proteoglycans in micellar aggregates, or the ability of the proteoglycans to interact with liposomes or with hydrophobic chromatography matrices. Direct evidence, such as the

identification of a transmembrane core protein domain, with a hydrophobic amino acid sequence, is lacking.

Various cell types can specifically bind heparin and highly sulfated heparan sulfate (Kraemer, 1977; Kjellén et al., 1977). The number of specific binding sites per cell appears high (up to 4×10^6), but binding is of low affinity (Kd = 10^{-7} M) (Kjellén et al., 1980).

Binding does not appear to result in the internalization and degradation of the receptor-bound proteoglycan. The origin, fate and significance of these "periferal" membrane proteoglycans are not clear.

Heparan sulfate proteoglycans also occur in the fibrillar structures of stromal matrices and basement membranes (Hassell et al., 1980; David and Bernfield, 1978, 1981), often in apparent codistribution with collagens, fibronectin and laminin (Hedman et al., 1982). Still others, of which the heparin proteoglycan is the prototype, are components of intracellular storage granules. It should be noted that, in the instance of membrane and basement membrane proteoglycans, the core protein appears to be the determinant which is responsible for the occurrence of the proteoglycan at these subcellular locations (Höök et al., 1984b).

4. Proteoglycan structure

Structural analyses have been complicated by the relative scarcity and poor solubility of these components, and by the lack of appropriate tools. Comparative data of proteoglycans derived from a variety of tissues suggest great diversity in structure, which stems both from differences in the structure of the core proteins, and from variations in the number and length of the heparan sulfate chains (Hassell et al., 1986). In general, the latter appear to be clustered within one region of the core protein. Core protein molecular weight estimates have varied from 10,000 (Fujiwara et al., 1984) to more than 350,000 dalton (Fenger et al., 1984). Structural heterogeneity also exists amongst proteoglycan cores isolated from a single tissue (Deboeck et al., 1987; Lories et al., 1987), and is also exemplified by the existence of distinct hydrophobic membrane intercalated and non-hydrophobic periferally membrane-bound and secreted forms (see above). Studies with polyclonal antibodies suggest the existence of epitopes that are unique, others common to distinct matrix proteoglycans (Dziadek et al., 1985), but absence of cross reactivity, so far, between basement membrane and cell surface-associated proteoglycans. Thus, at least two structurally distinct proteoglycan cores, with different distribution, appear to be synthesized by a variety of cells.

5. Molecular interactions and functions of the heparan sulfate proteoglycans

By virtue of the glycosaminoglycan chains they carry, the heparan sulfate proteoglycans may, besides the binding to cell surfaces as described above, interact and bind to extracellular components such as fibronectin (Laterra et al., 1980; Stamatoglou and Keller, 1982; Yamada et al., 1980), fibronectin-collagen complexes (Johansson and Höök, 1980), fibrillar collagen (Koda and Bernfield, 1984) and laminin (Sakashita et al., 1980). Through their glycosamino-glycan chains, heparan sulfate proteoglycans may also self-associate (Fransson et al., 1983). Fibronectin, heparan sulfate proteoglycans and collagen fibers codistribute (Hedman et al., 1982) and the interactions of these matrix components with heparan sulfate proteoglycans are thought to be crucial to the normal **assembly of the extracellular matrix**. Indeed, the gelation of basement membrane extracts of constant composition (matrisomes) depends on their heparan sulfate proteoglycan component (Martin et al., 1984). Poor basal lamina assembly by epithelial cells producing basement membrane proteoglycan with shortened and under-sulfated heparan sulfate chains (David et al., 1981; David and Bernfield, 1982; David and Van den Berghe, 1983; David et al., 1987), supports this hypothesis. Yet, heparinase digestion of

the matrix of human fibroblasts causes no fibronectin release or disruption of the fibrillar structure of the matrix (Hedman et al., 1984) suggesting that intact heparan sulfate proteoglycans are not required for general **maintenance of matrix integrity**. In contrast, the anchorage of collagen-tailed acetylcholinesterase to the basement membrane, a specialization of the matrix at synaptic sites, appears mediated by a heparan sulfate proteoglycan (Brandan et al., 1985). Through their cell-, matrix- and self-interactions, heparan sulfate proteoglycans may allow or enhance **cell attachment to the matrix and intercellular adhesion**. This is substantiated by the ability of free heparin or heparan sulfate chains or anti heparan sulfate proteoglycan antibodies to modulate or interfere with the attachment of myeloma cells to collagen (Stamatoglou and Keller, 1983), of rat hepatocytes (Johansson and Höök, 1984) and of chinese hamster ovary cells (Schwartz and Juliano, 1985) to fibronectin, and of neural retina cells to 'adherons' and to each other (Cole et al., 1985; Schubert and La Corbière, 1985). The restriction of heparan sulfate proteoglycans to the sinusoidal domain of the plasma lemma of hepatocytes (Stow et al., 1985) and the concentration of heparan sulfate proteoglycans in fibroblast adhesion sites (Lark and Culp, 1984; Woods et al., 1984) are also suggestive for a role in adhesive processes. Interestingly, the membrane-intercalated heparan sulfate proteoglycan form also interacts with F-actin, suggesting the possibility that membrane heparan sulfate proteoglycans may connect the endo- and exo-cytoskeleton and perhaps be transmembrane mediators of **cell shape** changes that are dictated by (changes in) the extracellular matrix (Rapraeger and Bernfield, 1982). The inhibition of the spreading of mouse 3T3 cells on fibronectin substrata by exogenous heparin (Laterra et al., 1983) and the codistribution of actin and heparan sulfate proteoglycans in spreading rat fibroblasts (Woods et al., 1984) as detected in immunofluorescence would be consistent with this putative function. A direct or indirect role for heparan sulfate proteoglycans in **cell growth** has been proposed. For example, heparan sulfate proteoglycans appear to be able to control smooth muscle cell proliferation (Castellot et al., 1984) and cell migration (Majack and Clowes, 1984; Terranova et al., 1985). Heparin (Lander et al., 1982) and antibodies against the heparin binding domain of laminin (Edgar et al., 1984) inhibit neurite outgrowth, while cell surface heparan sulfate proteoglycans derived from dorsal root ganglion neuronal cells stimulate Schwann cell proliferation (Ratner et al., 1985). The ability of heparin to influence cellular growth appears to depend on its sulfation characteristics, requiring both N- and O-sulfate, and a 3-0-sulfated, N-sulfated glucosamine in particular (Castellot et al., 1984, 1986). Interestingly, heparin molecules can be isolated which have antiproliferative activity but no anticoagulant properties (Castellot et al., 1984). **Anticoagulation,** due to the acceleration of antithrombin III interaction with thrombin and factors Xa and IXa, depends mainly on the occurrence of a unique antithrombin-binding site formed by a specific pentasaccharide sequence (Lindahl et al., 1984) which is shared by heparin and the chains of some heparan sulfate proteoglycans, e.g. those of endothelial cells (Marcum and Rosenberg, 1985; Marcum et al., 1986) and of the basement membrane (Pejler and David, 1987). By (similarly) accelerating other proteinase-inhibitor interactions, e.g. those between protease nexin I and arginine-specific serine proteases (Low et al., 1981) the heparan sulfate proteoglycans may **modulate pericellular proteinase activity** in more general ways. Perhaps the best documented function of heparan sulfate proteoglycans is that of acting as a **charge barrier** and to contribute to the selectivity of the glomerular filtration apparatus (Kanwar and Farquhar, 1979). By analogy, cell surface heparan sulfate proteoglycans may screen off the cell surface and create a cellular micro-environment. Finally, the core protein of some skin fibroblast cell surface heparan sulfate proteoglycans consists of 2 disulfide bonded subunits

-each of Mr 80-100,000- (Cöster et al., 1983)- with a structure and functional properties resembling those of the **cellular receptor** for transferrin. Indeed, this skin fibroblast heparan sulfate proteoglycan appears to bind -via its core protein- to apotransferrin in a pH-dependent, and to holotransferrin in a pH-independent fashion, that mimics the binding of these ligands to their receptor (Fransson et al., 1984). This suggests that some proteoglycans may be involved in the transduction of specific messages or ligands across cell membranes.

6. Heparan sulfate proteoglycans of transformed cells

Both quantitative and qualitative changes in glycosaminoglycans associated with the surface of transformed cells have been reported (Höök et al., 1984). Of particular interest are indications that the heparan sulfate synthesized by virally transformed cells (Underhill and Keller, 1975; Winterbourne and Mora, 1981; Shanley et al., 1983) and by transformed cell lines (David and Van den Berghe, 1983) are undersulfated compared to the untransformed counterparts. Similarly, the heparan sulfate isolated from a hepatoma was shown to have a lower degree of sulfation than the heparan sulfate isolated from normal liver (Nakamuro et al., 1978). Poor extracellular matrix formation (David and Van den Berghe, 1983), altered self association of the heparan sulfate proteoglycans (Fransson et al., 1981) and reduced cellular adhesion (Stamatoglou and Keller, 1983) are possible functional correlates. Reduction of the pericellular heparan sulfate content may also result from extracellular cleavage of heparan sulfate chains by endoglycosidase secreted by tumor cells (Kramer et al., 1982). Heparinase activity of malignant cells may also directly contribute to the destruction of extracellular matrices during the local invasion, intravasation and extravasation of metastasing tumor cells (Thorgeirsson et al., 1985).

References

1. Brandan, E., Maldonodo, M., Garrido, J. and Inestrosa, N.C. J. Cell. Biol. 101, 985-992, 1985.
2. Castellot, J.J., Beeler, D.L., Rosenberg, R.D. and Karnovsky, M.J. J. Cell Physiol. 102, 315-320, 1984.
3. Castellot, J.J., Choay, J., Lorneau, J.C., Pesiton, M., Sache, E. and Karnovsky, M.J. J. Cell Biol. 102, 1979-1984, 1986.
4. Cole, G.J., Schubert, D. and Glaser, L. J. Cell Biol. 100, 1192-1199, 1985.
5. Cöster, L., Malmström, A., Carlstedt, I. and Fransson, L.-Å. Biochem. J. 215, 417-419, 1983.
6. David, G. and Bernfield, M.R. Proc. Natl. Acad. Sci. USA 76, 786-790, 1979.
7. David, G. and Bernfield, M.R. J. Cell Biol. 91, 281-286, 1981.
8. David, G. Van der Schueren, B. and Bernfield, M.R. J. Natl. Cancer Inst. 67, 719-728, 1981.
9. David, G. 'Basal lamina formation by normal and transformed mammary epithelial cells : proteoglycan processing and turnover'. Ph.D. Thesis, 1981.
10. David, G. and Bernfield, M.R. J. Cell Physiol. 110, 56-62, 1982.
11. David, G. and Van den Berghe, H. J. Biol. Chem. 258, 7338-7344, 1983.
12. David, G. and Van den Berghe, H. J. Biol. Chem. 260, 11067-11074, 1985.
13. David, G., Nusgens, B., Van der Schueren, B., Van Cauwenberghe, D., Van den Berghe, H. and Lapière, Ch. Exp. Cell Res. In press, 1987.
14. Deboeck, H., Lories, V., David, G., Cassiman, J.J and Van den Berghe, H. Biochem. J. in

press 1987.

15. Dziadek, M., Fujiwara, S., Paulsson, M. and Timpl, R. EMBO Journal 4, 905-912, 1985.

16. Edgar, D., Timpl, R. and Thoenen, H. EMBO Journal 3, 1463-1468, 1984.

17. Fenger, M., Wewer, U. and Albertsen, R. FEBS Letters 173, 75-79, 1984.

18. Fransson, L.-Å., Carlstedt, I., Cöster, L. and Malmström, A. J. Biol. Chem. 258, 14342-14345, 1983.

19. Fransson, L.-Å., Carlstedt, I., Cöster, L. and Malmström, A. Proc. Natl. Acad. Sci. USA 81, 5657-5661, 1984.

20. Fransson, L.-Å., Sjöberg, I. and Chiarugi, V.P. J. Biol. Chem. 256, 13044-13047, 1981.

21. Fujiwara, S., Wiedemann, H., Timpl, R., Lustig, A. and Engel, J. Eur. J. Biochem. 143, 145-157, 1984.

22. Gallagher, J.T. and Walker, A. Biochem. J. 230, 665-674, 1985.

23. Gallagher, J.T., Lyon, M. and Steward, W.P. Biochem. J. 236, 313-325, 1986.

24. Goldberg, B.D. and Burgeson, R.E. J. Cell Biol. 95, 752-756, 1982.

25. Hassell, J.R., Gehron Robey, P., Barrach, H.J., Wilczek, J., Rennard, S. and Martin, G.R. Proc. Natl. Acad. Sci. USA 77, 4494-4498, 1980.

26. Hassell, J.R., Kimura, J.M. and Hascall, V.L. Ann. Rev. Biochemistry 55, 539-567, 1986.

27. Hedman, K., Johansson, S., Vartio, T., Kjellén, L., Vaheri, A. and Höök, M. Cell 28, 663-671, 1982.

28. Hedman, K., Vartio, T., Johansson, S., Kjellén, L., Höök, M., Linker, A., Salonen, E.-M. and Vaheri, A. EMBO Journal 3, 581-584, 1984.

29. Höök, M., Kjellén, L., Johansson, S. and Robinson, J. Ann. Rev. Biochem. 53, 847-867, 1984a.

30. Höök, M., Couchman, J., Woods, A., Robinson, J. and Christner, J. in 'Basement Membranes and Cell Movement'. Ciba Foundation, Symposium 108 (eds. Porter K. and Whelan J.) pp. 44-50, Pitman London 1984b.

31. Johansson, S. and Höök, M. Biochem. J. 187, 521-524, 1980.

32. Johansson, S. and Höök, M. J. Cell Biol. 98, 810-817, 1984.

33. Kanwar, Y.S. and Farquhar, M.G. Proc. Natl. Acad. Sci. USA 76, 1303-1307, 1979.

34. Kjellén, L., Oldberg, Å., Rubin, K. and Höök, M. Bioch. Biophys. Res. Commun. 74, 126-133, 1977.

35. Kjellén, L., Oldberg, Å. and Höök, M. J. Biol. Chem. 255, 10407-10413, 1980.

36. Kjellén, L., Pettersson, I. and Höök, M. Proc. Natl. Acad. Sci. USA 78, 5371-5375, 1981.

37. Koda, J.E. and Bernfield, M.R. J. Biol. Chem. 259, 11763-11770, 1984.

38. Kraemer, P.M. Biochem. 10, 1437-1445, 1971.

39. Kraemer, P.M. Biochem. Biophys. Res. Commun. 78, 1334-1339, 1977.

40. Kramer, R.H., Vogel, K.G. and Nicolson, G.L. J. Biol. Chem. 257, 2678-2686, 1982.

41. Lander, A.D., Fujii, D.K., Gospodarowicz, D. and Reichardt, L.F. J. Cell Biol. 94, 574-585, 1982.

42. Lark, M.W. and Culp, L.A. J. Biol. Chem. 259, 6773-6782, 1984.

43. Laterra, I., Ausbacher, R. and Culp, L.A. Proc. Natl. Acad. Sci. USA 77, 6662-6666, 1980.

44. Laterra, I., Silbert, J.E. and Culp, L. J. Cell Biol. 106, 112-123, 1983.

45. Lindahl, U., Thunberg, L., Bäckström, G., Riesenfeld, J., Nordling, K. and Bjöck, I. J. Biol. Chem. 259, 12368-12376, 1984

46. Lindahl, U., Höök., M., Bäckström, G., Jacobsson, I., Riesenfeld, J., Malmström, A., Rodén, L. and Feingold, D.S. Fed. Proc. Am. soc. Exp. Biol. 36, 19-24, 1977.

47. Lindahl, U., Feingold, D.S. and Rodén, L. TIBS 11, 221-225, 1986.

48. Lories, V., David, G., Cassiman, J.J. and Van den Berghe, H. Eur. J. Biochem. 158, 351-360, 1986.

49. Lories, V., Deboeck, H. David, G., Cassiman, J.J. and Van den Berghe, H. J. Biol. Chem. 262, 854-859, 1987.

50. Low, D.A., Baker, J.B., Koonce, W.C. and Cunningham, D.D. Proc. Natl. Acad. Sci. USA 78, 2340-2344, 1981.

51. Majack, R.A. and Clowes, A.W. J. Cell Physiol. 118, 253-256, 1984.

52. Marcum, J.A. and Rosenberg, R.D. Biochem. Biophys. Res. Commun. 126, 365-372, 1985.

53. Marcum, J.H., Athe, D.M., Fritze, L.M.S., Nawroth, P., Stern, D. and Rosenberg, R.D. J. Biol. Chem. 261, 7507-7517, 1986.

54. Martin, G.R., Kleinman, H.K., Terranova, V.P., Ledbetter, S. and Hassell, J.R. in 'Basement Membranes and Cell Movement', Ciba Foundation Symposium 108, (eds. Porter, R. and Whelan, J.) pp. 197-209, Pitman London 1984.

55. Nakamuro, N., Hurst, R.E. and West, S.S. Biochem. Biophys. Acta 538, 445-457, 1978.

56. Oldberg, Å, Kjellén, L. and Höök, M. J. Biol. Chem. 254, 8505-9510, 1979.

57. Pejler G. and David, G. Biochem. J., in press, 1987.

58. Rapraeger, A.C. and Bernfield, M.R. in 'Extracellular Matrix' (eds. Hawkes, S. and Way J.L.) pp.265-269, Academic Press N.Y., 1982.

59. Rapraeger, A.C. and Bernfield, M.R. J. Biol. Chem. 258, 3632-3636, 1983.

60. Ratner, N., Bunge, R.P. and Glaser, L. J. Cell Biol. 101, 744-754, 1985.

61. Rodén, L. in : The biochemistry of glucoproteins and proteoglycans (ed. Lunanz, W.J.) pp. 267-37, Plenum Publishing corporation, N.Y. 1980.

62. Sakashita, S., Engvall, E. and Ruoslahti, E. FEBS Letters 116, 243-246, 1980.

63. Schwartz, M.A. and Juliano, R.L. J. Cell Physiol. 124, 113-119, 1985.

64. Schubert, D. and La Corbière, M. J. Cell Biol. 100, 56-63, 1985.

65. Shanley, D.J., Cossu, C., Boetliger, D., Holtzer, H. and Pacifici, M. J. Biol. Chem. 258, 810-816, 1983.

66. Stamatoglou, S.C. and Keller, J.M. Biochem. Biophys. Acta 719, 90-97, 1982.

67. Stamatoglou, S.C. and Keller, J.M. J. Cell Biol. 96, 1820-1823, 1983.

68. Stow, J.L., Kjellén, L., Unger, E., Höök, M. and Farquhar, M.G. J. Cell Biol. 100, 975-980, 1985.

69. Terranova, V., DiFlorio, R., Lyall, R.M., Hic, S., Friesel, R. and Maciaq, T. J. Cell. Biol. 101, 2330-2334, 1985.

70. Thorgeirsson, U.P., Turpeenniemi-Hujassen, T. and Liotta, L.A. Int. Rev. Exp. Pathol. 27, 203-234, 1985.

71. Thunberg, L., Bäckström, G. and Lindahl, U. Carbohydrate Res. 100, 393-410, 1982.

72. Woods, A., Höök., M., Kjellén, L., Smith, C.G. and Rees, D.A. J. Cell Biol. 99, 1743-1753, 1984.

73. Yamada, K.M., Kennedy, D.W., Kimata, K. and Pratt, R.M. J. Biol. Chem. 255, 6055-6063, 1980.

David, G. : <u>Heparan sulfate proteoglycans and their role in the extracellular matrix</u>

Spier : I was intrigued that you missed virus attachment sites out, with all these sort of nice juicy charged groups and specific carbohydrates don't you think that viruses would be particularly attractive to these molecules ?

David : I do not know of any specific example where this has been documented. To my knowledge, not like the influenza receptor type with acidic residues. I do not know of any examples for these compounds.

Hooghe : Recently, somebody from Leuven claimed that heparin disappeared with binding of the virus.

David : Yes, I think for several components we will be able, with high doses of heparin, to obtain conformational states and influence binding properties. So, I think we have to compare also activity levels. Let me talk about the effects of heparin on the acceleration of the reaction between antithrombin and thrombin. We speak about very low concentration levels really, catalytic effects. You can knock off LDL from its receptors with high doses of heparin and yet there is no quite known details of these interactions and there is no scheme for heparin in this reaction and yet we can produce such effects at high concentration. So, not all effects of heparin are on binding interactions.

Horaud : There is some evidence in the 1960's that sulfated poly-saccharide influenced picornaviruses attachment and also some old papers published about dextran sulfate and the inhibitory reaction of dextran sulfate on the picornavirus

and if I remember some, but not all enveloped viruses. This can be significant now because we do not understand in the case of picornavirus, why some picornavirus are non infectious in some tissue culture and why they are able to induce a toxicity infection in other cultures and we know that the receptors for the replication of poliovirus are present on every cell. However, poliovirus does not replicate in the lung.

David :   That's interesting. I must declare myself incompetent on this issue I do not know about these studies.

Spier :   Carrying on the picornavirus story and to be heretical at the same time, I was looking for the receptor for Foot and Mouth disease virus and in fact all the studies that I did indicated that there wasn't a specific receptor. The more virus I added to the cell, the more bound to the cell. There was a straight line graph. So, I am just a little bit worried about whether or not there are specific receptors for the picornaviruses and this is why I was wondering whether or not these heparin-type proteins could be involved in virus attachment. They seem to be very active and have quite a lot of highly charged groups that would give you the non specific interactions that I was looking for ...

David :   Could be non-specific but still it should be limited because there is a well defined number of these residues at the cell surface. If it is really these components that act as receptors and if it is specific, the binding should be saturable.

Spier :   Well, how many molecules would you say would exist on the cell surface ?

David :   I still have to calculate that ... It depends what cell
            type. If you take liver cells for example, where there is
            a fraction which is itself not integrally imbedded in the
            membrane, the number of receptors for these molecules - at
            least if the cell is really loaded up with heparin sulfate,
            is an order of more than two millions copies per cell
            which is a huge number.

ROLE PLAYED BY GAP JUNCTIONS IN CELL-CELL COMMUNICATION

Dieter F. HÜLSER

Abt. Biophysik, Biologisches Institut,
Universität Stuttgart, Pfaffenwaldring 57
D-7000 Stuttgart 80

Intercellular communication is a necessary prerequisite for the evolution of complex multicellular organisms. Without information exchange neither plants nor animals would have developed. The different possibilities of intercellular communication can be classified as three independent signal pathways: 1.) indirect signal transfer by secreted molecules, 2.) direct signal transfer by plasma membrane bound molecules (receptors) and 3.) direct signal transfer from cell to cell by channels which span the plasma membranes of adjacent cells. Whereas in the first two cases the signal flow is unidirectional (hormone, neurotransmitter; sperm-egg-binding, immune system), in the latter case no rectification is observed. These cell to cell channels - called gap junctions - not only allow an intercellular pathway for the transfer of information between cells but also connect adjacent cells mechanically. They must, therefore, well be discriminated from desmosomes which anchor cells together to form structural or functional units as well as from tight junctions which seal membranes of epithelial cells to each other so that the paracellular path becomes impermeable to molecules and a polarity of apical and basolateral surface is maintained.

The communicating gap junctions consist of protein channels (connexons) to which both contacting cells contribute (10, 25). Up to several hundred channels are assembled to form a typical gap junction plaque (1). The density of these channels in the plasmamembrane may reflect their state of activity (2, 11, 19, 24).

Gap junctions are ubiquitously found in the animal kingdom from mesozoa to vertebrates. Not only regulation of embryonic development, cell differentiation and growth control depend on the existence of gap junctions but also the synchronous beating of heart muscle cells and the coordinated contractions of smooth muscle cells in the intestine.

Gap junctions are very often found in permanently growing cell cultures, which adhere to the surface of plastic petridishes. These monolayer cultures facilitate investigations of gap junction properties since individual cells can easily be discriminated unter lightmicroscopical observation. Cells with certain enzymatic defects will not proliferate under cell culture conditions without a substitution for the missing enzyme or its product. Defective cells, however, survive and proliferate when they are coupled via gap junctions (9). This metabolic cooperation first has been described by Subak-Sharpe and coworkers (34), who found that Chinese hamster cells lacking the enzyme hypoxanthine:guanine phosphoribosyltransferase (HGPRT⁻) are unable to incorporate exogenously administered hypoxanthine into their nucleic acid when they grew isolated, but as soon as they made contact with wild type cells (HGPRT⁺) they incorporated hypoxanthine. In this case, phosphoribosylpyrophosphate spreads as a signal via gap

A. O. A. Miller (ed.), Advanced Research on Animal Cell Technology, 25–37.
© *1989 by Kluwer Academic Publishers.*

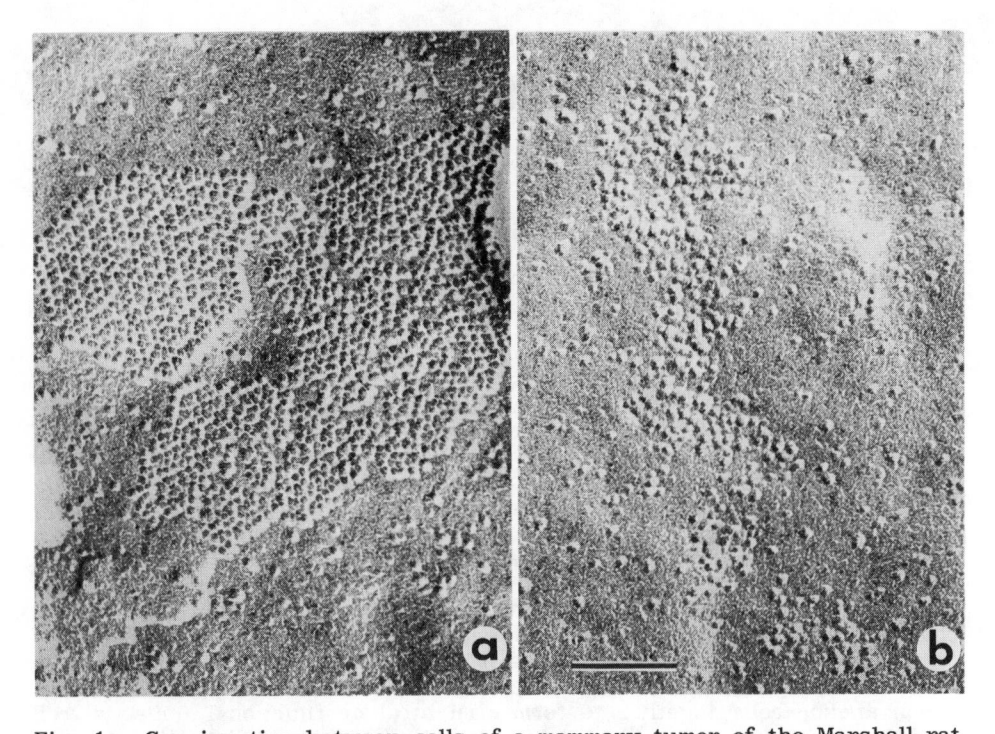

Fig. 1: Gap junction between cells of a mammary tumor of the Marshall rat
(BICR/M1R$_k$). An inactive (a) and an active (b) state is shown.
These unfixed culture cells were rapidly frozen in liquid propane,
fractured and replicated in a Balzers 301 instrument and the
pictures were taken in a Zeiss EM10 electron microscope.
Bar: 0,1 μm

junctions into the HGPRT$^+$ cells which therefore synthesize more nucleotides which are distributed via gap junctions into the HGPRT$^-$ cells (29).
Thus the enzyme block is bypassed and the defective cells can survive.
Using fluorescent dyes of different sizes it has been detected that molecules of up to M_r = 900 can pass intercellularly via gap junctions (27). In
insect cells the pore size seems to be wider, enabling the passage of
larger molecules up to M_r = 1800 (30). Since in most cases the intercellular
signal which is exchanged between the cells is not known, artificial signals
are used for the demonstration of open gap junctions. Very often the
fluorescent tracer Lucifer yellow (M_r = 457) is injected iontophoretically by
glassmicropipettes into a cell and its spreading into adjacent cells can be
observed under an epifluorescent microscope (33).

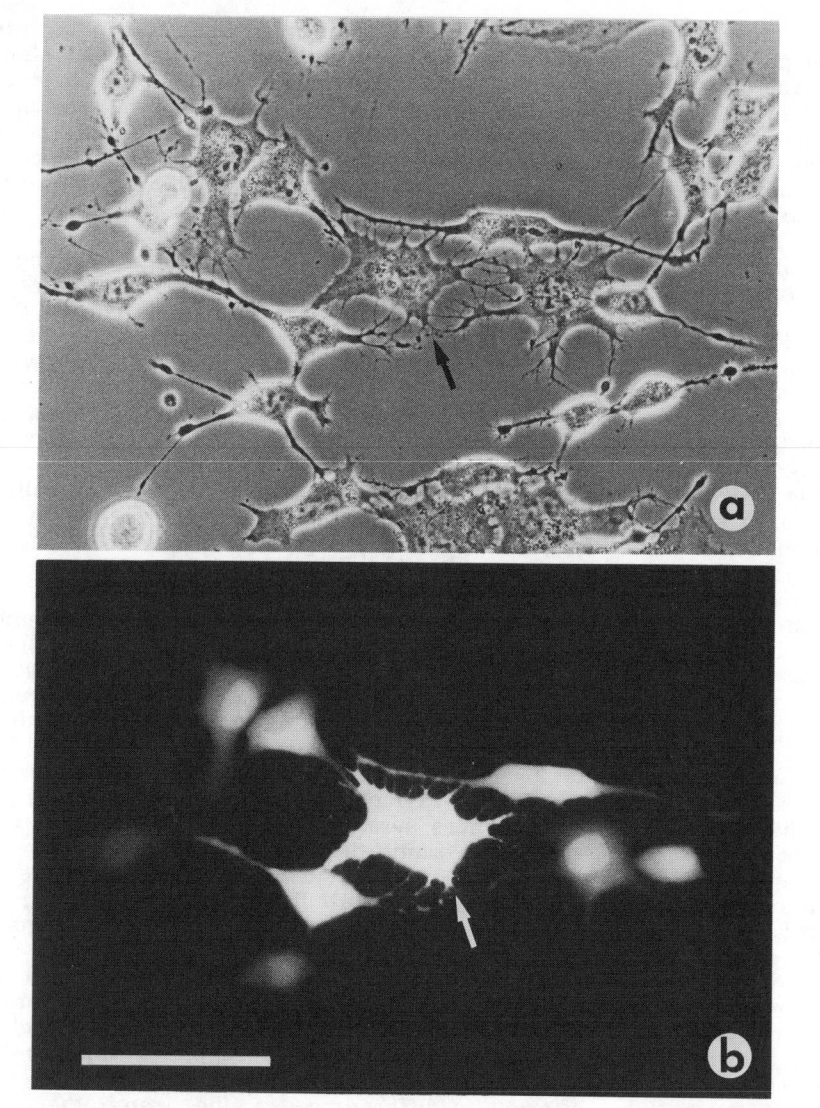

Fig. 2: Demonstration of gap junctional spreading of Lucifer yellow from an injected cell (arrow) into several coupled cells. The cells of a mammary tumor (BICR/M1R$_k$) of the Marshall rat were cultivated as monolayer cells. Lucifer yellow was injected as a 4 % solution in 1M LiCl.
(a) phase contrast picture,
(b) fluorescent picture, 3 min after injection of the tracer.
Bar: 100 μm

Interestingly, different cell types from different species can be interconnected by gap junctions (heterologous coupling), as has been shown for several cell cultures (18,27). Since human cells can be coupled to chicken cells and form heterologous gap junction channels the outer region of the connexon must be considered as very conservative. Both endogenous signals such as action potentials from chicken heart muscle cells and exogenously generated injected signals such as Lucifer yellow could be passed into human (FL, RD) and rodent (BICR/M1R$_k$, BT3C, BT5C, C6, B14, BHK-21) cells (27).

Measuring intercellular communication with endogenously or exogenously generated signals and demonstrating the gap junctions' structure at the same cells in the electron microscope does not necessarily prove that gap junctions mediate this kind of information exchange. With antibodies directed against the cytoplasmic end of the gap junction protein, however, this correlation was demonstrated (13). By injecting this type of antibodies into cells of xenopus embryos (8-cell stage), not only an inhibition of dye transfer but also severe damages in their further development were observed (36). This underlines the importance of gap junctions for developing organisms. The inhibition of dye transfer into or out of cells which had been injected with antibodies against gap junction proteins can also be demonstrated with some permanently growing cell culture lines, but most of them do not react with these antibodies (4). This indicates rather variable sites at the cytoplasmic ends of the connexons.

Isolation of gap junctions and biochemical analysis of their subunits resulted in some controversial findings: the reported figures for the molecular weight of a subunit range from 16k, 21k, 26k, 27k up to more than 45k (for review see 26). Each subunit spans the plasma membrane four times so that both the amino-terminal-region and carboxyl terminus are located in the cytoplasm (37). The amino acid sequences of the 21k and 26k-proteins are up to 87 % homologous, which is contributed mostly by the membrane embedded and the extracellular parts of the proteins, whereas the cytoplasmic parts are less homologous. Nevertheless, 21k and 26k antibodies are crossreactive in many tissues such as pancreas, stomach, kidney, ovary, heart, brain and uterus (12). This explains not only why heterologous coupling can be observed in culture but also why intracellular regulation of gap junction permeability seems to be differently regulated in different cell types.

The opening and closing of gap junction channels is influenced by two endogenous signal pathways, using as second messengers cAMP or diacylglycerol (8, 21). The cAMP signal pathway opens gap junction channels, whereas the diacylglycerol signal pathway downregulates the gap junction permeability. Besides a possible regulation by phosphorylation, the fine tuning may well be different in different cells, since many other substances have been described with regulatory effects (31). Whereas substances such as pH, Ca^{++} and retinoic acid may well represent physiological interactions, many other substances (Heptanol, Oxonol, Glutaraldehyd) react in a non physiological manner and always lead to a closing of gap junction pores.

The opening and closing characteristics of a single gap junction channel can be followed by patch clamp measurements, where the current across a channel can be determined. Depending on the type of experiment (whole cell recording with a cell pair or cell attached measurements) only open and closed states have been determined (22,32,35) or several states have been described for connexons (23,28). In the latter case, the final prove for a gap junction channel - the test with blocking antibodies - is still missing.

Cells in culture - either freshly prepared from biopsies or permanently growing as cellines - may well exist in a state where the permeability of gap junctions is completly upregulated. This is the case for the mammary tumor cells BICR/M1R$_k$ which remain permanently in the fully open state when they are cultivated as monolayer cells. However, as soon as they grow as threedimensionally organized multicell spheroids their coupling status changes considerably (14). With increasing age and size these multicell spheroids reduce their junctional permeability without reducing the amount of gap junction plaques. In case of stress, e. g. treatment by ionizing radiation - as is the case for cancer therapy - these gap junctions may open again and thus enable a better metabolic cooperation (5). This explains, why tumor cells with gap junctions must be treated with a higher radiation dose when their survival rate should be the same as for tumor cells without gap junctions (6).

With this threedimensional cell culture model more complex questions can be answered than is possible for monolayer cells (20). Confronting multicell spheroids of normal tissue with malignant cells (15,17), showed that coupling competent tumor cells occupied and progressively replaced the normal tissue within a few days (3). This invasive behaviour of tumor cells is only found for tumor cells which communicate via gap junctions with the host tissue. Non-coupled HeLa cells, for instance, do not show this invasive activity (3). These cervical cancer cells, however, are arranged to a densely packed external layer of epithelial morphology, demonstrating that in threedimensionally growing multicell spheroids a reorganisation and differentiation is reestablished, which cannot be obtained in monolayer cultures. Under these conditions not only tumor cells display the typical characteristics of solid tumor growth in the organism, but also normal cells regain their organotypic synthetic capacity (16). Multicell spheroids can be formed with many normal cells which are then structured similar as in the intact organs as has been shown for liver, brain, stomach, heart and kidney. Heart cell spheroids regularly contract, indicating the presence of active gap junctions (3).

At a first glance, studying the role of gap junctions in intercellular communication seems not to be of any relevance for the improvement of cell culture conditions. However, by comparing the regulatory functions of gap junctions under in vivo conditions, with primary cell cultures and with permanently growing cellines, the in vitro growth conditions have been improved. These studies helped to demonstrate the capability of threedimensionally growing multicell spheroids and may lead to successful mass cultivations of mammalian cells with high yields of cell production.

30

LITERATUR

1.) Bennett, M. C. L. and Goodenough, D. A.:
Gap junctions, electrotonic coupling, and intercellular communication
In: Neurosciences Research Program Bulletin 16, 372 - 486,
MIT Press, 1978

2.) Berdan, R. C. and Caveney, S.:
Gap junction ultrastructure in three states of conductance
Cell Tissue Res. 239, 111 - 122, 1985

3.) Bräuner, Th.:
Interzelluläre Kommunikation und invasives Wachstum maligner Zellen.
Konfrontation von Normalgewebe mit Tumorzellen in Monolayer- und
Multizell-Sphäroid-Kultur.
Dissertation, Biologisches Institut, Universität Stuttgart, 1987

4.) Brümmer, F.:
Interzelluläre Kommunikation   Regulation der Gap-Junction-Öffnungs-
weite
Dissertation, Biologisches Institut, Universität Stuttgart, 1987

5.) Dertinger, H. and Hülser, D.:
Increased radioresistance of cells in cultured multicell spheroids.
I. Dependence on cellular interaction
Radiat. Environ. Biophys. 19, 101 - 107, 1981

6.) Dertinger, H. and Hülser, D. F.:
Intercellular communication in spheroids
In: Recent Results in Cancer Research 95, 67 - 83,
Springer-Verlag Berlin - Heidelberg, 1984

7.) Flagg-Newton, J. L., Simpson, I. and Loewenstein, W. R.:
Permeability of the cell-to-cell membrane channels in mammalian cell
junctions
Science 205, 404 - 407, 1979

8.) Flagg-Newton, J. L., Dahl, G. and Loewenstein, W. R.:
Cell junction and cyclic AMP: I. Upregulation of junctional membrane
permeability and junctional membrane particles by administration of
cyclic nucleotide or phosphodiesterase inhibitor
J. Membrane Biol. 63, 105 - 121, 1981

9.) Gilula, N. B., Reeves, O. R. and Steinbach, A.:
Metabolic coupling, ionic coupling and cell contacts
Nature 235, 262 - 265, 1972

10.) Goodenough, D. A. and Revel, J. P.:
A fine structural analysis of intercellular junctions in the mouse liver
J. Cell Biol. 45, 272 - 290, 1970

11.) Greule, J. and Hülser, D. F.:
Comparison of different freeze-fracture methods
Eur. J. Cell Biol. 43, Suppl. 17, 20, 1987

12.) Hertzberg, E. L.: personal communication, 1987

13.) Hertzberg, E. L.:
Antibody probes in the study of gap junctional communication
Ann. Rev. Physiol. 47, 305 - 318, 1985

14.) Hülser, D. F. and Brümmer, F.:
Closing and opening of gap junction pores between two- and three-
dimensionally cultured tumor cells
Biophys. Struct. Mech. 9, 83 - 88, 1982

15.) Laerum, O. D., Mork, S. and De Ridder, L.:
Morphology of malignant rat glioma cells during invasion into chick
heart fragments in vitro
In: Tumour Progression and Markers, Proceedings of the Sixth Meeting
of the European Association for Cancer Research, 25 - 31,
Kugler Publications Amsterdam, 1982
16.) Landry, J., Bernier, D., Ouellet, C., Goyette, R. and Narceau, N.:
Spheroidal aggregate culture of rat liver cells: histotypic reorgani-
zation, biomatrix deposition, and maintenance of functional activities
J. Cell Biol. 101, 914 - 923, 1985
17.) Mareel, M. M. K.:
The use of embryo organ cultures to study invasion in vitro
In: Tumor invasion and metastasis, (eds.: L. A. Liotta, J. R. Hart)
207 - 230, Martinus Nijhoff Publishers, Den Haag-Boston-London 1982
18.) Michalke, W. and Loewenstein, W. R.:
Communication between cells of different type
Nature 232, 121 - 122, 1971
19.) Müller, A., Blanz, W. E., Laub, G. and Hülser, D. F.:
Pattern analysis of freeze-fractured gap junctions
Eur. J. Cell Biol. 36, Suppl. 7, 46, 1985
20.) Mueller-Klieser, W.:
Multicellular spheroids - A review on cellular aggregates in cancer
research
J. Cancer Res. Clin. Oncol. 113, 101 - 122, 1987
21.) Murray, A. W. and Fitzgerald, D. I.:
Tumor promoters inhibit metabolic cooperation in coculture of epidermal
and 3T3 cells
Biochem. Biophys, Res. Comm. 91, 395 - 401, 1979
22.) Neyton, J. and Trautmann, A.:
Single-channel currents of an intercellular junction
Nature 317, 331 - 335, 1985
23.) Paschke, D. and Hülser, D. F.:
Voltage dependent behaviour of a large membrane channel with sub-
conductance states
Eur. J. Cell Biol. 43, Suppl. 17, 41, 1987
24.) Peracchia, C.:
Gap junctions - Structural changes after uncoupling procedures
J. Cell Biol. 72, 628 - 641, 1977
25.) Revel, J. P. and Karnovsky, M. J.:
Hexagonal array of subunits in intercellular junctions of the mouse
heart and liver
J. Cell Biol. 33, C7 - C12, 1967
26.) Revel, J. P., Yancey, S. B. and Nicholson, B. J.:
The gap junction proteins
TIBS 11, 375 - 377, 1986
27.) Schmid, A. and Hülser, D. F.:
Intercellular communication in co-cultures of permanently growing cell
lines
Eur. J. Cell Biol. 39, Suppl. 12, 31, 1986
28.) Schwarze, W. and Kolb, H.-A.:
Voltage-dependent kinetics of an anionic channel of large unit
conductance in macrophages and myotube membranes
Pflügers Arch. 402, 281 - 291, 1984

32

29.) Sheridan, J. D., Finbow, M. E. and Pitts, J. D.:
Metabolic interactions between animal cells through permeable inter-
cellular junctions
Exp. Cell Res. 123, 111 - 117, 1979
30.) Simpson, I., Rose, B. and Loewenstein, W. R.:
Size limit of molecules permeating the junctional membrane channels
Science 195, 294 - 296, 1977
31.) Spray, D. C., White, R. L., Campos de Carvalho, A., Harris, A. L.
and Bennett, M. V. L.:
Gating of gap junction channels
Biophys. J. 45, 219 - 230, 1984
32.) Spray, D. C., Ginzberg, R. D., Morales, E. A., Gatmaitan, Z. and
Arias I. M.:
Electrophysiological properties of gap junctions between dissociated
pairs of rat hepatocytes
J. Cell Biol. 103, 135 - 144, 1986
33.) Stewart, W. W.:
Functional connections between cells as revealed by dye-coupling with
a highly fluorescent naphthalimide tracer
Cell 14, 741 - 759, 1978
34.) Subak-Sharpe, H., Bürk, R. R. and Pitts, J. D.:
Metabolic co-operation between biochemically marked mammalian cells in
tissue culture
J. Cell Sci. 4, 353 - 367, 1969
35.) Veenstra, R. D. and DeHaan, R. L.:
Measurement of single channel currents from cardiac gap junctions
Science 233, 972 - 974 1986
36.) Warner, A. E., Guthrie, S. C. and Gilula, N. B.:
Antibodies to gap-junctional protein selectively disrupt junctional
communication in the early amphibian embryo
Nature 311, 127 - 131, 1984
37.) Zimmer, D. B., Green, C. R., Evans, W. H. and Gilula, N. B.:
Topological analysis of the major protein in isolated intact rat liver
gap junctions and gap junction-derived single membrane structures
J. Biol. Chem. 262, 7751 - 7763, 1987

Hulser, D.F. : <u>Role played by gap junctions in cell-cell communication</u>

Preat :   Did you study the effects of tumour promoter on your
            culture system ?

Hulser :   Tumour promoters have been studied because they interact
            with the diacylglycerol signal pathway and therefore you
            always get a downregulation of the permeability. But this
            effect of tumour promoters is really controversial. There
            was just a conference on gap junctions in Asilomar and I
            think six groups reported on tumour promoters and the
            effects of gap junctions. In 50 % of the cases there was a
            down regulation, while in the other cases there was no
            such an effect.

Vandevoorde : The same question with relation to the action of Tumour
            Necrosis Factor ?

Hulser :   Sorry, I am not aware that somebody did it.

David :   There has been a recent report that gap junctions could be
            induced in liver cultures with heparin.

Hulser :   These were liver cells or hepatocytes or other cells ?

David :   Liver cells.

Hulser :   There is certainly more involved than gap junctions as you
            saw from the control because for the control, we had
            quiescent liver cells and heart cells. They make contact
            and there is no doubt that they don't even grow on each
            other. That means the cells must make contact not only by
            gap junctions but also there must be something else in
            between. It might well be that you have conditions in

your tissue culture where the cells do not make contacts. I know that for growing cultivated hepatocytes, you need special media otherwise they won't proliferate and when they proliferate or when they survive, then they make gap junctions within three days. This could only be that then, you find the gap junctions a little bit earlier. I do not know. Normally you find gap junctions in hepatocyte cultures.

David : No molecular interactions are known between connexons or heparin-like sequences ?

Hulser : Not that I know.

Chowdhury : In the whole liver the gap junctions seem to be down regulated during cholestasis. Have you looked at the effect of cholestatic agents like ethylene estradiol on your liver cell cultures ?

Hulser : No, what we know is that if you bring the liver cells into culture, then they obviously up regulate so that if you add dibutyryl CAMP or so you cannot increase the amounts of gap junctions or the amounts of coupling because obviously, everything is up regulated and so it does not work. This is our interpretation.

Miller : There is one thing which puzzles me. One of the aim in growing animal cells on microbeads is to reach high cell densities in order to promote what is called metabolic cooperation. Now, what I have seen astonishes me. It seems as if, when the cells grow as a flat monolayer, that the gap junctions are open so that they can more or less freely exchange molecules with a molecular weight of approximately 900. Now, when they are in three-dimen-

sions, your morulae-type aggregates, the gap junctions look as if they are closed. So, on microbeads, I would expect them also to be closed but we know the higher the cell density, the better the cells exchange, so, there is something I do not understand.

Hulser : I would consider microbeads in a first approximation as a monolayer and as you might remember the table I showed, not all of these cell lines we investigated were completely up regulated or the cell permeability of the gap junctions was not completely open because there were only 100 % for ionic coupling but only 10 or 15 % for the luciferin yellow dye, which means you can always have cells in tissue culture, growing as monolayers, where an up regulation is still possible. The one example I showed you concerns mammary tumor cells which were completely up-regulated and therefore always completely open. As soon as they grew in three dimensions, they started to downregulate.

Anonymous : Do you know whether there is a good model to measure gap junctions in vivo like in liver or is it supposed to happen only in vitro ?

Hulser : Normally, you take part of the liver out and then measure this but then it becomes an in vitro system. I know that measurements have been made on the kidney but it is rather complicated because every movement of the animal will be registered by the electrode.

Merten : Did you say that in aggregates the gap junctions were closed or not ?

Hulser : In vitro they are closed.

Merten :   If I suppose for instance that they are open, they could
              be very interesting for the aggregate cultures because if
              you make aggregates normally, you have some nutritional
              problems in the center of the aggregates. But if perhaps
              you have open gap junctions, there could be diffusion of
              the products not between the cells but within the cells
              too, to the center of the aggregate.

Hulser :   Yes, there are some in an animal or in an organ but these
              gap junctions must not always be open. We all have a
              telephone but we are not continuously talking on the
              phone. I think that the cells, if you grow them three-
              dimensionally, or if they grow in an organ, then, if
              necessary, they can open these gap junctions.   Otherwise
              they might be closed and as soon as we take cells out of
              the animal, then it is a completely different situation.

Merten :   But this could be very interesting for a cell culture in
              capsules such as in the Damon process. They have capsules
              of maximum 500 microns and normally in the center, it is
              full of cells. You have problems with the nutrition but
              if there were open gap junctions, then perhaps there could
              be some nutrients reaching the center.

Hulser :   We still have these mini tumors and this is the reason why
              they have been used. You get these necrotic areas and
              even if the cells are able to get gap junctions, you do
              not find them, so, the distance where this appears might
              be so or 100 microns as a living shell outside but I do
              not think you are able to completely supply 1 mm spheres
              with nutrients.

Cassiman :   Can I just state for clarity, when you say you measured
              patch clamping and gap junctions by patch clamping, that
              surprises me that you would look for a junction that is

intercellular with patch clamping. Can you say something about that ?

Hulser :   These were connexons, these were the semi-channels. If two cells make contact, then both provide a hemi-channel and you must know where they make contacts and then hope that if you patch here, that there are some channels on the track to the gap junction.

Miller :   I presume that the cDNA has been made for these proteins. Would it be experimentally feasible to transfect HeLa cells with the cDNA from such gap junctions proteins and make these HeLa cells which you said were lacking the gap junctions now respond to such things ?

Hulser :   Indeed, I forgot to mention that LOEWENSTEIN's group has done similar experiments with the messenger RNA and was able to induce gap junctions in cells which did not have gap junctions. With cDNA this type of experiment has not yet been done.

FIBRONECTIN-MEDIATED CELLULAR (INTER)ACTIONS

B. DE STROOPER*, M. JASPERS, F. VAN LEUVEN, H. VAN DEN BERGHE AND
J.-J. CASSIMAN
*Aspirant NFWO
Centrum voor Menselijke Erfelijkheid
Campus Gasthuisberg O&N6, Herestraat
B-3000 Leuven

I. Fibronectins : multifunctional glycoproteins

1. Biological role of fibronectin in vitro and in vivo

1.1. Localization of fibronectin in situ

Fibronectins are large glycoproteins (440 kDa), present in many tissues and in most body fluids. The soluble form, usually referred to as "plasma fibronectin", is present in rather high
concentrations (300 µg/ml) in blood (Mosesson and Umfleet, 1970). The insoluble, multimeric
form, called "cellular fibronectin", is found in connective tissue and basement membranes
and is organized in short fibrils, in amorphous granules, or in thick cross-striated fibrils consisting of collagen I and II decorated with fibronectin (Fyrand, 1979; Couchman et al., 1979;
Fleishmajer and Timpl, 1984). Fibronectin containing matrices are particularly prominent in
vivo at sites of tissue remodelling and cell migration (e.g., wound repair, neural crest cell
migration). In vitro, fibronectin localizes in extracellular fibrillar structures and in the basement membranes of most cultured cells (Yamada, 1978; Hedman et al., 1978). In these structures, fibronectin rather than collagen is the major structural component (Chen et al., 1978).
Extensive accumulation of collagen fibrils appears to be a subsequent event and the distribution seems to depend on the integrity of the fibronectin matrix as shown by inhibition studies
with anti-fibronectin antibodies (Mc Donald et al., 1982).

1.2. Functions of fibronectin

Both in vitro and in vivo experiments have shown the variety of biological processes in which
fibronectins are involved. In Table 1, the major biological activities as deduced from in vitro
assays are schematically presented. Of particular importance is the role of fibronectins in
adhesion, migration and differentiation during embryogenesis, wound healing and cancer.

a. Fibronectin during embryogenesis

Fibronectin containing fibrils are present in the early blastula stage of amphibian development (Boucaut and Darribere, 1983). These fibrils develop progressively into a network on
the inner surface of the blastocyst roof. Antifibronectin antibodies or synthetic peptides containing the Arg-Gly-Asp (RGD) sequence, which is recognized by the fibronectin receptor (cfr.
infra) interfere with normal gastrulation, but not with the formation of the neural plate

A. O. A. Miller (ed.), Advanced Research on Animal Cell Technology, 39–61.
© 1989 by Kluwer Academic Publishers.

(Boucaut et al., 1984a,b). In chick embryo, fibronectin appears just prior to the formation of the primitive streak (Duband and Thiery, 1982). Its appearance correlates well with the onset of the first active movement of the cells.

TABLE 1 : BIOLOGICAL ACTIVITIES OF FIBRONECTIN

```
Cell-substrate adhesion
      Cell attachment to collagen and gelatin
      Cell-attachment to fibrin
      Cell adhesion and spreading on glass and plastic


Cell-cell adhesion


Cell-migration
      Stimulation of cell motility
      Haptotaxis or chemotaxis


Cell-morphology
      Maintenance of flattened and spread cell shape with
      few cell surface microvilli
      Alignment of confluent fibroblasts into parallel arrays
      Stimulation of actin microfilament bundle organization


Stimulation or inhibition of cytodifferentiation


Stimulation of growth or cytokinesis


Nonimmune opsonic activity for macrophages
```

Another example for the role of fibronectin is observed during neural crest cell migration. Most crest cells remain in direct contact with fibronectin-rich fibrils throughout their migratory period (Duband et al., 1982). Arg-Gly-Asp containing decapeptides, injected in vivo, inhibit this migration. In vitro, neural crest cells bind preferentially to fibronectin coats (Rovasio et al., 1983) and their speed of locomotion on these coats ($\pm$ 40 μm/hr), is close to the in vivo estimate (Thiery et al., 1984).

A direct role for fibronectin in cell differentiation has also been suggested. Both stimulatory and inhibitory effects are observed. In neural crest cells, exogeneously added fibronectin promotes an adrenergic phenotype (Sieber-Blum et al., 1981; Loring et al., 1982). It is not clear if this is due solely to selection by the presence of fibronectin of a subpopulation commited to form adrenergic cells, or by a direct effect on catechol amine synthesis. In other cell systems, fibronectin functions as an inhibitory molecule for differentiation, e.g., myoblast fusion (Chiquet et al., 1979; Podleski et al., 1979), adipocyte differentiation (Spiegelmann and Ginty, 1983) and regulation of the chondrogenic phenotype (Pennypacker et al., 1979; West et

al., 1979). Again it is not clear whether fibronectin has direct effects on biosynthetic processes in these cells or whether the inhibition is a consequence of the reorganization of the cytoskeleton, induced by fibronectin.

b. Fibronectin and wound healing

A full thickness wound in the skin is almost immediately filled by a blood clot composed of platelets trapped in a fibrin meshwork. While the clot forms, fibronectin is crosslinked to fibrin by factor XIIIa (Mosher et al., 1979). The role of fibronectin in clot formation remains unclear, but Dixit et al. (1985) showed that a monoclonal antibody recognizing a 70 kDa fragment of fibronectin containing the cell binding site, could inhibit platelet aggregation in vitro.

In all the further steps of the woundhealing process fibronectin appears somehow to be involved (Grinnell, 1984). A point of particular importance is that the fibronectin content of the wound matrix (granulation tissue) is much higher than that of the adjacent tissue (Grinnell et al., 1981).

Following hemostasis, invasion of the wound region by neutrophils and monocytes, which control infection and remove tissue debris, occurs. Fibronectin is involved in neutrophil and monocyte mobility, chemotaxis and adhesion to endothelial cells (Wall et al., 1982; Norris et al., 1982; Bevilaqua et al., 1980; Czop et al., 1985). Neutrophils can also proteolytically modify fibronectin. This cleaved or "inflamed" (Vercellotti et al., 1983) fibronectin allows neutrophils to enhance their own adherence to endothelial cells.

The ingrowth of fibroblasts and endothelial cells from the subdermis and the formation of granulation tissue depends again on the presence of fibronectin. The adhesion of fibroblasts to fibrin requires fibronectin, and maximal adhesion requires that fibronectin is crosslinked by factor XIII to fibrin (Grinnell et al., 1980). Endothelial cell adhesion and chemotaxis is also promoted by fibronectin (Bowersox and Sorgente, 1982).

Finally, the granulation tissue is covered by a neo-epidermis formed by keratinocytes, which migrate in from the wound edges, probably again guided by fibronectin (Stenn et al., 1983). While laminin also mediates epidermal cell adhesion, it has been shown that in granulation tissue, laminin is absent, while fibronectin is available (Stanley et al., 1981).

A similar role for fibronectin in wound healing of the cornea has been shown (Fuyikawa et al., 1981).

c. Fibronectin and transformation

A great deal of the initial interest in fibronectin was due to the observation that cell surface associated fibronectin was significantly decreased in transformed cells (Pearlstein et al., 1980; Ruoslahti et al., 1981; Ruoslahti, 1984). This was especially impressive in experimental systems with cells transformed with temperature sensitive mutant viruses. At the nonpermissive temperature, these cells express a normal phenotype and have normal quantities of fibronectin at their surface. At the permissive temperature, however, the cells become transformed and concomitantly a 5-6 fold decrease in fibronectin is observed (Adams et al., 1977; Gahnberg et al., 1974; Hynes and Wyke, 1975; Rieber and Romano, 1976; Stone et al., 1974; Vaheri and Ruoslahti, 1974). Exogeneously added fibronectin, isolated from normal cells, can partially restore the morphology, adhesiveness and contact inhibition of such transformed cells (Yamada et al., 1976).

The correlation with malignancy in vivo is less clear. Tumor cells in vivo, at least in some cases, also express reduced amounts of matrix fibronectin (Asch et al., 1981; Labat-Robert et

42

al., 1981; Stenman and Vaheri, 1981). This seems to reflect a generalized lack of matrix components, which correlates well with the degree of morphological differentiation, rather than a specific deficiency in fibronectin accumulation. Possibly for similar reasons, cells, isolated from metastases (Smith et al., 1979), have less fibronectin than cells isolated from the primary tumors. Since metastatic cells have a high affinity for the extracellular matrix of endothelial cells (Kramer et al., 1980; Vlodavsky et al., 1982) and since these interactions are inhibited by anti-fibronectin antibodies, the lack of fibronectin could permit metastatic cells to leave the primary tumor and settle in other fibronectin containing regions of the organism. The difficulty to find firm correlations between the fibronectin system and tumor invasion or metastasis is likely due to the diversity in phenotype of the malignant cells <u>in vivo</u>. Given the striking effects of fibronectin on cell adhesion <u>in vitro</u>, however, it is likely that the lack of fibronectin influences the invasive and metastatic capacity. Particularly interesting is the observation in nonpathological conditions that the ability of cells to become translocated along the neural crest pathway correlates with the lack of cell surface fibronectin (Erickson et al., 1980). Indeed, the phenotype of the migrating cell appears to be that of lacking cell surface fibronectin.

2. The modular organization of fibronectin

The diverse functions of fibronectin are reflected in its structural organization. Most work in that regard has been done on plasma fibronectin, but the cellular fibronectins have largely the same structure.
As shown in fig. 1, fibronectin contains two monomers, cross-linked by two disulfide bridges located at the carboxyl-terminus. This highly asymmetric location permits the molecule to span large distances. The two subunits are slighty different in sequence and arise from alternative mRNA splicing.

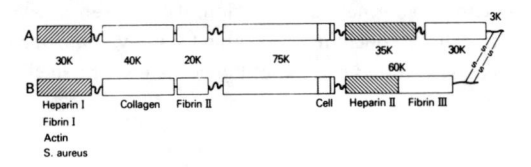

Fig. 1 : Current map of the structural and functional domains of human plasma fibronectin.

The amino-terminus is at the left and the carboxy terminus is at the right. The rectangles represent the protease-resistent domains described in the text. The two subunits designated A and B, are very similar in organization. They differ in a region toward the carboxy terminus as indicated. The binding activities of each domain are listed below the B chain (taken from Yamada et al., 1985).

Electron microscopy reveals plasma fibronectin to resemble a V-shaped structure. Each arm measures approximatively 60 nm x 2 nm (Engel et al., 1981; Erickson et al., 1981; Odermatt et al., 1982; Price et al., 1982). Under physiological conditions, the molecule appears to be more compact (Rocco et al., 1984) and this could prevent circulating fibronectin from binding to a number of different ligands if the molecule must open for binding. Actually, this might be the case for most of its interactions with cells, as fibronectin adsorbed to plastic coats or beads, or fibronectin added together with heparin, heparan sulfate or collagen is much more

effective in binding to cells than fibronectin in solution (Johansson and Höök, 1984).

A particular successful approach used to unravel the structure-function relationships in the fibronectin molecule has been to cleave the molecule in functional domains by limited proteolysis and to isolate the fragments by affinity chromatography on different ligands. This approach localized specific binding interactions of fibronectin to precise regions in the molecule.

2.1 The different domains of fibronectin (fig. 1)

The **amino-terminal domain** binds to a rather surprising number of ligands. It is a low affinity binding site for heparin and low sulfated heparan sulfate (Sekiguchi and Hakamori, 1980; Sekiguchi et al., 1983). The binding can be modulated by physiological Ca^{2+}-concentrations (Hayashi and Yamada, 1981). Other ligands include actin (Keski-Oja and Yamada, 1981), fibrin (Sekiguchi et al., 1981; Hayashi and Yamada, 1982), Staphylococcus Aureus and some Streptococci (Mosher 1980). Recently the receptor of Staphylococcus Aureus for this region was isolated (Fröman et al., 1987) and cloned (Flock et al., 1987). Cross-linking of fibronectin with fibrin by transglutaminase (factor XIII_a) occurs in the same region (Mosher et al., 1980). Since in blood clots fibronectin is associated with fibrin, the serum concentrations of fibronectin are 20 to 50% lower than those in plasma.

The **collagen-binding domain** is strongly conserved in evolution (Ruoslahti et al., 1979) and binds collagen of type I, II and III. It has, however, a higher affinity for gelatin (denatured collagen), which is used as affinity ligand to isolate this domain. The region is highly glycosylated and rich in disulfide bridges. It remains intact after digestion with many proteases (Ruoslahti et al., 1979; Hahn and Yamada, 1979a,b; Gold et al., 1979) but becomes much more susceptible to proteolytic degradation if glycosylation is prevented by tunicamycin (Bernard et al., 1982).

The **fibrin binding site II** is only a weak binding site and is easily destroyed by proteases such as trypsin, suggesting that it is not as compact structured as the other binding regions of fibronectin.

The **cell-binding domain** is of crucial importance for the adhesive function of fibronectin. The smallest proteolytic fragment, retaining full biological activity as evaluated by cell binding assays is one of 75 kDa (Hayashi and Yamada, 1983). The group of Pierschbacher and Ruoslahti (1981) succeeded in isolating a shorter sequence (11.5 kDa) with an adhesion inhibiting monoclonal antibody. This fragment has a more variable biological activity, probably because the tertiary configuration is not as stable as that of the 75 kDa fragment. This fragment (108 amino acid residues) was sequenced (Piersbacher et al., 1982). By testing various synthetic peptides, a tetrapeptide with the sequence Arg-Gly-Asp-(Ser) or RGD(S) was identified as the active site, recognized by cells adhering to fibronectin (Piersbacher and Ruoslahti, 1984a; Yamada and Kennedy, 1984). The arginine, glycine and aspartic acid are essential for activity, while the serine residue can be replaced by a limited number of other amino acids (Piersbacher and Ruoslahti, 1984b). Synthetic peptides containing the RGD sequence mediate cell attachment directly when presented to cells as an insoluble substrate, whereas in solution the peptides inhibit cell attachment to fibronectin.

Meanwhile it has become clear that the RGD-sequence is of crucial biological importance. This sequence plays not only a role in the adhesion of cells to fibronectin, but also in the adhesion or binding of cells and platelets to many other adhesive glycoproteins, as vitronectin (Suzuki et al., 1985), Von Willebrand factor, fibrinogen (Gartner and Bennett, 1985; Plow et

al., 1987), osteopontin (Oldberg, 1986), and perhaps other proteins.

Synthetic peptides containing the RGD sequence interfere with processes as platelet aggregation (Santoro, 1987), embryogenesis (cfr. supra), blastocyst implantation (Armant et al., 1986) and metastasis (Humphries et al., 1986). The RGD sequence occurs in proteins of several primitive organisms. The λ-receptor of E-coli contains an RGD sequence and can promote adhesion of mammalian cells. Surface proteins of yellow fever virus and foot-and-mouth disease virus also contain RGD sequences. It has been suggested that RGD interactions could play a role in the binding of these pathogens to their hosts. Finally, the RGD sequence is involved in the aggregation of Dictyostelium, a primitive slime mold (Gabius et al., 1985) and in the embryogenesis of Drosophila (Naidet et al., 1987).

The **heparin binding domain II** has the highest affinity for heparin and heparan sulfate (Sekiguchi et al., 1983; Hayashi and Yamada, 1982; Yamada et al., 1980). The binding is divalent cation insensitive.

Interest in the **fibrin binding domain III** follows especially from the fact that this domain shows differences between the A and the B monomer of plasma fibronectin (Richter and Hörmann, 1982; Hayashi and Yamada, 1983) Such a specific site could be a recognition signal during the assembly of the fibronectin monomers in the dimeric mature form.

The **carboxy-terminal domain** contains the disulfides which cross-link the two fibronectin monomers together.

2.2. The fibronectin gene (fig. 2)

Partial primary structure data revealed highly conserved amino acid sequences both between the plasma and the cellular fibronectin and among fibronectins from different species : bovine, rat, chicken, human and xenopus fibronectins are highly homologuous (Petersen et al., 1983; Pande, 1981; Kornblihtt et al., 1983, 1984; Schwarzbauer et al., 1983).

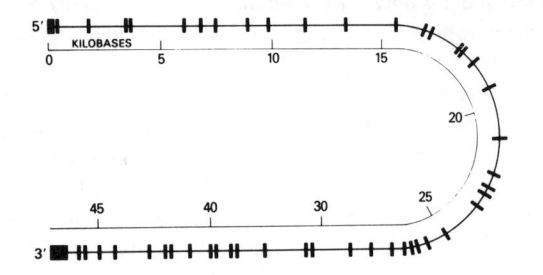

Fig. 2 : The fibronectin gene

The vertical bars indicate coding regions (exons), which are separated by intervening sequences (introns). The first and last exons at the 5' and 3' are larger, but the remaining exons are similar in size (see Hirano et al., 1983).

All these data confirm that the basic fibronectin polypeptide has three different types of internal repeats (I, II, III, respectively about 40, 60 and 90 amino acids long) (Skorstengaard et al., 1984; Petersen et al., 1983). This structural organization is reflected in the fibronectin gene, which has been isolated in chicken (Hirano et al., 1983). The entire gene spans 48 kb with a total exon length of 8 kb (fig. 2). The exons are of regular size (147 bp ±37) and are interrupted by introns of variable length (8 bp to 2648 bp). The length of the predicted peptides encoded by these exons matches fairly well with the length of the three internal repeats in the

fibronectin gene, which are apparently composed by either one or two of these units (Petersen et al., 1983). This structural organization suggests that the fibronectin gene has evolved by means of a massive series of duplications of conceivably three independent genes (Petersen et al., 1983), which have subsequently undergone evolutionary modification to the functional domains of present-day fibronectin.

The different monomers of plasma fibronectin and cellular fibronectin are encoded by one single gene (Akiyama and Yamada, 1985; Hirano et al., 1983; Kornblihtt et al., 1983; Yamada, 1985). The differences are produced by a mechanism of alternative splicing of the mRNA which is unprecedented (Hynes et al., 1984). The restricted difference regions of these mRNA's are located between the cell binding domain and the heparin binding domain II and between the heparin binding domain II and the fibrin binding domain II. At least ten different mRNA's can be generated by this process of alternative splicing (Kornblihtt et al., 1985). Interestingly the extra domain (ED) segment seems to be present in mRNA from fibroblasts, while it is absent in the liver hepatocyte mRNA's which are the source of plasma fibronectin (Kornblihtt et al., 1985). In rat fibronectin, the expression of a second RGD sequence appears also to be controlled by alternative mRNA splicing (Schwarzbauer et al., 1983) raising the possibility that some fibronectin molecules may have two distinct cell attachment sites.

Finally, in addition to this splicing-process, many posttranslational modifications are known to occur in fibronectin, so that fibronectins, although coded for by one single gene, are extremely versatile proteins, adapted to fulfill multiple tasks.

II. The fibronectin receptors : members of the Integrin family of cell surface receptor

1. Cellular receptors for fibronectin

As discussed above, adhesion of most cells to fibronectin in vitro, can be inhibited by Arg-Gly-Asp (RGD) containing peptides. Perturbation experiments in vivo of gastrulation and neural crest cell migration (cfr. supra) show the importance of this sequence in in vivo situations. Consequently much interest has gone to cellular receptors recognizing the RGD cell recognition signal (Ruoslahti and Piersbacher, 1986). On the other hand, evidence is accumulating that interactions of cells with fibronectin are more complex than their binding to this short recognition signal only. For instance, while fibroblasts indeed attach and spread on immobilized fibronectin through binding to the RGD sequence, an additional association with either one of the two heparin-binding domains in fibronectin is required for the subsequent organization of the intracellular actin into fibers terminating in focal adhesions (Woods et al., 1986). Membrane bound heparan sulfate proteoglycan may be responsible for the interaction with these fibronectin domains (Woods et al., 1984).

Another fibroblast-fibronectin interaction results in the polymerization of fibronectin into fibrils in the extracellular matrix and involves an unidentified "matrix assembly receptor" (Mc Keown-Longo, 1985). This receptor interacts with a region at the aminoterminal end of fibronectin, distinct from the cell attachment site.

A number of other surface molecules have been proposed to mediate the binding of cells to fibronectin, including glycopeptides (Huang, 1978), gangliosides (Kleinmann et al., 1979; Yamada, 1983) and glycoproteins in the 40-60 kDa molecular weight range (Aplin et al., 1981; Hughes et al., 1981; Oppenheimer-Marks et al., 1982, 1984; Urushihara and Yamada, 1986).

These possible receptors are less well characterized than the RGD receptors but are potentially important to explain the specificity in the interactions of cells with fibronectin.

2. The RGD-fibronectin receptor

Two main approaches have independently led to the identification of a complex of 140 kDa
glycoproteins as the cellular receptor for fibronectin. An immunological approach was based
on the assumption that antibodies raised against surface antigens only interfere with adhesion of cells to substrata if they react with molecules directly involved in the adhesive process
(see fig. 3). The breakthrough came when two groups identified a monoclonal antibody
(CSAT and JG22) which disrupts the adhesion of chicken cells to fibronectin (Neff et al., 1982;
Greve and Gottlieb, 1982).

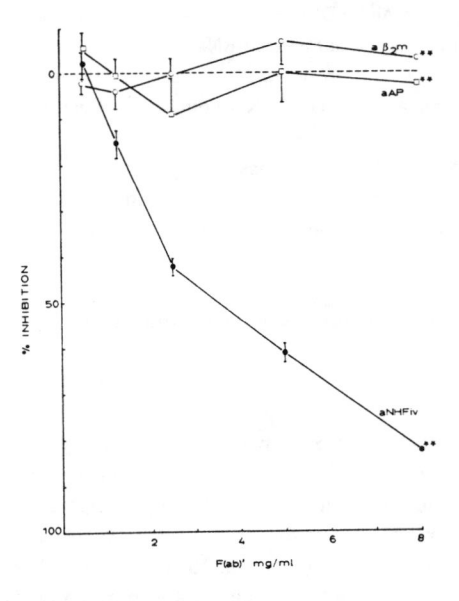

**Fig. 3 : Only antibodies which are directed against antigens directly involved in the adhesive process, are interfering
with cellular adhesion.**

Adhesion of fibroblasts to fibroblast cell layers was measured after 20 minutes in the presence of increasing concentrations of Fab' fragments prepared from different antibody preparations against surface antigens. aNHFiv is a polyclonal antiserum raised against fibroblasts. It has been shown that the inhibitory effect of this antiserum is partially
caused by antifibronectin antibodies and partially by antifibronectin receptor antibodies (Saison et al., 1983; Saison,
1986). aβ_2m and aAp are polyclonal antisera against respectively β_2-microglobulin and aminopeptidase M (Verlinden et al., 1981).

A similar monoclonal antibody, specific for BHK cells was identified by Brown and Juliano
(1985, 1986). The proteins recognized by these monoclonals are immunological related
(Brown and Juliano, 1986) but the functional effect of the monoclonal antibodies is species-
specific and no such antibody has been described for human cells. In human cells an alternative approach has been particularly successful. Pytela et al. (1985b, 1986) isolated the

fibronectin receptor of MG63 osteosarcoma cells on an affinity matrix, made of a 120 kDa fragment of fibronectin, containing the cell binding domain. Specific elution of the receptor was possible with a RGD-containing peptide. The same approach was subsequently used to isolate a whole series of extracellular matrix receptors including the vitronectin receptor (Pytela et al., 1985a), the collagen receptor (Dedhar et al., 1986) and the fibrinogen receptor of platelets (Pytela et al., 1986).

Although the human fibronectin receptor is clearly related to the avian receptor isolated with monoclonal antibody CSAT, some intriguing differences remain unexplained. For instance the avian receptor is a complex of three molecules, while the human receptor is clearly a heterodimer (Knudsen et al., 1985; Pytela et al., 1986) The avian receptor interacts with both laminin and fibronectin (Horwitz et al., 1985), while the human receptor is specific for fibronectin (Pytela et al., 1986). It is possible that the trimeric complex isolated with the CSAT monoclonal antibody in fact represents a mixture of two heterodimeric receptors, one for laminin and the other for fibronectin, which would be possible if these putative laminin and fibronectin receptors shared a common subunit recognized by the CSAT monoclonal antibody.

Formal pharmacological criteria for the fibronectin receptor status of the 140 kDa complexes in chicken and mammalian cells have not been met until now (Brown and Juliano, 1986). These criteria would include demonstration of specific ligand binding and correlation of receptor occupancy and biological function. Nevertheless, the evidence implicating the 140 kDa proteins as fibronectin receptors is impressive. First, there is the specific blocking of adhesion to fibronectin (and laminin) but not to other ligands by antibodies recognizing this complex (Decker et al., 1984; Horwitz et al., 1985; Brown and Juliano, 1986; Takada et al., 1987). Further, the complex binds to immobilized fibronectin (Pytela et al., 1986; Akiyama et al., 1986) and to fibronectin in solution (Horwitz et al., 1985; 1986). Third, after incorporation of the purified complex in liposomes, these liposomes specifically bind to fibronectin substrata (Pytela et al., 1986). Finally, immunolocalization studies both in vitro and in vivo suggest the involvement of the 140 kDa complex in cell-fibronectin interactions during cell adhesion and spreading and embryonic development (Chen et al., 1985, 1986; Damsky et al., 1985; Duband et al., 1986; Krotoski et al., 1986). The antibody CSAT interferes with neural crest cell migration (Bronner-Fraser, 1985).

Interaction of the fibronectin receptor with cytoskeletal components (talin) was shown in vitro (Horwitz et al., 1986), which established the RGD-receptor as the transmembrane link between fibronectin and the cytoskeleton.

3. Structural features of the fibronectin receptor

The α and the β subunit of the human fibronectin receptor have been cloned and sequenced (Argraves et al., 1986; Tamkun et al., 1986; Hynes, 1987). Both subunits are integral membrane proteins with a small cytoplasmic domain and with the aminoterminus in the extracellular space (fig. 4).

The β subunit, which has the lowest molecular weight, contains a remarkable four fold repeat of a cystein-rich motif. The function of these repeats is not known but their potential compact structure could explain at least the protease-insensitivity of this subunit which is lost after reduction of the disulfide bonds (Giancotti et al., 1985). The cytoplasmic domain contains two tyrosines residues, one of which is localized in a sequence homologous to the

48

autophosphorylation site of the epidermal growth factor receptor (Tamkun et al., 1986). This site is also the binding place for talin (cfr. supra). Hirst et al. (1986) have shown that the β subunit can indeed be phosphorylated on a tyrosine residue after viral transformation of cells and that this is accompanied by a reorganization of the receptor complex in the membrane of the transformed cells.

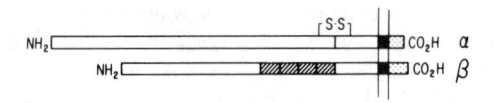

Fig. 4 : Structure of the fibronectin receptor

In the β subunit there are four repeats of a cysteine-rich motif (cross hatched area). In the α subunit the posttranslational cleavage site and the disulfide bridge are indicated (see text for more details).

The α chain is posttranslationally cleaved, and has a small ($\pm$ 20 kDa) and a large ($\pm$ 150 kDa) subunit, which are linked by a disulfide bridge (Argraves et al., 1986) The small subunit contains the transmembrane domain and is the membrane anchoring region of the molecule. In the extracellular domain there appears to be also a repeated structural motif with homology to the Ca^{++}-binding domain of calmodulin.

4. The Integrin family

The structural characteristics of the fibronectin receptor place it in a broad family of adhesive glycoproteins (see Table 2), the so called Integrin Family (Tamkun et al., 1986; Hynes, 1987). All the members of this family are heterodimers and function in cell migration during embryogenesis, in thrombosis, in lymphocyte help and killing, and in other adhesive processes. Deficiencies lead to severe immunological (Kishimoto et al., 1987a) or bleeding disorders (Glanzmann disease, see George et al., 1984).

There is an extensive homology between the different Integrins, especially for the β subunits. Biochemical, immunological and sequencing data have led to evidence for only three different β subunits (β_1, β_2, β_3) which can combine with at least ten different α subunits (see Table 2). All three β subunits have a similar overall structure and are about $\pm$ 45% homologous at the amino acid level. All fifty-six cysteine residues, including the four cysteine-rich motifs are conserved. The tyrosine in the homology region with the epidermal growth factor receptor autophosphorylation site is conserved in β_1 and β_3 (Fitzgerald, 1987) but not in β_2 (Kishimoto et al., 1987b; Law et al., 1987) The β_1 and β_2 subunits were localized in the human genome (Young et al., 1987; Marlin et al., 1986; Akao et al., 1987) respectively on chromosome 10 and 21.

The α subunits are also homologous to each other, but to a lesser extent. The current thought is that the β subunits fulfil an, as yet unidentified, common function, for instance transmission of binding signals to the cytoskeleton or regulation of expression of the Integrins. The α subunits would be responsible for the selectivity of the ligand binding. This is, however, not yet established.

TABLE 2 : MEMBERS OF THE INTEGRIN RECEPTOR FAMILY

β subunit	group	receptor	probable subunit composition	known ligands	functions and pathology
β_1	VLA-family	fibronectin receptor	$\alpha_F\beta_1$	FN ,	adhesion to fibronectin, cell migration, cytoskeletal interaction
		chicken fibronectin receptor	$\alpha_0\beta_1$	FN,L	
		VLA-1	$\alpha_1\beta_1$	-	-
		VLA-2	$\alpha_2\beta_1$	-	-
		VLA-3	$\alpha_3\beta_1$	-	-
		VLA-4	$\alpha_4\beta_1$	-	-
		VLA-5	$\alpha_5\beta_1(\alpha_F\beta_1?)$	-	- .
β_2	LFA-family	LFA-1	$\alpha_L\beta_2$	ICAM-1	leucocyte adhesion, T lymphocyte help, lymphocyte cytotoxicity
		Mac-1	$\alpha_M\beta_2$	C3bi	C3b receptor, monocyte and neutrophil adhesion
		p150-95	$\alpha_X\beta_2$	C3bi	neutrophil adhesion (LAD-deficiency)
β_3	cytoadhesins	glycoprotein IIb/IIIa	α_{IIb}/β_3	FN,FB,VN,VWF (TSP, collagen?)	platelet adhesion and aggregation (Glanzmann disease)
		vitronectin receptor	α_{VN}/β_3	VN	adhesion to vitronectin
		others ?			

The Integrin family

The different integrins are classified according to their β subunits. Their probable subunit composition is indicated as in Hynes (1987). The α subunits are denoted by the first letter of their ligands, when their ligands are well known. The others are denoted by the first letter of the original cell type (α_L : leucocyte, α_M : macrophage), or, where no simple designation exists, by the numbers or letters used by the authors, who first described them.

FN : fibronectin, LM : laminin, VN : vitronectin, FB : fibrinogen, VWF : Von Williebrands factor, TSP : Trombospondin, C3bi : inactivated form of C3b, a component of complement, VLA : very late antigens. For more details : see text and Hynes (1987).

50

III. Conclusion

The interactions of cells with fibronectin are extremely complex and versatile, both at the level of the fibronectin molecule in the extracellular matrix as well as at the level of the receptors in the membrane. Recent evidence even suggests the existence of tissue specific RGD-fibronectin receptors, as important differences appear to exist between the receptors of different cells both in affinity (Johansson et al., 1987) and in expression (Cardarelli et al., 1987). The picture becomes even more complex as cells seem to contain more than one RGD-fibronectin receptor at a time, and as these receptors appear to bind one, two or more different ligands (Giancotti et al., 1987; Pytela et al., 1986; Hemler et al., 1987; Takada et al., 1987).

The fibronectin/fibronectin receptor system is very conserved in evolution as demonstrated by studies in Drosophila (Naidet et al., 1987; Leptin, 1987). Fibronectin itself is a very ancient molecule, present in very primitive multicellular organisms such as the sponge Microciona prolifera (Akiyama and Johnson, 1983)

All these lead us to the conclusion that cell-fibronectin interactions are of fundamental biological importance. Insight in these interactions will provide clues to a better understanding of cellular organization in multicellular organisms and may eventually lead to the treatment of cancer, a disease of cellular disorganization.

Literature cited

Adams, S.L., Sobel, M.E., Howard, B.H., Olden, K., Yamada, K.M., de Crombrugge, B., and Pastan I. Proc. Natl. Acad. Sci. USA 74, 3399-3403, 1977

Akao, Y., Utsumi, K.R., Naito, K., Ueda, R., Takahashi, T., Yamada, K. Somat. Cell Mol. Genet. 13, 273-278, 1987

Akiyama, S.K., and Johnson, M.D. Comp. Biochem. Physiol. 76B, 687-694, 1983

Akiyama, S.K., and Yamada, K.M. J. Cell Biochem. 27, 97-107, 1985

Akiyama, S.K., Yamada, S.S., and Yamada, K.M. J. Cell Biol. 102, 442-448, 1986

Aplin, J.D., Hughes, R.C., Jaffe, C.L., and Sharon, N. Exp. Cell. Res. 134, 488-494, 1981

Argraves, W.S., Pytela, R., Suzuki, S., Millian, J.L., Piersbacher, M.D., Ruoslahti, E. J. Biol. Chem. 261, 12922-12924, 1986

Armant, D.R., Kaplan, H.A., Mover, H., and Lennarz, W.J. Proc. Natl. Acad. Sci. USA 83, 6751-6755, 1986

Asch, B.B., Komet, B.R., and Burstein, N.A. Cancer Res. 41, 2115-2125, 1981

Bernard, B.A., Yamada, K.M., and Olden, K. J. Biol. Chem. 257, 8549-8554, 1982

Bevilaqua, M., Mosesson, M.W., and Bianco, C. Clin. Res. 28, 340, 1980 (abstr.)

Boucaut, J.C., and Darribere, T. Cell. Diff. 12, 77-83, 1983

Boucaut, J.C., Darribere, T., Poole, T.J., Aoyama, H., Yamada, K.M., and Thiery, J.P. J. Cell. Biol. 99, 1822-1830, 1984a

Boucaut, J.C., Darribere, T., Boulekbache, H., and Thiery, J.P. Nature 307, 364-367, 1984b

Bowersox, J.C., and Sorgente, N. Cancer Res. 42, 2547-2551, 1982

Bronner-Fraser, M. J. Cell Biol. 101, 610-617, 1985

Brown, P.J., and Juliano, R.L. Science 228, 1448-1451, 1985

Brown, P.J., and Juliano, R.L. J. Cell Biol. 103, 1595-1603, 1986

Cardarelli, P.M., and Piersbacher, M.D. J. Cell Biol. 105, 499-506, 1987

Chen, A.B., Mosesson, M.W., and Solish G.I. Am. J. Obstet. Gynecol. 125, 958-961, 1978

Chen, W.T., Hasegawa, E., Hasegawa, T., Weinstock, C., and Yamada, K.M. J. Cell. Biol. 100, 1103-1114, 1985

Chen, W.T., Weng, J., Hasegawa, T., Yamada, S.S., and Yamada, K.M. J. Cell Biol. 103, 1649-1661, 1986

Chiquet, M., Puri, E.G., and Turner, D.C. J. Biol. Chem. 254, 5475-5482, 1979

Couchman, J.R., Gibson, W.T., Thorn, D., Weaver, A.C., Rees, D.A., and Parish, W.E. Arch. Dermatol. Res. 266, 295-310, 1979

Czop, J.K., Kadish, J.L, Zeph, D.M., and Austen, K.F. Immunology 54, 407-417, 1985

Damsky, C., Knudsen, K.A., Bradley, D., Buck, C.A., and Horwitz, A.F. J. Cell Biol. 100, 1528-1539, 1985

Decker, C., Greggs, R., Duggan, K., Stubbs, J., and Horwitz, A. J. Cell. Biol. 99, 1398-1404, 1984

Dedhar, S., Ruoslahti, E., Piersbacher, M.D. J. Cell Biology 104, 585-593, 1987

Dixit, V.M., Haverstick, D.H., O'Rourke, K., Hennessy, S.W., Brackelmann, T.J., Mc Donald, J.A., Grant, G.A., Santoro, S.A., and Frazier, W.A. Proc. Natl. Acad. Sci. USA 82, 3844-3848, 1985

Duband, J.L., and Thiery, J.P. Dev. Biol. 94, 337-350, 1982

Duband, J.L., Rocher, S., Chen, W.T., Yamada, K.M. and Thiery, J.P. J. Cell Biol. 102, 160-178, 1986

Engel, J., Odermatt, E., Engel, A., Madri, J., Furthmayr, H., Rohde, H.H, and Timpl, R. J. Mol. Biol. 150, 97-120, 1981

Erickson, C.A., Tosney, K.W., and Weston, J.A. Dev. Biol. 77, 142-156, 1980

Erickson, H.P., Carrell, N., and McDonagh, J. J. Cell. Biol. 91, 673-678, 1981

Fitzgerald, L.A., Steiner, B., Rall, S.C.Jr., Lo, S.S., and Phillips, D.R. J. Biol. Chem. 262, 3936-3939, 1987

Fleischmajer, R., and Timpl, R. J. Histochem. Cytochem. 32, 315-321, 1984

Flock, J.I., Fröman, G. Jönsson, K., Guss, B., Signäs, C., Nilson, B. Raucci, G., Höök, M., Wadström, T., and Lundberg, M. EMBO 6, 2351-2357, 1987

Fröman, G., Switalski, L.M., Speziale, P. Höök, M. J. Biol. Chem. 262, 6564-6571, 1987

Fuyikawa, L.S., Foster, C.S., Harrist, T.J., Laningan, J.M., and Colvin, R.B Lab. Invest. 45, 120-129, 1981

Fyrand, O. Br. J. Dermatol. 101, 263-270, 1979

Gabius, H.-J., Springer, W.R., and Barondes, S.H. Cell 42, 449-456, 1985

Gahnberg, C.G., Kiehn, D., and Hakomori, S.I. Nature 248, 413-415, 1974

Gartner, T.K., and Bennett, J.S. J. Biol. Chem. 260, 11891-11894, 1985

George, N.J., Nurden, A.T., Phillips, D.R. N. Eng. J. Med. 311, 1084-1095, 1984

Giancotti, F.G., Tarone, G., Knudsen, K., Damsky, C., and Comoglio, P.M. Exp. Cell. Res. 156, 182-190, 1985

Giancotti, F.G., Languino, L.R., Zanetti, A., Peri, G., Tarone, G., and Dejano, E. Blood 69, 1535-1538, 1987

Gold, L.I., Garcia-Prado, A., Frangione, B., Franklin, E.C., and Pearlstein, E. Proc. Natl. Acad. Sci. USA 76, 4803-4807, 1979

Greve, J.M., and Gottlieb, D.I. J. Cell. Bioch. 18, 221-229, 1982

Grinnell, F., Feld, M., and Minter, D. Cell 192, 517-525, 1980

52

Grinnell, F., Billingham, R.E., and Burgess, L. J. Invest. Dermatol. 76, 181-189, 1981

Grinnell., F. J Cell. Biochem. 26, 107-116, 1984

Hahn, L.H.E., and Yamada, K.M. Proc. Natl. Acad. Sci. USA 76, 1160-1163, 1979a

Hahn, L.H.E., and Yamada, K.M. Cell 18, 1043-1051, 1979b

Hayashi, M. and Yamada, K.M. J. Biol. Chem. 256, 11292-11300, 1981

Hayashi, M. and Yamada, K.M. J. Biol. Chem. 257, 5263-5267, 1982

Hayashi, M., and Yamada, K.M. J. Biol. Chem. 258, 3332-3340, 1983

Hedman, K., Vaheri, A., Wartiovaara, J. J. Cell. Biol. 76, 748-760, 1978

Hemler, M.E., Huang, C., and Schwartz, L. J. Biol. Chem. 262, 3300-3309, 1987

Hirano, H., Yamada, Y., Sullivan, M., De Crombrugghe, B., Pastan, I., and Yamada, K.M. Proc. Natl. Acad. Sci. USA 80, 46-50, 1983

Hirst, R., Horwitz, A., Buck, C., and Rohrschneider, L. Proc. Natl. Acad. Sci., USA 83, 6470-6474, 1986

Horwitz, A., Duggan, K., Buck, C., Beckerle, M.C., Burridge, K. Nature 320, 531-533, 1986

Horwitz, A., Duggan, K., Greggs, R., Decker, C., and Buck, C. J. Cell. Biol. 101, 2134-2144, 1985

Huang, R.T.C. Nature 276, 624-626, 1978

Hughes, R.C., Butters, T.D., and Aplin, J.D. Eur. J. Cell. Biol. 26, 198-207, 1981

Humphries, M.J., Olden, K., and Yamada, K.M. Science 233, 467-470, 1986

Hynes, R.O., and Wyke, J.A. Virology 64,492-504, 1975

Hynes, R.O., Schwarzbauer, J.E., and Tamkun, J.W. Pitman, London (Ciba Foundation Symposium 108) p. 75-92, 1984

Hynes, R.O. Cell 48, 549-554, 1987

Johansson, S., and Höök, M. J. Cell. Biol. 98, 810-817, 1984

Johansson, S., Forsberg, E., and Lundgren, B. J. Biol. Chem. 262, 7819-7824, 1987

Keski-Oja, J., and Yamada, K.M. Biochem J. 193, 615-620, 1981

Kleinman, H.K., Martin, G.R., and Fishman, P.H. Proc. Natl. Acad. Sci. USA 76, 3367-3371, 1979

Kishimoto, T.K., Hollander, N., Roberts, T.H., Anderson, D.C., and Springer, T.A. Cell 50, 193-202, 1987a

Kishimoto, T.K., O'Connor, K., Lee, A., Roberts, T.M., Springer, T.A. Cell 48, 681-690, 1987b

Knudsen, K.A., Horwitz, A.F., and Buck, C.A. Exp. Cell. Res. 157, 218-226, 1985

Kornblihtt, A.R., Vibe-Pedersen, K., and Baralle, F.E. Proc. Natl. Acad. Sci. USA 80, 3218-3222, 1983

Kornblihtt, A.R., Vibe-Pedersen, K., and Baralle, F.E. EMBO J. 3, 221-226, 1984

Kornblihtt, A.R., Umezawa, K., Vibe-Pedersen, K. and Baralle, F.E. EMBO J. 4, 1755-1759, 1985

Kramer, R.H., Gonzalez, R., and Nicolson, G.I. Int. J. Cancer 26, 639-645, 1980

Krotoski, D.M., Domingo, C., Bronner-Fraser, M. J. Cell Biol. 103, 1061-1071, 1986

Labat-Robert, J., Birembaut, P., Robert, L., and Adnet, J.J. Diagn. Histopatol. 4, 299-306, 1981

Law, S.K.A., Gagnon, J., Hildreth, J.E.K., Wells, C.E., Willis, A.C., and Wong, A.J. EMBO J. 6, 915-919, 1987

Leptin, M., Aebersold, R., and Wilcox, M. EMBO 6, 1037-1043, 1987

Loring, J., Glimelius, B., and Weston, J.A. Dev. Biol. 90, 165-174, 1982.

Marlin, S.D., Morton, C.C., Anderson, D.G., and Springer, T.A. J. Exp. Med. 164, 855-867, 1986

McDonald, J.A., Kelley, D.G., and Brockelmann, T.J. J. Cell. Biol. 92, 485-492, 1982

McKeown-Longo, P.J., and Mosher, D.F. J. Cell Biol. 100, 364-374, 1985

Mosesson, M.W., and Umfleet, R.A. J. Biol. Chem. 245, 5728-5736, 1970

Mosher, D.F. Prog. Hemostasis Thromb. 5, 111-151, 1980

Mosher, D.F., Schad, P.E., and Kleinman, H.K. J. Clin. Invest. 64, 781-787, 1979

Neff, N.T., Lowrey, C., Decker, C., Tovar, A., Damsky, C., Buck, C., and Horwitz, A. J. Cell. Biol. 95, 654-666, 1982

Naidet, C., Sémériva, M., Yamada, K.M., and Thiery, J.P. Nature 325, 348-350, 1987.

Norris, D.A., Clark, R.A.F., Swigart, L.M., Huff, J.C., Weston, W.L., an Howell, S.E. J. Immunol. 129, 1612-1618, 1982

Odermatt, E., Engel, J., Richter, H., and Hörmann, H. J. Mol. Biol. 159, 109-123, 1982

Oldberg, A. Franzén, A. Heinegaard, D. Proc. Natl. Acad. Sci USA 83, 8819-8823, 1986

Oppenheimer-Marks, N., and Grinnell, F. J. Cell. Biol. 95, 876-884, 1982

Oppenheimer-Marks, N., and Grinnell, F. Exp. Cell Res. 152, 467-475, 1984

Pande, H., Corkill, J., Sailor, R., and Shively, J.E. Biophys. Res. Commun. 101, 265-272, 1981

Pearlstein, E., Gold, L.I., and Garcia-Pardo, A. Mol. Cell. Biochem. 29, 103-128, 1980

Pennypacker, J.P., Hessell, J.R., Yamada, K.M. and Pratt, R.M. Exp. Cell Res. 121, 411-415, 1979

Petersen, T.E., Thøgersen, H.C., Skorstengaard, K., Vibe-Pedersen, K., Sahl, P., Sottrup-Jensen, L., and Magnusson, S. Proc. Natl. Acad. Sci. USA 80, 137-141, 1983

Pierschbacher, M.D., Hayman, E.G., and Ruoslahti, E. Cell 26, 259-267, 1981

Pierschbacher, M.D., and Ruoslahti, E. Nature 309, 30-33, 1984a

Pierschbacher, M.D., and Ruoslahti, E. Proc. Natl. Acad. Sci. USA 81, 5985-5988, 1984b

Pierschbacher, M.D., Ruoslahti, E., Sundelin, J., Lind, P., and Peterson, P.A. J. Biol. Chem. 257, 9593-9597, 1982

Plow, E.F., Piersbacher, M.D., Ruoslahti, E., Marguerie, G.A., and Ginsberg, M.H. Proc. Natl. Acad. Sci. USA 82, 8057-8061, 1985

Podleski, T.R., Greenberg, I., Schlessinger, J., and Yamada, K.M. Exp. Cell Res. 123, 104-126, 1979.

Price, T.M., Rudee, M.L., Pierschbacher, M., and Ruoslahti, E. Eur. J. Biochem. 129, 359-363, 1982

Pytela, R., Pierschbacher, M.D., and Ruoslahti, E. Proc. Natl. Acad. Sci. USA 82, 5766-5770, 1985a

Pytela, R., Pierschbacher, M.D., and Ruoslahti, E. Cell 40, 191-198, 1985b

Pytela, R., Piersbacher, M.D., Ginsberg, M.H., Plow, E.F., and Ruoslahti, E. Science 231, 1559-1562, 1986

Richter, H., and Hörmann, H. Hoppe-Seyler's Z. Physiol. Chem. 363, 351-364, 1982.

Rieber, M., and Romano, E. Cancer Res. 36, 3568-3573, 1976

Rocco, M., Aresu, O., and Zardi, L. FEBS Letters 178, 327-330, 1984

Rovasio, R.A., Delouvee, A., Yamada, K.M., Timpl, R., and Thiery, J.P. J. Cell Biol. 96, 462-473, 1983

Ruoslahti, E. Cancer Metastasis Reviews 3, 43-51, 1984

Ruoslahti, E., Engvall, E., and Hayman, E.G. Coll. Res. 1, 95-128, 1981

Ruoslahti, E., Hayman, E.G., Kuusela, P., Shively, J.E., and Engvall, E. J. Biol. Chem. 254, 6054-6059, 1979

Ruoslahti, E., and Piersbacher, M.D. Cell 44, 517-518, 1986

54

Saison, M., Van Leuven, F., Cassiman, J.J., and Van den Berghe, H. Exp. Cell Res. 143, 237-245, 1983

Saison, M. Adhesion of human skin fibroblasts, isolation and characterization of a plasma membrane protein, involved in fibroblast adhesion to fibronectin. Ph. D. Thesis, University of Leuven, 1986

Santoro, S.A., and Lawing, W.J. Jr. Cell 48, 867-873, 1987

Schwarzbauer, J.E., Tamkun, J.W., Lemischka, I.R., and Hynes, R.O. Cell 35, 421-431, 1983

Sekiguchi, K., Fukuda, M., and Hakomori, S.I. J. Biol. Chem. 256, 6452-6462, 1981

Sekiguchi, K., and Hakomori, S.I. Proc. Natl. Acad. Sci. USA 77, 2661-2665, 1980

Sekiguchi, K., and Hakomori, S.I. J. Biol. Chem. 258, 3967-3973, 1983

Sekiguchi, K., Hakomori, S.I., Funahashi, M., Masumoto, I., and Seno, N. J. Biol. Chem. 258, 14359-14365, 1983

Sieber-Blum, M, Sieber, F., and Yamada, K.M. Exp. Cell Res. 133, 285-295, 1981

Skorstengaard, K., Thørgensen, H.C., and Petersen, T.E. Eur. J. Biochem. 140, 235-243, 1984

Smith, H.S., Riggs, J.L., Mosesson, M.W. Cancer Res. 39, 4138-4144, 1979

Spiegelmann, B.M., and Ginty, C.A. Cell 35, 657-666, 1983

Stanley, J.R., Alvarez, O.M., Bere. E.W. Jr., Eaglstein, W.H., and Katz, S.I. J. Invest. Dermatol. 77, 240-243, 1981

Stenman, S., and Vaheri, A. Int. J. Cancer 27, 427-435, 1981

Stenn, K.S., Madri, J.A., Tinghitella, T., and Terranova, V.P. J. Cell. Biol. 96, 63-67, 1983

Stone, K.R., Smith, R.E., and Joklik, W.K. Virol. 58, 86-100, 1974

Suzuki, S., Oldberg, A., Hayman, E.G., Pierschbacher, M.D., and Ruoslahti, E. Embo J. 4, 2519-2524, 1985

Takada, Y., Huang, C., and Hemler, M.E. Nature 326, 607-609, 1987

Tamkun, J.W., De Simone, D.W., Fonda, D., Patel, R.S., Buck, C., Horwitz, A.F., and Hynes, R.O. Cell 46, 271-282, 1986

Thiery, J.P., Duband, J.L., Tucker, G., Darribere, T., and Boucaut, J-C Matrices and cell differentiation. Alan R. Liss, Inc., 150th Ave., New York. 187-198, 1984

Urushihara, H., and Yamada, K.M. J. Cell. Physiol. 126, 323-332, 1986

Vaheri, A., and Ruoslahti, E. Int. J. Cancer 13, 579-586, 1974

Vercellotti, G.M., Mc Carthy, J., Furcht, L.T., Jacob H.S., and Moldow, C.F. Blood 62, 1063-1069, 1983

Verlinden, J., Van Leuven, F., Cassiman, J.J., and Van den Berghe, H. FEBS Letters 123, 287-290, 1981

Vlodavsky, I., Ariav, Y., Atzom, R., and Fuks, Z. Exp. Cell Res. 140, 149-159, 1982

Wall, R.S., Cooper, S.L., and Kosck, J.C. Exp. Cell Res 140, 105-111, 1982

West, C.M., Lanza, R., Rosenbloom, J., Lowe, M., Holtzer, H., and Avdalovic, N. Cell 17, 491-501, 1979

Woods, A., Höök, M., Kjellén, L., Smith, C.G., and Rees, D.A. J. Cell Biol. 99, 1743-1753, 1984

Woods, A., Couchmann, J. , Johansson, S., and Höök, M. EMBO J. 5, 665-670, 1986.

Yamada, K.M. J. Cell. Biol. 78, 520-541, 1978

Yamada, K.M. Ann. Rev. Biochem. 52, 761-799, 1983

Yamada, K.M., Kennedy, D.W., Kimata, K., and Pratt, R.M. J. Biol. Chem. 255, 6055-6063, 1980

Yamada, K.M., Yamada, S.S., and Pastan, I. Proc. Natl. Acad. Sci. USA 73, 1217-1221, 1976

Yamada, K.M., and Kennedy, D.W. J. Cell Biol. 99, 29-36, 1984

Yamada, K.M., Humphries, M.Y., Hasegawa, T., Hasegawa, E., Olden, K., Chen, W.T., and Akiyama, S.K. In "The cell in contact", 303-332, A Neuroscience Institute Publication, John Wiley and Sons, New York, 1985, Editors Edelman, G.M., and Thiery, J.P.

Zhang, Y., Saison, M., Spaepen, M., De Strooper, B., Van Leuven, F., David, G., Van den Berghe, H., and Cassiman, J.J. Somat. Cell Mol. Genet., accepted for publication

De Strooper, B. : <u>Cell-fibronectin interactions</u>

Foidart :   You made an analogy between a receptor for fibronectin and
            for EGF concerning the phosphorylation of the tyrosine
            residues. Do you have any idea of the fate of the
            receptor once it is phosphorylated ?

De Strooper: No, there are not many studies being done. As far as I
            know there is only one report that I know of. We have
            done <u>in vitro</u> studies of the phosphorylation of the
            chicken fibronectin receptor in chicken fibroblasts which
            were transformed by several viruses and they show that it
            is the low molecular weight subunit which is
            phosphorylated. There is also some phosphorylation of one
            of the two alpha subunits. Phosphorylation happens in a
            domain which is also interacting with tallin on the
            cytoskeleton. It has been shown <u>in vitro</u>, that the
            chicken fibronectin receptor can interact with the
            cytoskeleton and maybe phosphorylation is important in
            inhibiting this interaction. But what is the fate of the
            receptor after phosphorylation, I do not know.

Remacle :   The fibronectin receptor is able to react with other
            components of the cytoskeleton than actin ?

De Strooper : The receptor interacts with tallin and this tallin
            interacts then with vinculin. The following steps are not
            clear.

Cassiman : I am surprised that the people who are working with
            viruses and who are looking for receptors are not reacting
            now.

De Strooper : There are RGD (Arg-Gly-Asp) sequences present in several
Foot and Mouth disease viruses but it is not clear about
their functional importance.

Cassiman : There is one such sequence in EBV.

Spier : What is the probability of any random protein having an
RGD sequence ?

De Strooper : Very high, very high indeed. Something like one to
twenty. There are lots of examples where RGD sequences are
important in vitro and in vivo and so it is surely a
cellular recognition signal. For instance, for collagen,
its presence has been shown on the bends, at specific
amino acid residues, so that there must be specific
tertiary conformation around this sequence. As a result
you can elute collagen receptors with such peptides. You
can do that with the RGD sequence, RGD-containing
sequences which were used to elute fibronectin.
So, it appears that there is the RGD sequence as the
recognition system but the tertiary conformation which is
given by the other amino acids around it, are important
for the final recognition as well and this explains also
why you need such high concentrations of peptides for
inhibition assays : only a fraction of the total amount of
peptides you add in your assay has a good conformation to
compete with the recognition of fibronectin and it has
been shown in addition that longer peptides are more
biologically effective in such in vitro assays.

Spier : Has anybody done any modelling of the amino acid sequences
that you are alluding to in order to ascertain the lowest
energy three-dimensional configuration which is likely to
be involved ?

De Strooper : I think this is going on. YAMADA has already published some results on the different possibilities and the different constrains on these peptides.

Spier : Can I add in that Richard LERNER at the Scripps Institute has been playing around with polypeptide antigens and he has learned to substitute the hydrogen bond with nitrogen-nitrogen bonds and in that way he rigidizes what would have otherwise been a very floppy structure. So, I think you know possibly, once you have actually decided what the three dimensional configuration is, there is a possibility of actually holding it in that particular form and that may get around the need to have such large concentrations and avoiding the effects you are talking about.

De Strooper : Yes.

Foidart : I was attending an EMBO workshop last week in Santa Margarita in Italy attended by E. RUOSHLATI who issued the state of the art. You really need dipeptide rather than a tripeptide and this is the minimal requirement that you have but then, in order to obtain a higher affinity, what they have done is to cyclisize the tetrapeptide and they can get up to a cyclic structure. At that stage you get a much more stable peptide which has a higher affinity and a better stability.
Now, if I may add a comment, I think that we have to look at the interactions between the extracellular matrix proteins and the plasma membrane the same way as we look at the interactions between antibody and antigens. What I mean by that is that all these interactions between large molecular weight proteins and the receptors that we just heard about, the integrin family receptors, are based on molecules which are closely related together, the type you

just described. They have alpha and beta subunits and as
you said, have many analogues in their amino acid sequences.
But everything appears like in the F_c region of the
antibody, all antibodies have an F_c region which is common
to all antibodies and what makes the specificity of the
antigen-antibody recognition is the F_{ab} fragment. It is
what probably happens in the cell matrix interaction. It
is probable that like the F_c region, the RGD fragment is
what causes the matrix to recognize the specific cell
receptor. What causes the specificity of the cell
receptor for fibronectin or laminin appears to be a
different amino acid sequence on each macromolecule and
very recently YAMADA and RUOSHLATI have isolated it. The
results are not yet published. I heard last week that
they isolated a twenty amino acid sequence on the fibrin
II region of the fibronectin molecule which is responsible
for the specificity of the fibronectin receptor. So, we
have really to look at cell-matrix interactions like we
look at the F_{ab}-F_c regions in antibody-antigens reactions.

De Strooper : There are some recent reports also, indicating that for
instance on hepatocytes, the receptor is very similar to
the receptor present on fibroblasts but has some differences in molecular weight and has a much higher affinity
for fibronectin than for instance the receptor on fibroblasts. I think that is really the image which is coming
up.

Foidart :   If I may add a final comment concerning laminin, there is
another peptide which is YIGSR. It has the main acid
sequence of the fibrin I-IV of laminin which is responsible for binding to laminin receptor which is completely
different from the fibronectin receptor.

De Strooper : Maybe, I should add also... In fibrinogen there is in the gamma chain also another peptide which is also recognized by IIB3E, the fibrinogen receptor and which is completely different of the RGD and it has been shown that there is competition between the RGD sequence and this peptide. But I did not want to go in details in these interactions because they are very complex.

Spier : I was fascinated by the first slide that you showed where you take a tumorigenic cell with a basically "demoralized" actin cytoskeleton, you add fibronectin to it and it becomes beautiful, looking just like you would wish. Can't this discovery be exploited in some way ?

De Strooper : These observations have been made between, seventy six and seventy seven and the question raised by them was : what was the connection between fibronectin and the cytoskeleton. During the following ten years, researches have sought for that receptor and now we can understand what is happening there and maybe then you can really manipulate this finding and use it to correct malforma-tion. The results presented were obtained in vitro. These are virally transformed cells and in humans most tumours are not virally induced. Also the phenotype of the cells shown above appears normal but they still continue growing and dividing like malignant cells. So it is not completely the transformed phenotype which is changed but it is merely an effect on the cytoskeleton itself and that is why I have also given the example of the phosphorylation because I think it is the phosphorylation which occurs in these transformed cells- in these virus transformed cells- which inhibits the interactions of fibronectin receptor with the cytoskeleton. Because of that, you loose fibro-nectin from the surface of these cells. If you add enough

fibronectin you can maybe immobilize these fibronectin receptors - not all the receptors will be phosphorylated - and you can maybe induce a normal cytoskeleton but this is speculation and may be too simple.

Part II.

"Normal" Versus "Abnormal" Cells

USE OF CELL LINES IN BIOTECHNOLOGY

JOHN C. PETRICCIANI
FORMER CHIEF OF BIOLOGICALS
WORLD HEALTH ORGANIZATION
GENEVA, SWITZERLAND

1. INTRODUCTION

The benefits and problems associated with the use of continuous cell lines (CCLs) for the production of various types of biological products have been discussed many times during the past few years (1, 2, 10). The purpose of this paper is to review past major events and more recent results which have led to the widening use of CCLs.

The acceptance of CCLs has been gradual; nevertheless, there has been a great deal of progress, so that at the present time most of the focus of attention is on the manufacturing process to show that it can eliminate and/or inactivate specific cellular contaminants of concern for a given product.

Discussions and decisions associated with the use of abnormal mammalian cells for the production of human biological products began in 1954 when the United States Armed Forces Board met to consider what cells would be an acceptable substrate to use for the production of an experimental live adenovirus vaccine (5). They basically had two choices: i) HeLa, a human cancer cell line; or ii) primary monkey kidney cells. They reasoned that there would be less risk in using a cell culture derived from the tissues of normal animals than if cells derived from a human tumor were used; and they therefore approved primary monkey kidney cell cultures. The significance of that decision cannot be overstated, because it essentially ruled out as unsafe the use of cells other than those derived from normal tissues of animals in good health.

Then in 1961 Hayflick and Morehead suggested that human diploid cells (HDCs) could serve as a useful and well-standardized cell system for the production of biological products. As reviewed by others (4, 5, 8), intense debate, disagreement, and dogmatism surrounded the question of safety of products derived from HDCs. Safety issues were raised and discussed which centered largely around the theoretical possibility that HDCs may contain a human oncogenic agent which would then contaminate vaccines produced in them. It is of some interest to note that real contaminants such as SV_{40}, B virus, and the Marburg agent found in primary monkey cells were considered of less significance by many than the theoretical agent that might be in HDCs. Little seemed to be resolved and the arguments, especially in the USA, continued to rage for years on the theoretical dangers of producing vaccines in HDCs. Eventually, however, vaccines derived from such cells were used in clinical trials, and ultimately they were licensed by national control authorities in many different countries. Experience showed that the theoretical risks had no basis in fact.

It was not until the mid-1970s that the next serious proposal to use other than primary or diploid cells was made when the Namalwa human lymphoblastoid cell line was proposed as a source of interferon (IFN) for

A. O. A. Miller (ed.), Advanced Research on Animal Cell Technology, 65–73.
© 1989 by Kluwer Academic Publishers.

human clinical studies (2). As in the case of the safety considerations of adenovirus vaccine in 1954 which could be produced in human cancer cells, IFN from malignant human lymphoid tissue posed serious questions of safety.

A conference was held at Lake Placid in 1978 to discuss the issues and to open a reconsideration of the entire area of cell substrates (3). A decade later it is easy to appreciate how important that IFN research and development effort was because it laid the groundwork for approaches which might be taken to ensure the safety of products derived from a variety of other CCLs including those now being used in biotechnology. Following the Lake Placid meeting, interest began to develop on the use of other CCLs for manufacturing other products.

2. BIOTECHNOLOGY

Recombinant DNA technology was developed shortly thereafter, and bacterial cells were selected as the first expression system. It quickly became evident, however, that bacterial cells were not able to synthesize adequate amounts of certain products such as hepatitis B surface antigen. In addition, the inability of bacterial cells to glycosolate was thought to be a potential drawback at least for some products in which glycosolation sites might be important for their proper biological functioning.

Because of the limitations of bacterial systems, manufacturers began to explore the use of yeast and several different mammalian cells as expression systems for the production of some new biologicals. Among major determinants in the choice of the cell substrate were the following: rapid growth, high level of expression, and the ability to grow to high cell density in suspension culture. Those requirements focused attention on CCLs. The development of hybridoma technology and the potential usefulness of monoclonal antibodies also led to heightened interest in the use of CCLs since all hybridomas fall into the general class of CCLs.

Once again, however, there was a great deal of concern relating to the safety of products derived from these various abnormal mammalian cells. A number of meetings and discussions tried to address the safety issues, and a substantial consensus began to grow that under certain conditions CCLs would be acceptable. In an effort to develop a more formal consensus on the use of CCLs, the United States Public Health Service (PHS) sponsored a workshop in July 1984 at the National Institutes of Health in Bethesda (1). The focus of attention at that meeting was on the risks that might be associated with the following cellular contaminants in a biological product: i) transforming proteins; ii) viruses; and iii) DNA. The outcome of that meeting was once again a clear agreement that CCLs were in principle acceptable as cell substrates.

3. WHO REVIEW OF THE ISSUES

Concerns remained, however, regarding the safety of products derived from CCLs, and they have impeded the more general acceptance of CCLs as a substrate for biologicals administered to humans. In response to that, WHO convened a Study Group in November 1986 in order to provide an independent international evaluation of the safety issues. The purpose of the Study Group was to undertake a general review of the acceptability of CCLs as substrates for biological products administered to humans, and to provide advice to WHO on certain specific points of concern, especially the long-term risk of malignancy represented by heterogenous

contaminating DNA, with an emphasis on potentially oncogenic or regulatory sequences.

There is already some evidence for the safety of certain CCLs as substrates for the production of biologicals. For example, inactivated foot and mouth disease vaccine has been prepared in the BHK-21 cell line, and it is estimated that more than 10^8 doses have been administered to cattle over a 20-year period. Carcass inspection has failed to discover any ill-effects attributable to the vaccine, suggesting that in the short term (2-4 years) this vaccine has been safe. Although human clinical experience with biologicals produced in CCLs has been more recent, and therefore more limited than with veterinary products, it also was noted that over 10^7 doses of inactivated polio vaccine produced in Vero cells already have been administered to children since 1983.

There was agreement that the major potential risks associated with the use of biologicals produced in CCLs fall into the same three categories identified in the 1984 PHS meeting: transforming proteins, viruses, and heterogeneous contaminating DNA.

The Study Group's report (12) concluded that, in general, CCLs are acceptable as substrates for the production of biological products, but that differences in the nature of the products derived from CCLs and the specifics of the manufacturing processes must be taken into account in making a decision on the acceptability of each given product. There is, therefore, no reason to exclude CCls from consideration as substrates for biological products. In this regard, there was agreement with the actions taken to date in approving the use of products derived from CCLs, after having been satisfied that a given manufacturing process yields a product with no detectable risk attributable to the cell substrate.

The importance of validating the efficiency with which various steps in a manufacturing process inactivate and/or eliminate unwanted material such as cellular DNA and viruses was emphasized. Validating the ability of a process to yield a product with certain specifications and to establish the consistency of that process are essential elements in providing the basis for an acceptable biological derived from CCLs. Once a process has been validated and consistency of production has been established, limited tests appropriate for each product should suffice, as has been the usual practice with biologicals in the past.

4. MAJOR ISSUES
4.1 Transforming Proteins

A number of proteins have now been identified which can induce the proliferation of many different types of cells, but most of them are active only intracellularly. Such proteins are encoded by cellular transforming genes or by an altered form of those genes.

Although transforming proteins have been a part of safety considerations, they have not loomed large as realistic problems. The major reason for the lesser importance attached to oncogene-encoded proteins is that only those which act extracellularly pose even a theoretical risk. That in turn means the consideration is restricted to growth factors which might be secreted by cells. But even then, several aspects concerning the known growth factors lead to the conclusion that they are trivial considerations with respect to safety of products derived from CCLs.

First of all, they do not appear to be oncogenic in and of themselves, and their growth promoting effects are transient and reversible. In addition, they do not replicate and many of them are rapidly inactivated

when administered in vivo. And finally, they are secreted in such small quantities that it would be necessary to concentrate them along with the product during the manufacturing process in order for them to begin to express any biological activity in vivo. In view of in vivo data (9) in which 0.3 ug/gm body weight of transforming growth factor alpha was required to induce a biological effect in highly susceptible mice, it is reasonable to assume that even mg amounts of a growth factor would be needed to begin to observe an effect in humans. Such large amounts of a contaminating protein are extraordinarily unlikely in biotechnology products, and the Study Group therefore did not consider that the presence of contaminating known growth factors in the concentration at which they are ordinarily to be found constitutes a serious risk in the preparation of biologicals. The risk was considered to be so remote that no testing was recommended.

4.2 Viruses

Viral contaminants from cells and tissues used to produce biologicals have been a problem for manufacturers, the medical profession, and the recipients of biologicals ever since the early days of vaccine production in cell cultures, and viral agents continue to be discovered. It is interesting to note, however, that in each case where a virus of potential pathogenicity for humans was discovered (e.g. SV_{40}), it was primary cell cultures or primary tissues in which the viruses were found. In contrast, no endogenous virus has ever been isolated from any of the HDCs used for vaccine production. This fact underscores the point that cells drawn from a well characterized cell bank offer great advantages because with careful studies it should be possible to establish whether or not a viral agent is present in the cells, and if so whether or not it carries any risk for humans. In this regard, potential viral contaminants derived from human and nonhuman primates clearly pose the greatest risk, while avian viruses are probably the least worrisome.

In considering the potential for infecting man with viruses, the Study Group agreed that cells may be divided into three risk categories with respect to their potential for carrying viral agents pathogenic for man:

High risk: Any material derived from either human or primate blood or bone marrow cells, caprine or ovine cells, or hybridomas when at least one fusion partner is of human or nonhuman primate origin.

Medium risk: Any material derived from mammalian non-hematogenous cells such as fibroblasts or epithelial cells.

Low risk: Any material derived from either human diploid lines or avian tissues.

Taking into account the above classification, the Study Group agreed that different degrees of concern, and therefore testing, were appropriate for products manufactured from the various types of cells mentioned. Nevertheless, it was emphasized that when either diploid or continuous cell lines are used for production, a cell seed lot system must be used and the cell seed must be characterized as already provided in WHO Requirements.

The Study Group stressed the importance of validating the ability of a manufacturing process to eliminate and to inactivate those viruses which may pose a risk to humans when cells carrying viruses are proposed for use in the manufacture of human biologicals. In addition, cells from humans and animals with diseases of unknown origin should not be used, nor should animal cells which may contain "slow viruses" be used to produced biologicals for human administration.

4.3 DNA

In evaluating the theoretical risk of cellular DNA in products administered to humans, it is important to keep in mind the fact that normal cellular DNA has already been accepted as a contaminant in the sense that all products produced from cell culture systems up to the present have to some degree contained DNA, and it has neither frightened anyone nor has it apparently caused any adverse health effects. Concern about DNA from CCLs should therefore concentrate on those portions of the genome of CCLs which are different from the genome of normal cells. Only recently with the discovery of cellular oncogenes (c-onc) has there been significant progress in identifying whatever DNA differences do exist between normal and abnormal cells. Certainly oncogenes do not lead to a direct explanation for cancer or even to a solid understanding of the genetic changes which occur as a cell progresses from normal to abnormal to overtly tumorigenic. For example, strong promoters and gene amplification may be important in some cases, and chromosome rearrangements with subsequent activation of a normal unexpressed gene or inactivation of a gene which normally acts to supress an oncogene (or other genes) are possibilities worth consideration. But even though the precise role of cellular oncogenes in human cancer remains to be determined, they are useful models to consider in trying to come to grips with the biological meaning of contaminating DNA because they represent special sequences of some tumor cell DNAs which have transforming potential, and experiments using cellular oncogenes can be helpful in making judgements about the significance of various levels of DNA which might be present in products.

Although specific data are not available, it is reasonable to assume that contaminating DNA is heterogeneous and contains random portions of the genome, and that various sizes of DNA are present without selective elimination or concentration of any specific gene sequences.

Various approaches have been taken to calculate risk. One is to assume that at least some contaminating DNA sequences are large enough to retain the potential for biological activity, and that one or more c-oncs are present in the contaminating DNA. Although the size of c-oncs vary, the smallest are in the range of 1 kb which serves to illustrate the point. One copy of a 1 kb c-onc is equivalent to about 1×10^{-6} pg of DNA. A variety of studies have shown that under optimal experimental conditions $\underline{in}$ $\underline{vitro}$, purified c-onc DNA has a transformation efficiency of about 10^4 focus forming units per ug. Converted to pg, that means 100 pg of purified c-onc DNA are needed to give one focus of transformation in the 3T3 system. Going back to the cellular DNA with a c-onc, it becomes clear that there is a factor of about 10^{-8} (10^{-6}pg/100pg) between a minimally effective DNA transformation dose $\underline{in}$ $\underline{vitro}$ and the amount of c-onc DNA which theoretically might be in a productif it were originally present in only one copy per cell. If more realistic assumptions were made which relate to the $\underline{in}$ $\underline{vivo}$ situation, the safety factor becomes larger by at least several orders of magnitude. For example, if one does similar calculations based on the assumption that purified c-onc DNA has an $\underline{in}$ $\underline{vivo}$ transformation efficiency of less than one per 10 ug, then the 10^{-8} safety factor just mentioned becomes 10^{-13}. Those and similar calculations by others (6, 11) have suggested that if the DNA in a product is in the range of 100 to 10 pg per dose, then there is a safety factor of at least 10^{10}. A detailed description of these risk assessment calculations was presented recently (7).

The Study Group concluded that, based on the experimental data available to date, the probability of risk associated with heterogeneous contaminating DNA in a product derived from a CCL is so small as to be negligible when the amount of such DNA is 100 pg or less in a single dose given parenterally. The assessment of the safety of any product with respect to DNA should take into consideration: i) the elimination of the biological activity of DNA by various steps in the manufacturing process; and ii) the reduction in the amount of DNA during the purification of the product in the manufacturing process. A given product may be considered safe on the basis of reliable data from either of those two elements, or their combination.

The use of special DNA sequences such as viral regulatory sequences in the construction of recombinant cells is considered acceptable because there is no evidence to suggest that such sequences would impose any additional risk beyond that of heterogeneous contaminating DNA in general. Nevertheless, the manufacturing process should be validated to show that they are not concentrated in any detectable contaminating DNA.

5. SUMMARY AND CONCLUSIONS

Looking back at the history of issues and decisions that were made regarding the acceptability of various cell substrates leads one to wonder how much more rapidly products might have been developed if different decisions had been made. It is also strikingly evident that primary tissues and primary cells have been the source of whatever problems have existed in the past. This is not to say that everything is known about CCLs or that there are no surprises in store as more experience is gained with their use in biological production. It is difficult to argue, however, that one is not better off with a very well-tested cell system and a manufacturing process which can cope with even theoretically worrisome contaminants.

The prevention and treatment of human diseases will be greatly facilitated by products derived from CCLs. Clearly there is much at stake in terms of world health regarding the resolution of what might be termed the last of the cell substrate issues, and the Study Group's conclusions provide a framework for the international acceptance of a wide rage of cells in the manufacture of many different types of products.

One of the most important general conclusions made by the Study Group regarding the acceptability of CCLs was defining the point at which the decision is really made for any given product. They pointed out that when approval is given for clinical research in humans with a product derived from a CCL, the cell acceptability issue is basically settled. The final approval of the product will, of course, depend on establishing safety and efficacy data. But the cell acceptability issue cannot reasonably be deferred until the time when a decision for approving the product is made. Both biomedical and ethical considerations dictate that it must be made before the first humans receive the experimental product. In the words of the Study Group: "It would be inimical to the basic concepts of medicine knowingly to impose the risk of cancer on patients with serious diseases or on healthy people, particularly since alternatives exist to many of the products from CCLs now available or that are expected to be approved soon." In other words, more decisions have already been made on CCLs than might appear to be the case because many more products from CCLs have been approved for experimental use than have been approved for general use. For example, the USA FDA has already agreed to human clinical studies of products derived from Vero, CHO, and

several hybridomas. The National Control Authorities of other countries also have approved products from these cells as well as other CCLs.

Another important implication of the Study Group's report is that the phenotypic characteristic of tumorigenicity is irrelevant to the acceptability of a product derived from a CCL. The reason is simply that there are no biological data which suggest that tumorigenic CCLs would pose any greater risk than non-tumorigenic CCLs. As pointed out earlier, the results of all the in vitro and in vivo experiments reported to date are uniformly negative regarding the carcinogenic potential of heterogenous cellular DNA even when it is inoculated in mg amounts. It is probably just as artificial to argue that non-tumorigenic CCLs are safer than tumorigenic CCLs as it is to suggest that the transformation of 3T3 cells by c-oncs is related to the potential of c-oncs to cause normal human cells to become malignant. In both cases the "normal" cells are in fact grossly abnormal in the sense that they already have an infinite in vitro life potential. In addition, when 3T3 or cells such as Vero are tested for tumorigenicity in the most sensitive systems, their biological behavior is clearly consistent with that of the tumorigenic phenotype. The focus of attention has therefore now been placed on the manufacturing process to yield products pure enough to be safe to use regardless of the tumorigenic profile of the cell substrate. This obviously opens the door to the use of a wide variety of cells where the major determinant in selecting a substrate is efficiency of production. The practical result of this approach is illustrated by the evolving situation with hepatitis B vaccines (HBV).

The first such vaccine was derived from human plasma, and this "substrate" still provides a large proportion of all of the HBV currently used. The second generation of HBVs came onto the international market in 1986 with the introduction of rDNA vaccines produced in yeast. No regulatory authority, including the USA FDA, has concluded that HBV derived from plasma is now unacceptable even though an alternative is available which on the surface would seem to offer safety advantages over the plasma vaccine because the yeast substrate does not contain any of the worrisome human pathogenic components known to be potentially present in plasma. Plasma derived HBVs have remained on the market because regulatory agencies have taken into account reliable data showing that the various manufacturing processes now in use to produce the vaccines give final products in which risk has been reduced to undetectable levels. It therefore makes no scientific sense to disapprove plasma derived HBV. And from an ethical as well as a public health point of view it would be extremely unfortunate if plasma derived HBVs were forced off the international market for nonscientific reasons since they continue to provide the major means of control of HB disease at the lowest cost.

In the near future rDNA HBVs produced in CCLs are likely to be approved in several countries, and these products will provide a third alternative. And finally, early human experiments have begun with HBV derived from human hepatoma cells. The Study Group's report implies that each of these vaccines should be acceptable when reliable data are provided to establish the safety of the product when it is manufactured in such a way that risk factors such as contaminating cellular DNA and endogenous viruses are eliminated. In the view of the Study Group, the basic decision on the acceptability of the CCL needs to be reconsidered only if new data are developed which suggest unanticipated risks.

While this approach is scientifically sound and consistent with past

regulatory philosophy in many countries, it has nevertheless been of concern to some for a variety of reasons. Opening the door to the use of cells which had for decades been considered unacceptable inevitably leads to an unfocussed general apprehension that something bad is going to happen. In other words, there is a certain amount of anxiety (unfocussed) as opposed to fear (focussed). In addition, there are economic forces which complicate any of these discussions because manufacturers who use non-CCL systems (e.g. bacteria and yeast) would understandably like to take a marketing advantage of any doubts which might be stirred up, regarding the safety of their competitors' products. Ultimately, however, such strategies can only be self-defeating because it is probably unrealistic for any competitive manufacturer to restrict product development to non-CCL systems.

From the point of view of world health, it is extremely important to allow product development and marketing to proceed along lines which will result in safe products which can compete with one another so that acceptable products can be available at the lowest cost. No one with a sense of responsibility and conscience wants to see the approval of risky products, especially vaccines which will be used by the hundreds of millions of doses in children. On the other hand, social responsibility demands that arguments related to relative risk and safety of various manufacturing strategies be sharply focussed and supported by relevant data. The stakes are too high in terms of world health for anything else.

REFERENCES

1. Beale AJ: Choice of cell substrate for biological products. Adv. Exp. Biol. Med. 118, 83-97, 1979.

2. Hayflick L, Plotkin S, and Stevenson RE: History of the acceptance of human diploid cell strains as substrates for human virus vaccine production. Develop. Biol. Standard. 68, 1987.

3. Hilleman MR: Cell line saga: an argument in favor of production of biologicals in cancer cells. Adv. Exp. Biol. Med. 118, 47-58, 1979.

4. Hopps HE and Petricciani JC (eds.): Abnormal Cells, New Products, and Risk. Tissue Culture Association, Gaithersburg, MD, 1985.

5. Lowy DR: Potential hazards from contaminating DNA that contains oncogenes. In: Abnormal Cells, New Products, and Risk (HE Hopps and JC Petricciani, eds.) pp 36-40, Tissue Culture Association, Gaithersburg, MD, 1985.

6. Perkins FT and Hennessen W (eds.): Use of Heteroploid and Other Cell Substrates for the Production of Biologicals. Karger, Basel 1982

7. Petricciani JC, Hopps HE and Chapple PH (eds.): Cell Substrates. Plenum, New York, 1979.

8. Petricciani JC and Regan PJ: Risk of neoplastic transformation from cellular DNA: calculations using the oncogene model. Develop. Biol. Standard. 68, 1987.

9. Salk J: The spector of malignancy and criteria for cell lines as substrates for vaccines. Adv. Exp. Biol. Med. 118, 107-113, 1979.

10. Tam JP: Physiological effects of transforming growth factor in the newborn mouse. Science 229, 673-675, 1985.

11. Wahl G: Detection of adventitious agents and sensitivity of methods. In: Abnormal Cells, New Products, and Risk (HE Hopps and JC Petricciani, eds.) pp 50-56, Tissue Culture Association, Gaithersburg, MD, 1985.

12. World Health Organization Technical Report Series, No. 747: Report of the WHO study group on biologicals: Acceptability of cell substrates for production of biologicals, Geneva, 1987.

WHY NOT USE PRIMARY CELLS ?

C. REMACLE, B. AMORY, P. GILON, F. GREGOIRE, N. HAUSER, B. HERBERT, D. MASQUELIER, J.L. MOURMEAUX, B. REUSENS, B. SCHELLEN, L. SKA.

Université Catholique de Louvain, Laboratoire de Biologie Cellulaire, Place Croix du Sud, 5, B1348 Louvain-la-Neuve, Belgium

INTRODUCTION.

For the cell biologist, perhaps an ultimate goal is to explain how, from the unicellular zygote, the various cells of the organism proliferate, move, differentiate, and die. If the signals regulating these events are discovered, the complete developmental pattern would be explained. The program is currently under study in small animals, as *Caenorhabditis* and *Drosophila* (1,2). If one imagines a similar approach with more complex animals, man comprised, the key is found to take every desired cell at the good time and the good place, and then to govern its proliferation rate and differentiation. Needless to say that to have differentiated cells at work is a major biotechnological target, thanks to the variety of tests and products they can provide.

Unfortunately, we are facing the apparent antinomical problem : to use differentiated cells, and to keep them proliferating.

The problem appears in the technical approach of cell cultures: (3)
* Organ culture will preserve tissue architecture, cell interactions, histochemical and biochemical differentiation. But explants cannot be propagated and display great variations between replicates.
* Primary cultures issued after cell dispersion or migration from normal explant are often heterogeneous, depending on the pre-treatment. Primary cells have a low growth fraction, but retain specific functions more readily.
* The cell strain obtained after subculture of these cells allows the expansion, but may cause the loss of specialized cells unless care is taken to select out the correct lineage, or reinduce differentiated properties.
* The cell line issued from tumors can express at least partial differentiation, while retaining the capacity to divide.
* After several subcultures, the cell strain will either die out (crisis), or transform, showing greater growth rate and higher yield, but also genomic instability, divergence from the donor phenotype, and loss of tissue specific markers.

The crisis of cell strains is a very intriguing phenomenon. Human cells have a finite capacity for replication in vitro. Embryo derived fibroblasts -which have the greatest capacity for replication in culture- undergo about fifty population doublings before division ceases, in about one year (4). The events are divided into three phases (5): phase I represents the primary culture, phase II, the active replication period, and phase III, the period when cell replication slows, ceases, and ultimately ends with complete degeneration and

75

A. O. A. Miller (ed.), Advanced Research on Animal Cell Technology, 75–92.
© *1989 by Kluwer Academic Publishers.*

death. After having experimentally ruled out trivial explanations for these events, Hayflick put forth the notion that the phase III might represent aging at the cell level. Hundreds of studies concerned this phenomenon, which is still waiting an explanation.

Cultures of mouse embryo cells behave differently. It is a well-known fact that in conventional serum-supplemented media, the mouse cells rapidly lose proliferative potential, leading to growth crisis (6,7). This senescence is often followed by the appearance of genomically altered, immortalized cell lines. Depending on the age of the mouse embryo and culture conditions, the emergence of immortal variants can be avoided, and the culture stays in the senescent non-proliferating state (8).

Recently however (9), mouse embryo cells were seen to undergo the usual crisis when cultured in serum-supplemented media, but when cultured in serum-free synthetic medium containing EGF as a growth factor, cells grew exponentially without exhibiting growth crisis. Contrary to the immortal aneuploid variants, the cell line -or cell strain- which is obtained in serum-free medium do not display chromosomal aberrations. Moreover, evidences are presented that serum contains growth-inhibiting factors leading to the crisis of mouse cells, and that selection of cells unresponsive to these factors contribute to immortalization in conventional culture conditions.

Although this awaits further confirmation in other species, it is not the first time that the inhibitory activity of serum upon cells in culture is put in light (3,10). So that the problem of the proliferative crisis is perhaps in the way of being solved, namely by the use of specific factors in serum-free defined media.

In theoretical cell lineages, the differentiation process can be superimposed to mitoses (11). From the unicellular, totipotent, and virtually immortal zygote emerge two kinds of lineages : the germ line and the somatic cells. In the developing embryo, the somatic cells form pluripotent stem cell pools, the determination of which taking shape progressively. Up to now, determination has been almost impossible to study. No pure population -or pure interacting populations- of cells which carry out the program of determination were obtained. Consequently, the signals regulating the gene function, as well as the genes concerned, are still unknown.

By contrast, some aspects of the differentiation program are now being discovered. After the period of determination, the fate of the stem cell pools is quite varied. Certain pools are totally committed for differentiation during a brief developmental period of time, as is the case for neurons, while other tissues will keep stem cells in the differentiated part. These stem cells can divide symetrically to produce two identical stem cells, or asymetrically to produce one determined or fully differentiated cell, and one new stem cell. Unipotent stem cells provide one type of differentiated cell, as the basal cells of the epidermis. Pluripotent stem cells may produce two or more types of differentiated cells, as in hematopoiesis. The production of differentiated cells may involve additional steps of control.

In our laboratory, we are interested in the culture of normal differentiated cells, more specifically on their lineage and the signals regulating their proliferation and function. We intend to review now briefly three examples : precursors of adipose cells, insulin-secreting cells, and osteogenic cells, together with a look to the cell lines used in the same objective.

ADIPOCYTE PRECURSORS.

In normal adipocyte lineage, pluripotent mesenchymal progenitors proliferate and determine to become committed quiescent cells, which grow into adipocytes. Some factors influencing proliferation and differentiation were found : pancreatic polypeptide, cyclic AMP, estradiol, paracrine adipose tissue factor, pituitary growth factor, growth hormone, insulin (12). The activity of these factors was suspected in the course of studies performed on rat or human precursors of adipocytes, in primary cultures or during first subcultures.

But another set of results came from studies performed on cell lines, capable of accumulating triacyl-glycerols in culture, as cell lines cloned from 3T3 fibroblasts, which display a low (3T3-L1) or high (3T3-F442A) frequency of adipoconversion. Other lines were cloned from carcinoma, bone marrow, calvaria, or directly from adipose tissue (13). In Ob-17 cell line for example, a specific model of differentiation could be proposed (14), where stem cells -defined as adipoblasts- can become committed and acquire the susceptible state at any division during the growth phase. At confluence, these committed preadipocytes are present in clones of variable sizes, and are still morphologically indistinguishable from uncommited cells. Subsequently, postconfluent mitoses occur, and the differentiation program starts, leading to morphological and biochemical changes of the cells. Growth hormone, tri-iodo-tyronine, adipose conversion factor, and insulin are active along this line.

Doubts remain however, concerning the possible extrapolation of these data obtained on cell lines, to normal cells. Effectors of proliferation and adipoconversion of cloned cell lines cannot be expected to be necessarily effective in similar processes, in primary preadipocytes. Some examples are known that these cells react differently.

Unfortunately, primary cells and cell strains are more difficult to control, the main problem being the heterogeneity of the inoculated cells. The fat lobule consists of the stroma-vascular fraction and of mature adipocytes, which are liberated during usual collagenase isolation procedure. After light centrifugation, the adipocytes float and the cell pellet - the stroma-vascular fraction - contains an heterogeneous population of cells corresponding to blood cells, endothelium, mesothelium, as well as fibroblasts and presumed preadipocytes, the latter being unrecognizable with morphological criteria (16).

Various methods exist to purify this fraction to a certain extend : a) filtration trough gauze or nylon with progressively finer mesh retains undissociated blood capillaries and fragments of connective tissue; b) pre-culture and decantation after a few hours eliminates less-adherent epithelial cells; c) Gradient centifugation allows the sampling of a cell population of lower density, due to fat load, preexisting or stimulated during preculture (15-17). By using these methods, cells can be obtained where nearly general adipoconversion will occur later.

When such relatively pure cell populations are kept growing in standard medium containing fetal calf serum, they divide at a rate far lower than undifferentiated mesenchymal cells (Fig.1). The preadipocytes can be subcultured several times, but they loss progressively the capacity for adipoconversion. It could not be ascertained whether it was due to the overgrowth of less committed cells, or to a decrease in capacity of the whole population.

The modulators of cell proliferation are presently under study, the question being often to discriminate between the factors affecting really the cell cycle, and those slowing down mitoses in predifferentiated state.

First, the presence of extracellular organic matrix and the nature of the physical support have retained very little attention. However, they may affect significantly the proliferation rate, as shown by cell kinetics after inoculation of preadipocytes on two supports (Falcon TC plastic, and bacteriological dishes coated with a film of titanium) in serum-complemented medium (Fig.2).

Second, the presence of serum in culture media may evidently sustain the proliferation of non-confluent preadipocytes. On cell strains cultured in such media, the proliferation rate could be enhanced by adding factors originated from pituitary gland and collectively termed "mesenchymal growth factors" (12). Estradiol is also mitogenic, whereas pancreatic polypeptide suppresses the replication of cultured human adipocyte precursors (12).

To be efficient however, the study of specific factors modulating cell growth requires the use of defined media. Added to basal medium, insulin at high concentration, transferrin and FGF suffice to sustain the proliferation of the cell line 1246. The further supplementation with the presumed EFG binding protein is necessary for 3T3 and Ob17 lines, as well as stroma vascular cells of mice and rats (13,14). One may note that the mitogenic activity of insulin could be due to the well known interaction with IGF receptors, since IGF-1 and IGF-2 are far more potent than insulin in this respect (13,14). There is no doubt that the recent use of defined media will provide further information.

In appropriate conditions, preadipocytes will undergo adipoconversion. These may consist of a medium containing serum, lipids (0.5%), Tween-80 (0.1 mg/ml) and insulin at physiological concentration (16) (Fig.3).

When tested in serum-free or serum-poor media, the requirements for adipoconversion of cell lines were found quite variable. Among the factors, growth hormone, tri-iodo tyronine, dexamethasone, methyl-isobutyl xanthine, prostaglandins, fibrate drugs enhance adipoconversion, which remains poorly expanded however, so that undetermined serum factors must play a role (14,18). In particular, growth hormone seems important with respect to the onset of the differentiation program, enhancing the transcription of a specific gene (19). Insulin would act solely as a modulator in the phenotypic expression of Ob-17 cells (14), whereas in 3T3 cells, a dual role was shown : a mitogenic stimulation induce traverse of the cell cycle from commitment to expression stages; afterwards, insulin enhances the expression of adipocyte phenotype (20).

In this context, a recent paper (21) merits attention, as can be found therein unique evidence for a two-step process of terminal differentiation in 3T3-T mesenchymal stem cells. Before to accumulate lipids, 3T3 cells must first arrest growth in the G1 phase of the cell cycle, at a distinct state designed Gd, which differs from other restriction points (growth factor-dependent, and nutrient-dependent). Cells at Gd can remain quiescent, reinitiate proliferation, or undergo differentiation; in the latter case, both non-terminal and terminal phases are involved. The non-terminally differentiated state -Gd'- could be further dissected in two substeps. The first one can be induced to lose the differentiated phenotype when treated with retinoic acid and methyl-isobutyl xanthine, but not with phorbol

esters. The second one can do the same when treated with MIX and phorbol esters, but not with retinoic acid. These Gd' cells are non-terminally differentiated because they can both be induced to undergo proliferative responses when treated with fetal calf serum and insulin. By contrast, cells at the terminally-differentiated state are not responsive to any of these agents.

IN CONCLUSION, some specific signals are now being discovered, which will permit to control the proliferation and adipoconversion of precursors. The work is more advanced on cell lines, but the achievement of homogeneous cell strains will soon take advantage of these data.

INSULIN SECRETING CELLS.

During very early embryogenesis (see 22 for references), in association with ecto-endodermal delamination, certain cells acquire a special determination, marked by some neural characters. These cells will constitute the diffuse neuroendocrine system, a main component being the gastro-entero-pancreatic system. Nevertheless, the origin of islet cells from the neural crest is unlikely. Rather they arise from the primitive duct wall. Inversion of the polarity of the mitotic axis in certain epithelial cells of growing pancreatic ducts is currently viewed as the mechanical means to separate endocrine islets from the continuous duct epithelium. A mesenchymal induction has been suspected. The repetition of this process, and the proliferation of escaped cells lead to the formation of Langerhans islets, which become later irrigated and innervated. It remains to be established whether the various endocrine cell types secreting insulin, glucagon, somatostatin, and pancreatic polypeptide arise from separate populations of endodermal cells previously stemmed, or from pluripotent stem cells. During life, the pancreatic islets ensure hormone production, and display a slow turnover of the differentiated cells, which eventually become exhausted at the extreme life span, or before that time in pathological conditions such as diabetes (22).

Various methods were used to study islet cells in vitro, either issued from tumors or in primary cultures. The problem is that endocrine islets compose only about one percent of the pancreatic tissue, so that pure endocrine population seems difficult to obtain. Moreover, the endocrine islet comprises four cell types, the predominant one -about seventy percent- being the insulin-secreting B cell. Then, several groups use cell lines issued from rat, hamster and human tumors, as plentifull source of insulin-secreting cells (23-25). It is often evident that ultrastructural aspects of these cell lines do not correspond to normal B cells, namely regarding the weakness of insulin content. They do secrete insulin, but most of them respond barely to the usual secretagogues, as glucose and amino-acids.

On the other hand, the primary cultures of islet cells take on several forms, in addition to the complex organ culture started from massive explants. The vast majority of studies were performed in rat (26). Islets can be picked up individually after careful collagenase digestion of the pancreatic tissue, or collected in greater number using some recent post-isolation techniques of gradient centrifugation (27), or by adding seleno-methionine (28). These techniques are used mainly with adult material.

Cell dispersions of fetal and neonatal total pancreas can be inoculated into dishes. In the absence of corticosteroids, exocrine cells

disappear rapidly. As far as endocrine cells are concerned, two arrangements can then be obtained : a) either monolayers of endocrine cells,
which are growing among fibroblasts during one or two weeks, and
finally desintegrate; the fibroblastic growth can be slowed down by
certain means (26); b) spherization of endocrine monolayers into
neoformed islets (29-31) (Fig.4). This is a kind of sorting-out. The
fibroblasts expell endocrine cells from their support, providing
free-floating pure endocrine islets, composed of more than 90 % of B
cells, the other endocrine cell types being located at the periphery, as
in vivo (30-31). Another way of islet formation is a budding from
epithelial layer (30-31).

Finally, one step further to the purification was recently carried
out (32). On cell suspension from isolated islets, various parameters
can be selected for the isolation of one or more cell types. As B cells
are usually larger than the other cell types, centrifugal elutriation
separates them, according to the differences in sedimentation velocity.
Specific membrane antigens bound to fluorescent antibodies, or
metabolic characteristics leading to the production of fluorescent
endogenous molecules permit the separation of islet cells by use of
fluorescence-activated cell sorter. These preparations yield 90-95%
purification of islet cell types, which can survive in culture.

In this context, the influence of non-endocrine cells and extracellular matrices is very important. Endocrine monolayers grown on collagen will take a spherical shape when covered by another layer of
collagen (33). Spherical islets will arrange in monolayers, when
inoculated on osmotically disrupted fibroblasts (31,34). Moreover, cell
dispersion of fetal pancreas inoculated directly in collagen gel will
provide numerous unusual polarized structures within one week (35).
After a few days, fragments of pancreatic ducts have sealed their cut
ends, and epithelial cells surround a spherical lumen. Certain cells
feature endocrine differentiation. Later, the endocrine cells have
proliferated, forming large islets budded from epithelial layer. This
system allows for the study of undifferentiated epithelial cells -the
presumed stem cells-, differentiating and differentiated endocrine
cells, in the same preparation.

All these elements may indicate how flexible is the primary culture
of islet cells -especially B cells. Their purification from other
pancreatic cells is now solved.

The main problem with this culture consists in the poor proliferative capacity, so that subcultures were never achieved up to now.
Differentiated endocrine cells containing secretory granules may divide
in vitro (Fig.5), but their proliferative capacity is rapidly exhausted
in normal media. They may then survive for only a few weeks. Three
possible explanations can be proposed : a) a cytotoxic effect of serum.
Serum-free defined media were applied in only a few studies, but the
same exhaustion of growth was observed (36). The authors conceded
that its empirically defined composition may not be optimal. b)
Specific factors may be lacking, which would sustain mitoses of differentiated islet cells. c) B cells, as being subterminally differentiated,
might possess an internal limit to their mitotic capacity. There are
evidences in vivo that this could be so (37).

In any case, the mitotic rate of insulin cells can be modulated.
Schematically, two ways can be considered : metabolic, and receptor-
mediated.

Glucose is the major stimulus of adult-type secretory response. It is mitogenic also (37-39). It was found to stimulate B cell proliferation by regulating the number of cells entering the cell division cycle, as the cell cycle proceed in a similar manner during a period of about fifteen hours, irrespective of glucose concentration in the culture medium (38). Amino-acids are also mitogenic, especially on fetal material (40). At this stage, amino acids are precisely stronger stimulators of insulin secretion than glucose (30). The mediating role of cAMP is not yet clear, since reports have been divergent after use of various means to induce an increase of cAMP within B cells. For example, the phosphodiesterase inhibitors isobutyl-methyl xanthine (41) and theophylline (42) have been found respectively to stimulate and depress DNA replication. Recently, we noted an increased proliferation of B cells exposed to forskolin, a direct stimulator of the catalytic subunit of adenylate cyclase.

Receptor-mediated stimulation of B cell replication is complex also and requires further studies, especially with defined media which have been used only occasionnaly. Insulin-like Growth Factors markedly enhance B cell mitoses (43-44). In this respect, the B cell is quite special, as insulin at high concentration shows a cross-reaction with IGF receptors, which can explain the small mitogenic effect of insulin (44). As B cells release insulin in the extracellular spaces, the local concentration reached around the cells could be important in some kind of autocrine regulation of proliferation. Moreover, it was recently discovered that B cells themselve release detectable amounts of IGF (43,45). EGF is also able to stimulate B cell replication, but to a lesser extend than IGF (46).

The differentiated state of isolated endocrine islet cells can be easily maintained in culture (26). In contrast to endocrine cell lines, they retain their morphological characters (Fig.6), they continue to secrete insulin in the culture medium, in response to various stimuli, the main of which being glucose.

IN CONCLUSION, the major problem with this kind of primary endocrine cells no more concerns the purification procedure, nor the maintenance of differentiated state, but rather the proliferation in culture. Some possible solutions should be considered : a) the effects of serum-free defined media have not been largely explored; b) the mitotic capacity of the differentiated cells is perhaps heterogeneous, and could be subjected to external regulation. c) the culture of presumed stem cells seems possible now, and on the other hand, the regulation of the insulin gene expression begins to be better known (47).

OSTEOGENIC CELLS.

Despite numerous studies about embryonic bone development, namely in limb buds, the cell lineage of osteogenic cells is far from being understood. It is even complicated by the fact that co-exist various modes of bone formation, and also reparative processes as well as ectopic bone appearance. Multipotent mesenchymal stem cells are thought to be able to receive progressive commitment by a large series of factors. They transform to osteoprogenitor cells -or preosteoblasts- but the lack of a reliable morphological marker delays progress in defining the steps (48,49).

Using certain models in vivo, such as the periodontal ligament in rat, more precise pattern of osteoblast formation could be depicted (50). Self-renewing precursor cells synthetize DNA, and divide. Daughter cells may remain precursors, or become committed to osteoprogenitor cells. In presence of adequate stimuli, these cells increase in nuclear size, become G1 preosteoblasts, then progress through S-phase becoming G2 preosteoblasts. Under osteogenic circumstances, the latter enter mitosis and produce two osteoblasts. Hypothetical possibilities of cell death, cell regression back to progenitor cells, and input from more primitive stem cells were also suggested.

The osteoblast is concerned with making organic extracellular bone matrix. These cells are found wherever new bone matrix is being formed, and are often polarized at the bone surface. In vivo, the osteoblast apparently is not capable of cell division (48,49).

Finally, osteocyte is the mature cell that lie in lacunae, within the various types of bone. Osteocytes are probably osteoblasts that have been engulfed by the matrix, that they subsequently mineralize (48,49).

The committed cells can be recognized by several, more or less specific markers. Osteoblasts express distinct morphological phenotype, possess alkaline phosphatase, synthetize bone-specific type 1 collagen and other specific bone proteins such as osteonectin and osteocalcin (51). Antibodies were produced against these markers. New monoclonal antibodies are now appearing also, directed against cell-surface determinants of osteoblasts, osteocytes, or osteoclasts (53).

These markers are very important, as bone presents unique problems to the investigator attempting to isolate and culture cells (52,53). Particularly in the adult, bone tissue is relatively acellular and the cell composition is highly heterogeneous. Add to this a mineralized matrix and a hematopoietic marrow cell population, and you have a really complex challenge.

The most popular bone cell culture concerns the osteoblasts. Cell cultures characterized as osteoblast-like have now been established from both normal bone tissue and bone tumors, in a variety of species, ages, and anatomic locations. Cultures are initiated from cells isolated by enzyme digestion, explants, or others. Enrichment for cells expressing osteoblast markers is accomplished by mechanical means, differential digestion, or by cloning. Some permanent clonal lines have been started from normal bone, others are tumor-derived. Phenotypic drift towards less differentiated characters occurs in permanent lines, but over more passages and longer time periods than osteoblasts-enriched primary populations. In most cases, the phenotypic loss in these mixed populations was attributed to overgrowth of non-expressing cells (51,52).

As a matter of fact, contamination by non-osteoblast cells must be as lowest as possible. In the culture system we use in the laboratory, the starting material consists of calvaria of ten day-old rats. The sutures are cut off, so that frontal and parietal bones are free. The periosteum covering their surfaces is then peeled. In that manner, the contamination by poorly differentiated mesenchymal cells is markedly reduced. Small glass fragments are disposed on the endocranial surface of the bones. Osteoblasts migrate preferentiallyon glasswhich allows to isolate a nearly pure population. After five days of culture, the monolayer of osteoblasts covers the entire glass fragment. At that moment, the glass fragments are transfered into Petri dishes and

further incubated for two days. The cells are then trypsinized, counted and inoculated at known density into new Petri dishes (Fig.7). More than eighty percent of the inoculated cells are alkaline phosphatase-positive (Fig.8).

A few days later, ascorbic acid is added to the medium, in order to promote collagen synthesis, and its deposition as fibers and bundles in extracellular spaces. The organic matrix thickens progressively, forming a network which engulfs osteoblasts. In these conditions, gel electrophoresis of collagen secreted in vitro shows alpha-1 and alpha-2 bands characteristic of type I collagen that is bone specific. It is another mean to test the purity of the culture.

To get mineralization of the organic matrix, beta glycerophosphate is then added to the culture medium. The onset of mineralization takes two different forms : some kind of vesicle mineralization with the growing spots fusing together, or the deposition of spicular crystals along collagen fibers. The final bone tissue formed in culture is very similar to that found in vivo (Fig.8,9). Osteocytes are enclosed in the mineralized matrix, where X-ray dispersion analysis reveals calcium and phosphorus with the same ratio as in hydroxyapatite which is the mineral phase of bone.

IN CONCLUSION, such a system permits to isolate a nearly pure population of osteoblasts, which can proliferate, synthetize organic matrix, and mineralize it, under the influence of controlled factors of the culture medium. We cited only ascorbic acid and glycerophosphate, but a complex series of promoting factors is now the subject of intense research (54).

THE USE OF PRIMARY CELLS.

By these three short examples, we tried to remember some pitfalls and potentialities of the differentiated cell culture.

In first place, the good cells have to be selected in a population that is nearly always heterogeneous. Metabolic or antigenic markers are presently in development; they will allow to really recognize the cell, and to get more and more pure inoculates with the help of the cell sorting technology.

Then are concerned the apparently correlated problems of proliferation and differentiation. At present, we have no possibility to govern the stage of determination, but fortunately the animal or human embryo did the work in our place, to propose stem cell pools that we can select and handle.

Theoretically exist the postulates of internal limit to proliferation and of terminal differentiation, but in certain cases the first could be bypassed while keeping the genomic stability - that is without "transformation". On the other hand, terminal differentiation could be avoided by use of external signals while differentiated properties were fully expressed.

The tools that are now utilized are :
* the refining composition of defined media, where can be introduced selective enhancers of proliferation and differentiation, in the form of hormones, growth factors and other chemicals (55).
* the composition of physical support of cell cultures and extracellular matrices which are known to interplay with plasma membrane,

84

cytoskeleton, gene expression at the level of the nucleus or cytoplasm (56,57).

In the last years, a lot of fundamental researches has been, and continue to be allotted to the better knowledge of these factors. So that if a practical target is to obtain cells expressing extensively one or more specific characters (secretion, intracytoplasmic enzymes, and so on), together with security and reproducibility, then to play with primary cell strains should be soon considered, namely in place of introducing foreign gene assemblages into cells.

REFERENCES

1. Chalfie M, Horwitz HR, Sulston JE: Mutations that lead to reiterations in the cell lineages of *C. elegans*. Cell, 24, 59-69, 1982.
2. Gehring WS: Molecular basis of development. Sci. Amer., 253(4), 153-162, 1985.
3. Freshney RI(ed): Animal cell culture, a practical approach. Oxford: IRL Press, 1986.
4. Hayflick L, Moorhead PS: The serial cultivation of human diploid cell strains. Exp. Cell Res.,25, 585-621, 1961.
5. Hayflick L: Intracellular determinants of cell aging. Mech. Ageing Dev., 28, 177-185, 1984.
6. Todaro GJ, Green HJ: Quantitative studies on the growth of mouse embryo cells in culture and their development into established lines. J. Cell Biol., 17, 299-312, 1963.
7. Meek RL, Bowman PD, Daniel CW: Establishment of mouse embryo cells in vitro. Exp. Cell Res., 107, 277-284, 1977.
8. Van Gansen P: Le vieillissement cellulaire in vitro. Ann. Biol., 18, 147-177, 1979.
9. Loo DT, Fuquay JI, Rawson CL, Barnes DW: Extended culture of mouse embryo cells without senescence: inhibition by serum. Science, 236, 200-202, 1987.
10. Jakoby WB, Pastan IH (eds): Methods in enzymology, vol.53: Cell culture. New York: Academic Press, 1979.
11. Darnell J, Lodish H, Baltimore D: Molecular cell biology. New York: Scientific American Books Inc, 1986.
12. Roncari DAK: Pre-adipose cell replication and differentiation. Trends in Biochem. Sci., 486-489, Nov 1984.
13. Ailhaud G: Adipose cell differentiation in culture. Molec. Cell. Biochem., 49, 17-31, 1982.
14. Vannier C, Gaillard D, Grimaldi P, Amri EZ, Djian P, Cermolacce C, Forest C, Etienne J, Negrel R, Ailhaud G: Adipose conversion of Ob17 cells and hormone-related events. Internat.J.Obesity,9,s1, 41-53, 1985.
15. Dugail I, Quignard-Boulange A, Ardouin B, Brignant L: A method for separating cultured preadipocytes according to their density: application to stromal cells from overfed suckling rats. In Vitro Cell. Dev. Biol., 22, 375-280, 1986.
16. Grégoire F, Hauser-Gunsbourg N, Remacle C: Ultrastructural analysis of the in vitro differentiation of female rat preadipocytes. Biol. Cell, 56, 127-136, 1986.
17. Swierczewski E, Gaben-Cogneville AM: Characteristics of the rat adipoblast during differentiation in culture. Internat. J. Obesity, 9, suppl.1, 23-28, 1985.
18. Brandes R, Arad R, Bar-Tana J: Adipose conversion of cultured rat primary preadipocytes by hypolipidemic drugs. Biochim. Biophys. Acta, 877, 314-321, 1986.

19. Doglio A, Dani C, Grimaldi P, Ailhaud G: Growth hormone regulation of the expression of differentiation-dependent genes in preadipocyte Ob1771 cells. Biochem. J., 238, 123-129, 1986.
20. Gamou S, Shimizu N: Adipocyte differentiation of 3T3-L1 cells in serum-free hormone-supplemented media: effects of insulin and dihydroteleocidin. Cell Struct. Funct., 11, 21-30, 1986.
21. Weir ML, Scott RE: Regulation of the terminal event in cellular differentiation: biological mechanisms of the loss of proliferative potential. J. Cell Biol., 102, 1955-1964, 1986.
22. Hoet JJ, Remacle C:, Organization of the pancreatic islets, with special reference to diabetes. In DeGroot LS et al.(eds), Endocrinology, New York: Grune & Stratton, in the press.
23. Gold G, Gishizki ML, Chick WL: Contrasting patterns of insulin biosynthesis, compartmental storage, and secretion: rat tumors versus islet cells. Diabetes, 33, 556-562, 1984. ?
24. Lambert DG, Hughes K, Atkins TW: Insulin release from a cloned hamster B-cell line (HIT-T15): the effects of glucose, amino acids, sulfonylureas and colchicine. Biochem. Biophys. Res. Commun., 140, 616-625, 1986.
25. Thivolet CH, Chatelain P, Haftek M, Durand A, Pugeat M: Sucessful establishment of human insulinoma cell lines: morphological and functional characterization. Diabetologia, 29, 600a, 1986
26. Von Wasielewski E, Chick WL (eds): Pancreatic beta cell culture, Amsterdam: Excerpta medica ICS 408, 1976.
27. Lake SP, Anderson J, Chamberlain J, Gardner SJ, Bell PRF, James RFL: Bovine serum albumin density gradient isolation of rat pancreatic islets. Transplantation, 43, 805-808.
28. Huber W, Uhlschmid GK, Largiader P: Successfull transplantation of rat islets isolated by a new method. Transpl. Proc., 19, 924, 1987.
29. Hellerström C, Lewis NJ, Borg H, Johnson R, Freinkel N: Method of large-scale isolation of pancreatic islets by tissue culture of fetal rat pancreas. Diabetes, 28, 769-776, 1979.
30. Mourmeaux JL, Remacle C, Henquin JC: Morphological and functional characteristics of islets neoformed during tissue culture of fetal rat pancreas. Mol. Cell. Endocrinol., 39, 237-246, 1985.
31. Masquelier D, Amory B, Mourmeaux JL, Remacle C: Cell interactions during the in vitro neoformation of fetal rat pancreatic islets. Cell Differ., 18, 199-211, 1986.
32. Pipeleers D: Purified islet cells in diabetes research. Hormone Res., 23, 225-234, 1986.
33. Montesano R, Mouron R, Amherdt M, Orci L: Collagen matrix promotes reorganization of pancreatic endocrine monolayers into islet-like organoids. J. Cell Biol., 97, 935-939, 1983.
34. Meda P, Hooghe-Peeters EL, Orci L: Monolayer culture of adult pancreatic islet cells on osmotically disrupted fibroblasts. Diabetes, 29, 497-500, 1980.
35. Amory B, Mourmeaux JL, Remacle C: In vitro cytodifferentiation of perinatal rat islet cells within a tridimensional matrix of collagen. In vitro Cell. Dev. Biol., in the press.
36. McEvoy RC, Leung PE: Tissue culture of fetal rat islets: comparison of serum-supplemented and serum-free, defined medium on the maintenance, growth, and differentiation of A, B, ans D cells. Endocrinology, 111, 1568-1575, 1982.
37. De Clercq, Schmidt G, Delaere P, Remacle C: Nuclear events in B cells of young and senescent rat islets in organ culture. Cell Biol. Internat. Rep., 4, 817-827, 1980.

38. Swenne I: Role of glucose in the in vitro regulation of cell cycle kinetics and proliferation in fetal pancreatic B cells. Diabetes, 31, 754-760, 1982.
39. Reusens-Billen B, Remacle C, Daniline J, Hoet JJ: Cell proliferation in pancreatic islets of rat fetuses and neonates from normal and diabetic mothers: an in vitro and in vivo study. Horm. Metabol. Res., 16, 565-571, 1984.
40. Swenne I, Bone A, Howell SL, Hellerström C: Effects of glucose and amino acids on the biosynthesis of DNA and insulin in fetal islets maintained in tissue culture. Diabetes, 29, 686-692, 1980.
41. Ohgawara H, Carrol R, Hoffmann C, Takahashi C, Kikuchi M, Labrecque A, Hirata Y, Steiner DF: Promotion of monolayer formation in cultured whole pancreatic islets by 3-isobutyl-1-methylxanthine. Proc. Nat. Acad. Sci. USA, 75, 1897-1900, 1978.
42. Swenne I, Effects of cyclic AMP on DNA replication and protein biosynthesis in fetal rat islets of Langerhans maintained in tissue culture. Biosci. Rep., 2, 1867-1876, 1982.
43. Swenne I, Hill A, Strain AJ, Milner RDG: Growth hormone regulation of somatomedin C/insulin-like growth factor I production and DNA replication in fetal rat islets in tissue culture. Diabetes, 36, 288-294, 1987.
44. Rabinovitch A, Quigley C, Russel T, Patel Y, Mintz DH: Insulin and multiplication stimulating activity (an insulin-like growth factor) stimulate islet B-cell replication in neonatal rat pancreatic monolayer cultures. Diabetes, 31, 160-164, 1982.
45. Romanus JA, Rabinovitch A, Rechler MM: Neonatal rat islet cell cultures synthesize insulin-like growth factor I. Diabetes, 34, 696-702, 1985.
46. Chatterjee AK, Sieradzki J, Schatz H: EGF stimulates pro-insulin biosynthesis and ^{3}H-thymidine incorporation in isolated pancreatic rat islets. Horm. Metab. Res., 18, 873-874, 1986.
47. Nir U, Walker MD, Rutter WJ: Regulation of rat insulin 1 gene expression: evidence for negative regulation in nonpancreatic cells. Proc. Nat. Acad. Sci. USA, 83, 3180-3184, 1986.
48. Urist MR(ed): Fundamental and clinical bone physiology. Philadelphia: JB Lippincott Company, 1980.
49. Owen M: Lineage of osteogenic cells and their relationship to the stromal system. In Peck WJ(ed): Bone and mineral research, Amsterdam: Elsevier, vol.3, 1-25, 1985.
50. Roberts WE, Morey ER: Proliferation and differentiation sequence of ostoblast histogenesis under physiological conditions in rat periodontal ligament. Amer. J. Anat., 174, 105-118, 1985.
51. Majeska RJ, Rodan GA: Culture and activity of osteoblasts and osteoblast-like cells. In Butler WT(ed): The chemistry and biology of mineralized tissues. Birmingham: Ebsco Media Inc, 279-285, 1985.
52. Brand JS, Hefley TJ: Collagenase and the isolation of cells from bone. In Cell separation; methods ans selected applications. New York, Academic Press, vol.3, 265-283, 1984.
53. Nijweide PJ, Mulder RJP: Identification of osteocytes in osteoblast-like cell cultures using a monoclonal antibody specifically directed against osteocytes. Histochem., 84, 342-347, 1986.
54. Canalis E: Effect of growth factors on bone cell replication and differentiation. Clin. Orthoped. Rel. Res., 193, 246-263, 1985.
55. Sato G: this volume.
56. Foidart JM: this volume.
57. Watt FM: The extracellular matrix and cell shape. Trends in Biochem. Sci., 11, 482-485, 1986.

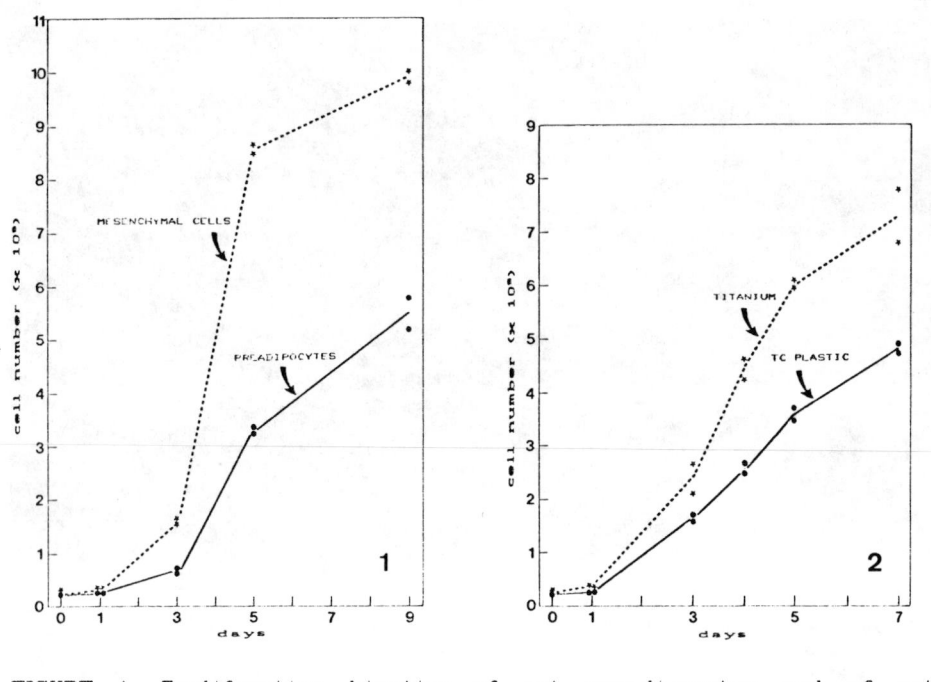

FIGURE 1. Proliferation kinetics of rat preadipocytes and of rat mesenchymal cells issued from muscle; 1st subculture; culture medium: DME, 10% fetal calf serum.

FIGURE 2. Proliferation kinetics of preadipocytes on two supports: Falcon Tissue Culture plastic, and titanium-coated bacteriological dishes; 1st subculture; culture medium : DME, 10% fetal calf serum.

FIGURE 3. Electron micrograph of differentiating preadipocyte in DME culture medium containing 10 % fetal calf serum, insulin and lipids. N: nucleus, L: lipid droplet. Bar=2 um.

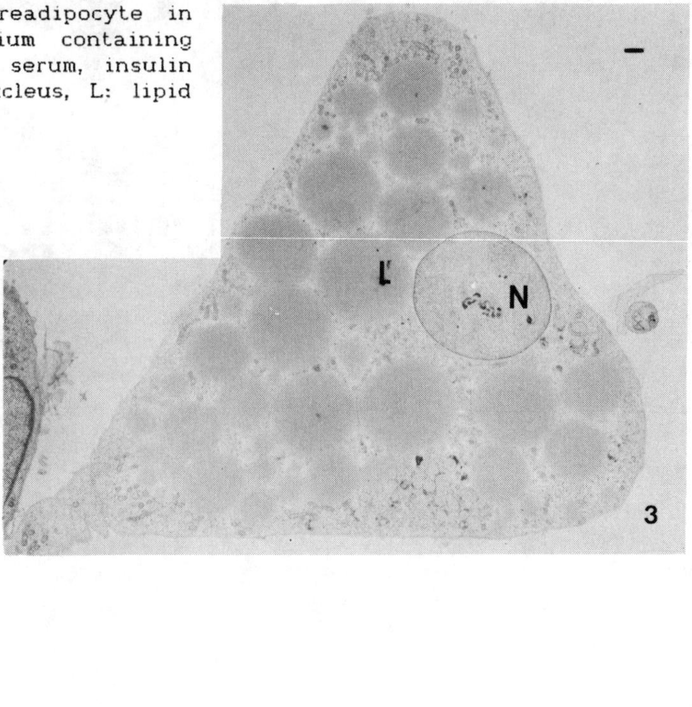

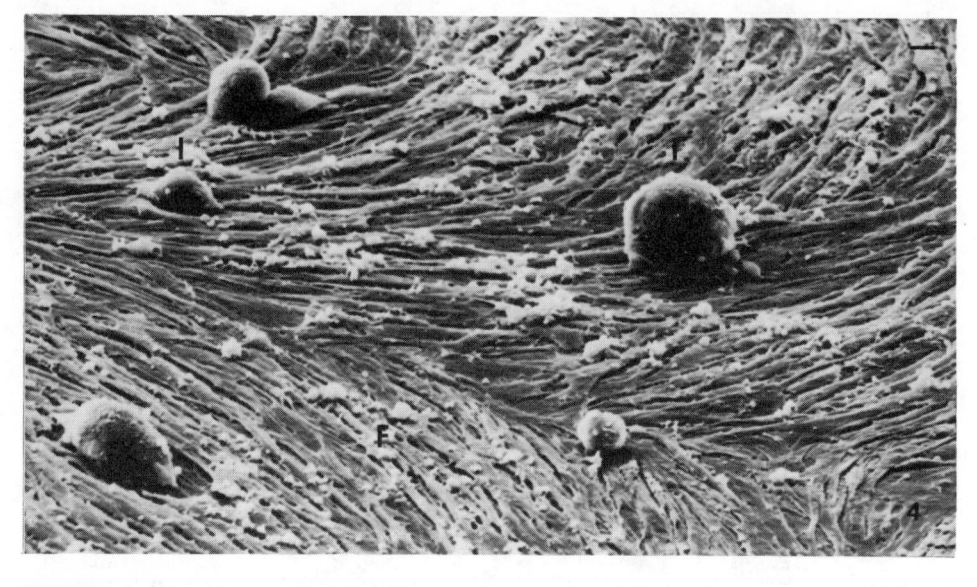

FIGURE 4. Scanning electron micrograph of islets neoformed from fetal rat pancreatic cell culture at day 7. I: endocrine islet; F: fibroblast layer. Bar=20 um.

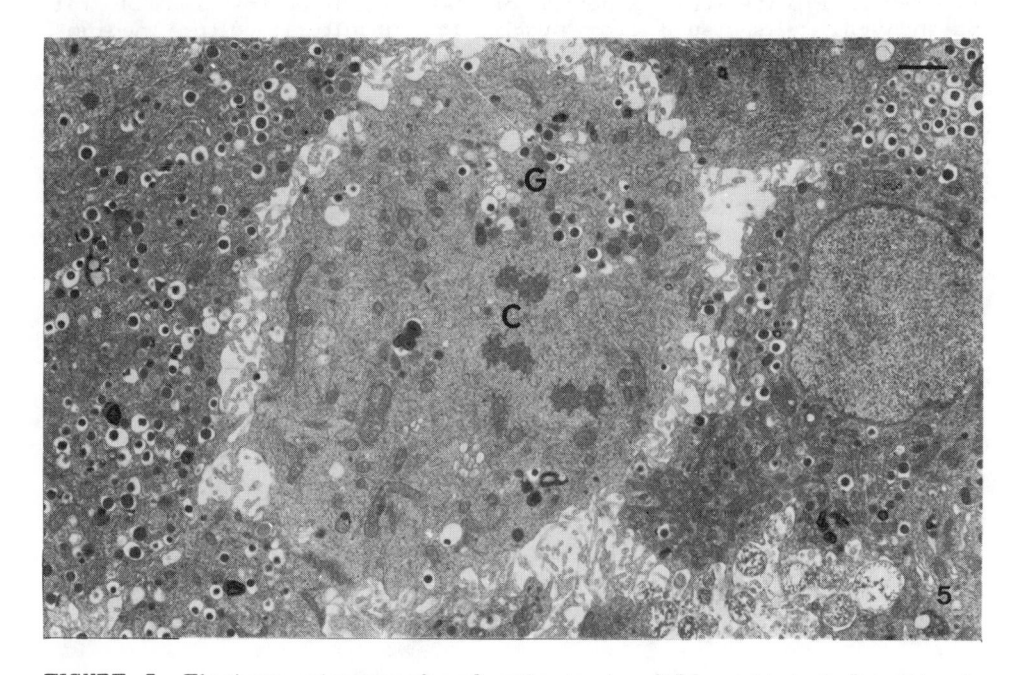

FIGURE 5. Electron micrograph of mitosis in differentiated B cell of adult rat islet cultured for 7 days. C: chromosomes; G: insulin granules. Bar=2 um.

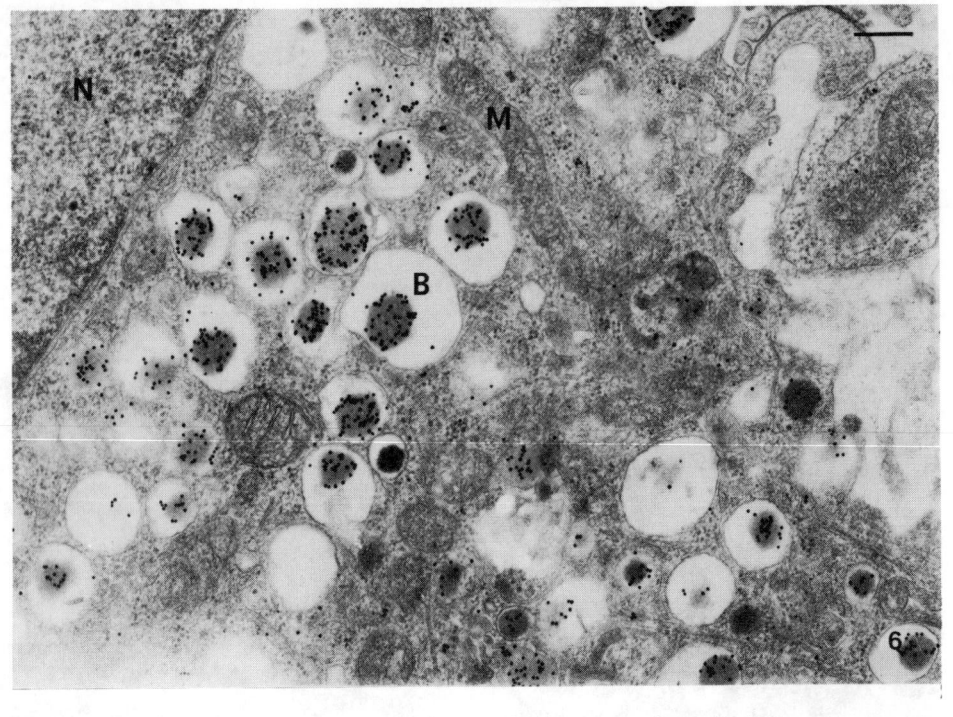

FIGURE 6. Electron micrograph of B cell in islet formed from pancreatic cell culture in tridimensional matrix of collagen. The grains of colloidal gold result from immunocytochemistry of insulin. N: nucleus; B: insulin granules; M: mitochondria. Bar=0.2 um.

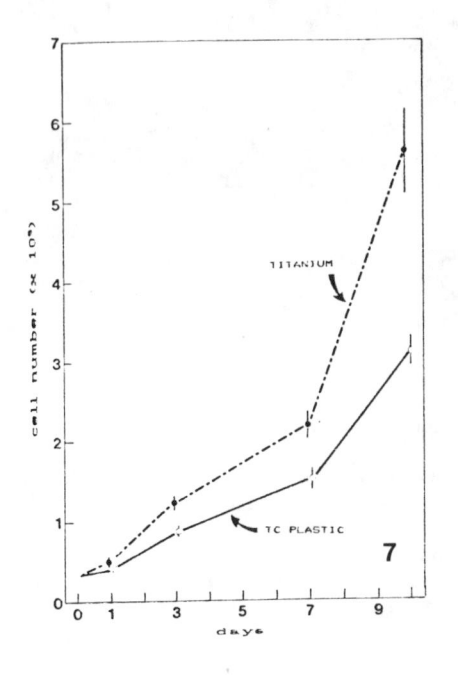

FIGURE 7. Proliferation kinetics of osteoblast cells cultured on two supports: Falcon Tissue Culture plastic, and titanium-coated bacteriological dishes. Culture medium: DME, 10% fetal calf serum.

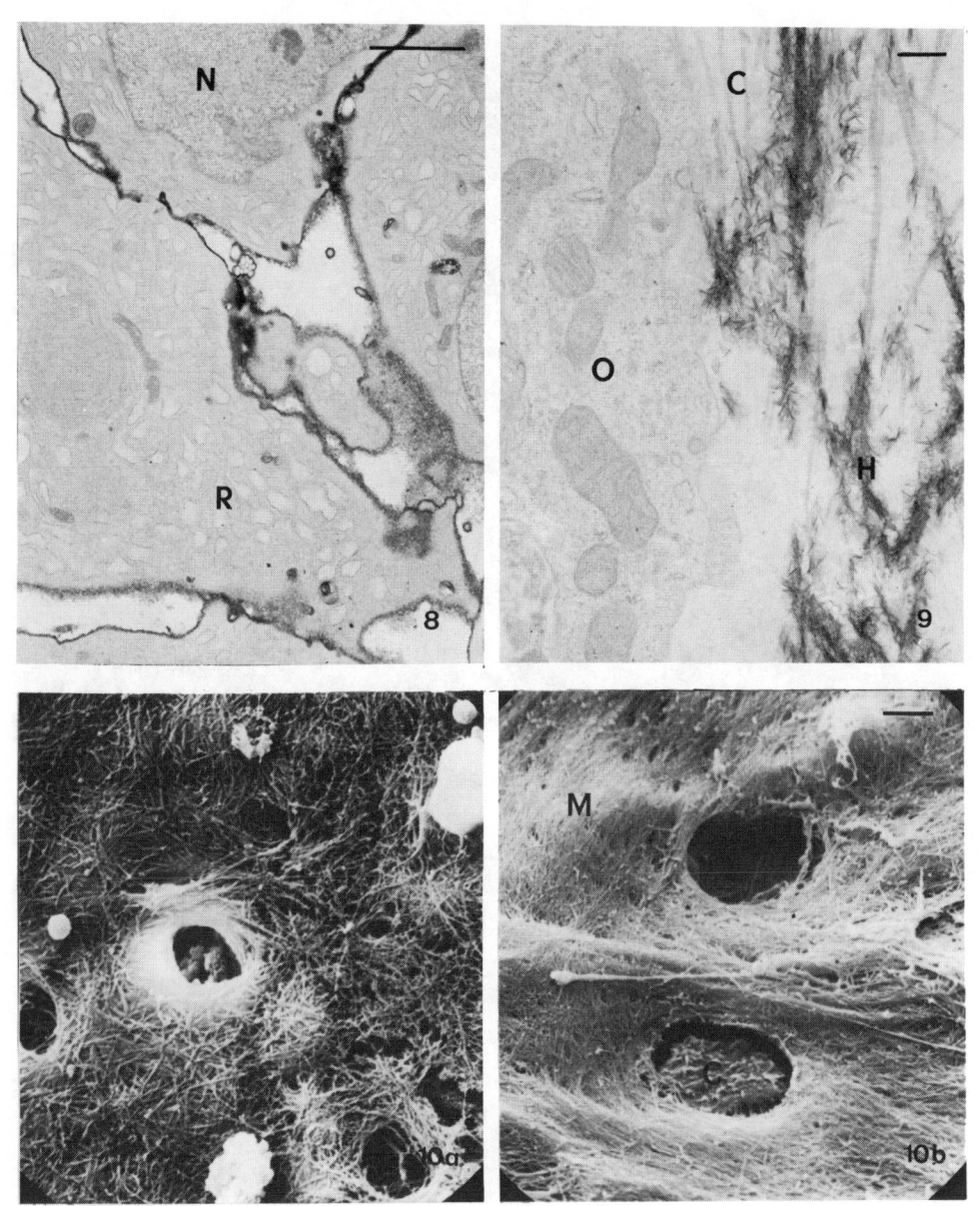

FIGURE 8. Electron micrograph of rat osteoblast primary culture, during proliferative state. The black border corresponds to the cytoenzymological reaction of alkaline phosphatase. N: nucleus; R: rough endoplasmic reticulum. Bar=2 um.

FIGURE 9. Electron micrograph of rat osteoblast primary culture, during matrix mineralization state. O: osteoblast; C: collagen fibers; H: hydroxy-apatite crystals. Bar=0.2 um.

FIGURE 10. Scanning electron micrographs comparing bone tissue formed in vitro (10a), and the endocranial surface of rat calvaria free of periosteum and osteoblasts (10b). M: bone matrix; C: cells. Bar=2 um.

Remacle, C. : <u>Why not use primary cells ?</u>

Miller :   It seems as if by playing carefully with the chemical
              composition of the medium, amongst which omitting the
              serum and using chemically-defined medium, the difference
              that John Petricciani mentioned between cells of finite
              life and CCLS would be blurred completely. So, does this
              difference really exist between the cells.   That is the
              first question and the second one is, does that mean that
              the serum contains factors which may promote senescence ?

Remacle :   Yes, maybe but I think that Dr Sato is far more competent
              in the field of defined media.   The work that I was just
              citing was published this year by LOO[1]. Their defined
              medium had a very simple composition. It contained only
              EGF as growth factor and the other usual increments of
              defined media. I can't remember its composition now but
              they really proved that the aneuploid cell line which is
              used of course has an abnormal chromosome number while the
              cell strain which has not undergone the senescence crisis
              has really a diploid chromosome number and no visible
              alterations.

              It is an essential work because it seems to go against the
              postulate of cell senescence.   There are also other
              indications namely in the group of SCOTT[2]. They have
              partially purified a protein isolated from serum which
              they called aproliferin and which inhibits the growth of

---

(1) Loo, D.T., Fuquay, J.I., Rawson, C.L. and Barnes, D.W. (1987). Extended culture of mouse embryo
cells without senescence : inhibition by serum. Science, <u>236</u>, 200-202.

(2) Weir, M.L., Scott, R.E. (1986). Regulation of the terminal event in cellular differentiation :
biological mechanism of the loss of proliferative potential. J. Cell Biol., <u>102</u>, 1955-1964.

various cells. They think that these kinds of proteins could be responsible in the serum-grown culture, of the senescence and of the proliferative crisis.

Baserga : There are a number of growth-regulation gene products that are either added into or secreted during the growth of cells and scientists have found that some of these genes are not expressed in primary cultures but become expressed in cells that have undergone the crisis, generating the interpretation that the cells have to adapt to a very hostile environment which is the environment in tissue culture and they do so by becoming hemizygous, expressing genes that are otherwise not expressed. That may explain why these cells become senescent.

Sherman : Just to add a point to Dr Remacle's discussion of factors in the serum which might suppress proliferations, Kurt HARRIS and his colleagues have shown that factors such as transforming growth factor β which in fact suppresses proliferation of cells - and I think that this also raises the general issue about how supposed transforming or proliferative growth factors - can under other circumstances have really quite opposite effects on cells depending on the composition of the medium.

Part III.

Large Scale Cultivation of Animal Cells

TECHNOLOGY OF MASS CULTIVATION : AN OVERVIEW

Raymond E. SPIER
Department of Microbiology - University of Surrey - Guildford, GU2 5XH, UK

1. Introduction

It is inevitable that in the development of a relatively new area of exper-
tise many of the basic concepts are yet in a state of some flux. We are
not clear about the units with which to define the performance of our
systems : or if we think that we know which units to use there is still
difficulty in relating figures couched in such terms to the problems that
the recipient of the information is grappling with. Again as a relative
newcommer to the biotechnology scene the proponents of the activity have
to contend with the many myths promulgated in the early days when is was
never clear why some experiments worked while others did not. As a result,
we have, in the field of animal cell biotechnology, a rich and diverse
array of alternative ways of tackling particular problems. This in itself
is both blessing and curse. On the one hand we can be very selective as
to which system we use to achieve a given end result; on the other hand we
can be easily bewildered by the wide choice available and put off decisions
until it has become more evident which system offers the prospects of productivity at the highest efficiency. More recently yet, it has become increasingly evident that the animal cell in culture has become the system
of choice for the expression of exogenous genes. For these reasons there
has been a burgeoning of the activity in this area as is obvious from the
number of conferences, books and articles dedicated to the use of animal
cells for the generation of commercially interesting materials.

2. The Units of Measurement

There are a number of techniques whereby products may be produced from
animal cells in culture. Each has its advantages and drawbacks. There is
as yet no consensus as to which of the in vitro technique is to be preferred and indeed the literature is replete with productivity data which have
little relevance to the "bottom line" which recounts the actual cost of
making a defined amount of material at a stated degree of purity with a

A. O. A. Miller (ed.), Advanced Research on Animal Cell Technology, 95–106.
© 1989 by Kluwer Academic Publishers.

quoted measure of biological activity per unit mass of material. At present we have a situation wherein the following parameters are quoted :

- the yield of product per unit animal cell, (or per unit or measure of biomass), tells the reader about the physiology of the cells and the degree to which those cells are performing in relation to some "animal cell biomass ultimate capability" ; it does not necessarily relate to economic or bioprocess performance,

- the yield of product per unit of medium used expresses one of a number of important factors which may be relevent to the overall economics of the process but which may also relate to the physiology of the cell, or when a recycle or make up system is used, it may relect a more materially efficient process ; clearly the two factors of overall raw material cost and the simplicity of the composition of the medium stream are some indication of the likely cost of the downstream processing operation,

- the yield of product per unit volume of bioreactor per unit of time is a measure of the efficiency of one of the key elements in the production system ; however, this factor can be misleading because in some bioreactors where the cells are held at high cell densities, (in excess of 10^8 cells/ml), either a large or small volume of medium may be made to flow through the cell containing part of the system ; under such circumstances it is more relevant to the determination of the space requirements of the bioreactor system to consider the volume of all the components which have been used to hold and feed the mass of animal cells in the system,

- the cost of a unit weight of the final product at a defined state of purity and activity is the most valuable determinant of the efficiency of a process, yet it is a parameter which is seldom determined and often difficult to acquire ; it is critically dependent on the performance of the cell and on medium utilisation efficiencies as well as on the nature of and performance of the downstream concentration and purification operation.

It is clearly necessary to arrive at a common way of evaluation of the alternative bioprocesses which are available and in a manner whereby those who make the decisions on the nature of the bioprocesses they chose to adopt are provided with figures they can rely upon.

3. Some myths relating to Animal Cell Biotechnology have been dispelled

It is the authors view that many of the supposed difficulties in the techniques of animal cell culture have resulted from observations made on cells

which were not in a prime physiological state.

- The Medium need not be expensive

It is held that the medium required for the growth of animal cells is expensive, complex and difficult to make with a high degree of reproductibility. The major cost of any defined culture medium (for such must be used in biotransformation activities) is in the labour and overhead costs which are similar for animal cell media as for media used to grow other classes of microbes. A second significant and equivalent cost factor is that incurred in the pre-use testing of the medium.

Whereas, in the past, the cost and the difficulty of obtaining serum batches with identical properties has led to problems, modern approaches to the "adapting or weaning" of cells on to both lower cost (by a factor of 10) and lower concentrations of serum (a factor of 2 to 5 is common) has reduced overall the cost of the medium considerably. Furthermore the use of modern techniques has also resulted in many production processes operating with media devoid of the serum component. The understanding of the need to pretest serum before use and to ensure that the production operation has adequate backup supplies of stored and pretested materials has aided the reproductibility of the operations considerably.

Further developments in the preparation of water (glass distilled and millicue treated) with its storage in a way which prevents the growth of contaminating and endotoxin producing microorganisms has removed yet another feature which has provided problems in the past.

- Almost any Cell can be grown

The difficulties which the animal cell technology pioneers experienced in obtaining cultures of the cell lines of their choice has been considerably allieviated. Current media supplemented with growth factors or immortalising agents specific for the required cell type have enabled most cells to become accustomed to the life in culture. Once this "transformation" has occurred, subsequent cultivation can be taken to ensure that during the adaptation to culture the specific properties for which the cell is needed are not lost.

The variability of the cells in culture has also been "tamed". This has been aided by off-line monitoring systems and by the present relative ease with which cells may be stored over liquid nitrogen and with which cells from the original population can be cloned ; a process which often results in the recovery of higher yielding cell populations.

- Animal Cells in Culture are not fragile

Much has been conjectured about the "fragility" of animal cells in culture. Those who have tried to determine the ability of animal cells to withstand shear fields have been surprised by the robustness of the cells rather than dismayed by their delicacy. Indeed when the cells are healthy physiologically it is difficult to disrupt them with a stirrer in a stirred tank reactor. Rotational speeds in excess of 600 rpm are required to observe such effects which could also be due to the phenomenon of vortexing which often occurs under such circumstances.

A second and related concept is that animal cells are destroyed by the air-liquid interface at the surface of a bubble such as in commonly used for the necessary culture aeration. This view is in need of modification consequent to some recent work which has shown that the cells are less prone to survive in the foam at the surface of the culture when that foam is unstable. A further factor is the statistical residence time of each cell of the culture in the zone of bubble disengagement as well as factors which relate to the physiological well-being of the cells.

- Contamination is avoidable

Contamination of the "rich" soup with which animal cells have to be provided is also alleged to be a disadvantage of the use of animal cells culture. Again the design of the modern fermenters, the care with which such fermenters are tested on installation and prior to use, and the clear definition of the operation procedures and the training of the fermentation technicians, are all paramount requirements in the operaion of a contamination-free system. Advances in all these areas has led to the routine operation of large-scale animal cell cultures with 2 % contamination rates in an industrial context. Also the new bioreactors have shown that culture lengths of several months can become commonplace even under circumstances in which fresh medium is provided to the system on a daily or weekly basis.

- Animal Cell Cultures can be scaled-up

That animal cell cultures can not be scaled-up is a commonly held view restricting developments. This contention is refuted by the 10.000 L scales of operation in the virus vaccine area and the larger units which have been designed to exploit genetically engineered recombinant animal cells. Also cells which require a surface for their growth can be grown in systems containing 1.000 L of medium or 10^{12} cells.

-- Animal Cell Cultures can operate at high biomass densities

While the early proponents of animal cell culture in the 1950s were impressed with their high density cultures expressing 10^5 cells/ml the culturists of the decades of 1960 and 1970 were comfortable with cultures whose cell concentrations ranged up to the low 10^6 cells/ml. Moreover, modern bioreactors designed to produce the equivalent of the secondary products of conventional microbial systems, operate for months at a time at cell concentations of 10^8 or higher cells/ml. This new development has to be seen in the light of the density of animal cells in the tissues of the body ($\pm$ 2 10^9 cells/ml at 12 microns cell diameter). This ability to operate at high biomass densities has destroyed the myth of animal cell systems necessarily representing low intensity operations. Clearly all the advantages of a high intensity system excreting products into the cell free compartment of the culture provides manufacturers with advantages in the smaller size and greater efficiency of the downstream concentration and purification processes.

An additional advantage consequent upon such high density culture systems is that the cells have much decreased requirements for medium fortified with animal serum. This enables further savings in the cost of medium and increases the ease with which final products can be prepared from crude culture supernatants.

-- Other Reasons for using Animal Cells in Culture

In addition to the above considerations there are also a number of practical and commercial factors which result in the promotion of the animal cell as a means of making products.

. In the event that one company has a patent based on genetically engineered prokaryote or yeast cell line which prevents other companies from making that product, then the use of a genetically engineered animal cell becomes an attractive alternative.

. The suspected contamination of a number of existing product materials derived from collected blood has led to the expenditure of effort to make the product materials from animal cells.

. The size, present and potential, of the market for materials which can only be made in animal cells in culture is expanding by leaps and bounds.

. The extension of our understanding in the processes of growth, replication and operation of animal cells in culture has led to increases in confidence in our ability to use such cells as components of reliable and

productive systems.

4. The reason that animal cells are the substrate of choice for the expression of exogenous genes

A combination of need and improvements in techniques and understandings has led to the present upswing in the level of activity in the use of animal cells in culture. The need was recognised in that the materials which can be made in genetically engineered prokaryotes did not perform as well as the native materials.

4.1. The post-translational processes

The linear polymer of amino acids which constitutes the back-bone of a protein molecule is but one of a number of features which defines the properties of that molecule in its biological context. The molecule coils, folds, pleats and loops according to the disulphide bridges which are formed and the way in which the environment, local to the newly formed molecule, influences the particular groups which are brought into juxtaposition. Yet not only is the final shape of the amino acid polymer determined by the environment in which it is formed but that folded molecule is then dressed up with an overcoat of added groups which are also crucial for the proper functioning of the finished product. Thus the nascent proteins are modified by :

- Glycosylation : a highly variable labelling device which determines the location and properties of the fully tailored molecule ; it also determines its antigenic properties and its longevity when injected into a recipient,

- Phosphorylation : many enzymes have been discovered recently which phosphorylate proteins on either tyrosine or threonine and serine residues ; the importance of these and other phosphorylation reactions is realised when such reactions seem to be implicated in the growth control mechanisms of the cell,

- Carboxylation, Sulphonation and Hydroxylation : reactions involved in bringing some of the blood clotting factors to an active state,

- Selenisation : an animal cell enzyme requiring step in the generation of the active form of the enzyme glutathione peroxidase,

- Signal peptide processing : determines the secretion, membrane insertion and intracellular location of the new molecule.

As both the enzymatic and immunogenic properties of the fully fledged protein molecules are determined by the three-dimensional structure of the

dressed-up molecule and the body of the animal cell is the natural site for such post-translational modifications, animal cells in culture have become the candidates of first choice for the expression of foreign genes.

5. Alternative Methods of Production of Materials from animal cells in culture

There are at present two main contenders for the preferred production system. One is based on the principle of providing each of the cells in the bioreactor with an environment which is completely homogeneous whereas the alternative technique allows the development of concentration gradients within the bioreactor such that some of the cells experience different conditions than their bioreactor cohabitants.

5.1. Homogeneous systems

Two bioreactor types vie for dominance in this area. The one is based on the mixing of the bioreactor contents using a stirring mechanism whereas the second system relies on the use of air bubbles to achieve the same effect and to aerate the medium at the same time. Both of these approaches tend to be characterized by short batch times, (figures of up to 10 days are given), even when some attempt is made to continually perfuse the system with fresh medium and the product streams are relatively dilute. There is also a greater tendency for such systems to be more demanding of medium containing added growth factors, be they whole serum or defined materials such as insulin, transferrin or interleukine 2. Where whole serum to be used in addition to the costs involved the protein load on the downstream processing operation leads to increases in cost per unit of material generated. In the event that a low protein plus specific growth factor route is chosen then the cost of the specialised materials becomes a significant part of the final bill for the product.

5.2. Inhomogeneous systems

There has presently become available a bewildering array of alternative systems for the cultivation and exploitation of hybridoma, (and other), types of animal cells in culture. In general the cells are immobilised either in containers or in the matrix of solid or semi solid materials. As they are not moving about the bulk of the bioreactor freely, the cells experience gradients in the supply of raw materials as well as in the concentration of waste materials and products. The kinds of alternatives available can be appreciated from the offered listing :

- Immured systems in which the cells are held within a container :

. Hollow microspheres made of polyornithine and polylysine,

. Immuration on the "shell side" of a hollow fibre bundle contained in an outer case,

. Immuration between sheets of parallel planar membranes,

- Entrapped systems wherein the cells are held within the interstices of a material or matrix of a gel :

. Entrapment within alginate, agarose or collagen particles which move freely through the body of the bioreactor,

. Entrapment within the crevasses of ceramic materials,

. Entrapment within an open sponge like matrix of polyester.

- Attached systems where the cells adhere to the surface of either the particles in a packed bed or to particles which can be distributed evenly throughout the bioreactor :

. The use of 3 mm glass spheres or other particulates, (stainless steel spings, diatomaceous earth, rashing rings, etc...),in columns has been used up to the 600 L scale :

. Microcarriers made from substituted dextrans, polypropylene, glass, cellulose and other materials have been used with varying degrees of success,

. Bundles of hollow fibres have also been used as surface dependent culture systems,

. Open ceramic matrices afford much surface for the growth of animal cells.

In each of the above cases run lengths of several months have been quoted and it is the experience of the investigators that there is little need for the complex or costly growth factors which compound the purification and economic problems. However, each of the systems mentioned above has draw-backs as well as advantages. Some of them are difficult to scale-up while others incur problems during the setting up phase. While the immured sys-tems are generally freely available from the equipment manufacturers the entrapped alternatives are often covered by patents which renders their use less free.

It is clear that from this welter of opportunities there are at present many options for the manufacture, on commercial scales, of proteins from animal cells.

6. Conclusions

6.1. An effective comparison between the alternative ways in which products

can be made from animal cells has yet to be made.

6.2. The use of high intensity bioreactors and simple defined media with long run lengths will improve the economics of bioprocesses based on animal cells to the extent that they can be regarded as competitive with those based on other microorganisms.

6.3. As experience accumulates many of the myths concerning the growth of animal cells in culture will recede into the mists of the past.

6.4. As the knowledge about the way in which gene expression in animal cells expands so will the biotechnologist be able to increase his/her ability to obtain from such cells products of value to the community of man.

Suggestions for further reading

1. "Animal Cell Biotechnology" (1985)

   Eds. R.E. Spier and J.B. Griffiths, Vol. 1 and 2 (3 in press July 1987), Academic Press, London.

2. "Large Scale Mammalian Cell Culture" (1985)

   Eds. J. Feder and W. Tolbert, Academic Press, London.

3. "Developments in Biological Standardisation"

   Vol. 42, 46, 50, 55, 60, 66

   Eds. W. Hennessen, R.E. Spier and J.B. Griffiths, Kager, Basle.

Spier, R. : <u>Technology of mass cultivation : an overview</u>

Horaud : Thank you very much Ray for this review which really underlines the basic problems of this technology, I mean, the choice. And the choice is not for me a matter of technology first of all because this equipment is extremely expensive and I have personal experience of being charged in Pasteur to organize such a laboratory. I am not as you know an expert in organizing a laboratory for large scale cultivation of animal cells. And when the person in charge of the project - who is present here, Otto Merten, started to present projects, all of us were flabbergasted by the price of this material. So ultimately you decide to proceed and to buy a piece of equipment of one million francs which is highly sophisticated having the piece of software, everything and then, two days later, you start to work and the propeller stops, the software is good but the propeller stops and the culture is ruined. This is exaggerated but I have some experience in this. Another problem is as follows : we are working expecially in the field of the vaccine production to transfer this product to other developing countries. In these countries it is not only a question of money, but of technology. And who knows, people working on genetic engineering on bacteria will be able to improve thermoresistance. Nobody knows. Because in science, the only thing that is predictable is the past and so, I think that the choice is really at the present time in large scale cultivation of animal cells. I was in the last meeting in Amsterdam of the European Congress of Biotechnology, twenty five companies were selling fermentors for large scale cultivation of animal cells and advertising that this particular piece of equipment they are selling is the best and well studied.

So, if you have specific questions on that particular subject, please, put them now.

Miller : You said that one use of animal cells was to turn around the patents if you are using bugs to produce biologicals. Do you think you really can bypass the Cohen Boyer patent ? I think you really have to pay. The patent covers the technique of DNA recombination wherever you use cells or bacteria. I do not think you can avoid that.

Spier : I think the concept of patent is - certainly in the UK system - that you cannot say you have got a patent until actually demonstrated by example a manifestation of the system. I do not believe that the Cohen Boyer patent actually included a manifestation of an animal cell system. But I think this is a technical point for the patent lawyers. I certainly think there is sufficient justification that if animal cells - even though they could have been mentioned in the Cohen-Boyer patent - can actually reproduce the method whereby that specific procedure could be demonstrated and exemplified to the point where somebody who is not expert in the field can actually take it on board and do it. In that case, that bit of the patent will not be valid.

Horaud : I would like to suggest that we do not discuss the problem of patent because we shall never get to the bottom of these problems. I can tell you that years ago, my director Jacques MONOD sent me to the United States to see a company making a product. So, when I came back, I pre-sented a report to Jacques and he said what is the busi-ness figure of that company ? It was, I don't remember fifty millions dollars to eighty millions of dollars and he asked me how many patents they were using. There are no

patents I said, this company has no patents at all. And the comments of Jacques MONOD was very nice : "We, here in Pasteur, we have thousands of patents and we have no money".

CELL CULTURE IN HOLLOW FIBRE BIOREACTORS

O.T. SCHÖNHERR

Cell Culture Dept.
R&D Diosynth
Oss, The Netherlands

1. INTRODUCTION

Mammalian cell technology deals with three interrelated cell culture aspects: cells and their characteristics, media composition and physical environment (see fig. 1).
· Cells are sensitive to shear forces, have a long generation time and are of the suspension or anchorage-dependent type.
· Culture medium is composed of a large number of chemically defined components, mostly supplemented with undefined protein-rich serum.
· Physical environment, like temperature, mass transfer and shear forces.

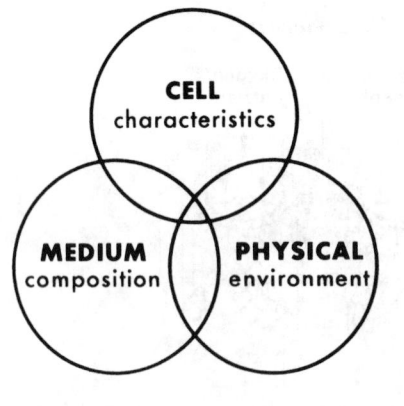

Fig. 1 Interrelated aspects of cell culture

A hollow fibre bioreactor aims to improve the conditions for cell culture, both on the level of physical environment and medium composition. Cells immobilized in the extra-capillary compartment of a hollow fibre bioreactor are much less exposed to shear forces than cells cultured in stirred or airlift fermentors. As firstly shown by Knazak (1972) cells can be maintained for several months in such a system. Nutrients and oxygen are supplied via the capillaries. Waste products and carbon dioxide are removed via the same membranes (see fig. 2).
The large surface area, provided by the hollow fibre membranes, makes the bioreactor a versatile system: both sus-

A. O. A. Miller (ed.), Advanced Research on Animal Cell Technology, 107–117.
© 1989 by Kluwer Academic Publishers.

pension and anchorage-dependent cells grow in these bioreactors. Membranes are available with various pore sizes. The mass transfer through the fibre wall of medium components and cell growth factors and cell products are influenced by the membrane characteristics, like wall thickness and cut-off values.

This communication shows that a large variety of cell lines can be grown in a culture medium without serum or other proteins at very high densities, if dialysis membranes are used. High concentrations of product are obtained and harvested on a semi-continuous basis.

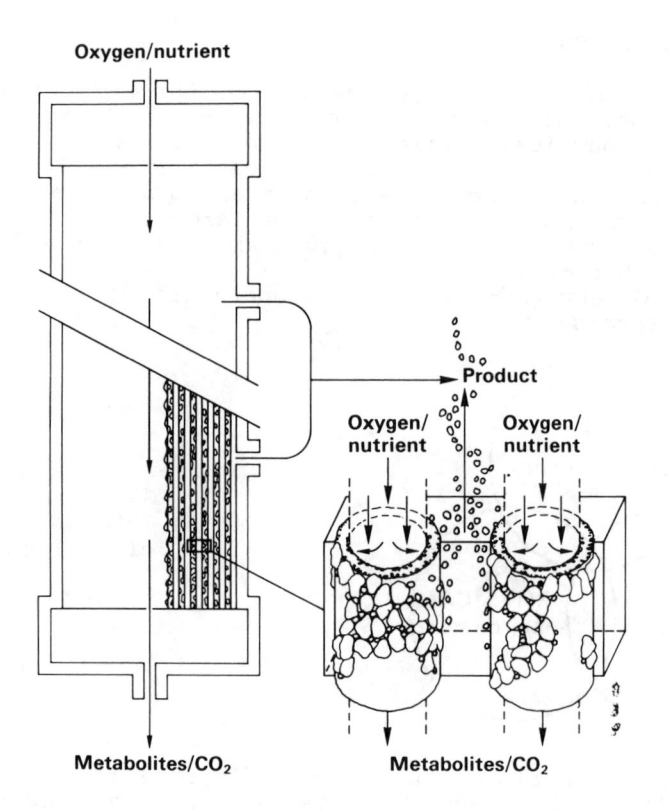

Fig. 2. Hollow fibre bioreactor. Cells are grown in the extra-capillary compartment of the bioreactor. Nutrients, wastes and gases are exchanged over the membranes. (Research within Akzo, 1987)

2. BIOREACTOR SYSTEMS

2.1. General
Hollow fibre reactors are used for haemodialysis of patients with renal failures and in laboratory and industrial processes for concentration, purification, dialysis and sterile filtration (Brock, 1983). Moreover, immobilization or compartmentalization has been applied for enzymes and microbial, plant and animal cells (Vick Roy, 1983).
Some physical aspects of bioreactors, like membrane characteristics, are of great importance.

2.2. Membranes
The main types of membranes in bioreactor are (modified) cellulose (acetate), polypropylene and polysulphone. Wall thickness varies between 5 μm of cellulose membranes in dialysis fibres (Cuprophan, Enka) to 100 μm of ultra- and microfiltration fibres.
The permeability of the membranes strongly influences the composition of culture medium in the cell compartment. Serum components and cell products may pass freely or are retained by the membranes (see table 1). The mass transfer of low molecular weight components, like glucose and amino acids, depends on the steepness of concentration gradients and characteristics of the membranes.
Oxygen and carbon dioxide may be exchanged as liquid soluble gases over the hydrophylic membranes or supplied by hydrophobic polypropylene or silicone rubber membranes.

MEMBRANES

Type/ Process	Dialysis diffusion	Ultra- filtration	Micro- filtration
Pore size (nm)	0,1-5	1-500	100-10.000
Mol.weight (Dalton)	10^3-10^4	10^4-10^6	10^5-
Permeability for	no	some	yes
- serum proteins			
- cell growth factors			
- cell products			
Clogging	no	yes	yes

Table 1: Characteristics of membranes used for hollow fibres

2.3. Cartridges
Commercially available haemodialysis cartridges (a.o. Organon Teknika, C.D. Medical) have been used for cell culture (see Schönherr et al., 1987b).
More expensive cartridges, especially designed for cell culture may be obtained from Enka or Endotronics. Ultrafiltration (Amicon, Queue) or microfiltration (Millipore, Enka) systems

have been developed for different purposes, but may be used
for cell culture as well. Table 2 gives an overview of some
commercially available cartridges.

Company	System	Membrane material	Extracapillary volume (ml)
Amicon	Vitafiber	polysulphone	2,5-250
CD Medical	Cell-pharm	cellulose (acetate)	75-90
Endotronics	Acusyst	cellulose acetate	20-100
Enka	dialyser	cellulose	20-1200
Organon Teknika	Allegro	cellulose	100-200
Queue/Monsanto	Hybrinet	polysulphone	45

Table 2: Some commercially available hollow fibre cartridges
and their characteristics

A slow transport of dissolved oxygen and carbon dioxide via
diffusion or ultrafiltration could lead of sub-optimal cell
culture conditions. A large membrane surface area (10 m^2 membrane per litre of extra capillary compartment), combined with
short distances (less than 100 µm) from fibres to cells are
provided by haemodialysis cartridges. This dense capillary
network resembles the situation in vivo, where the distance
between cells and blood capillaries never exceeds 100 µm
(Krogh radius). Several groups like Knazek et al (1972) and
Tharakan et al (1986) developed special prototypes of bioreactors with mixed fibres beds to improve gas transfer. However,
the construction of complex systems is expensive and the advantages of these bioreactors should firstly be demonstrated
with more experimental evidence.
Scaling up of cartridges can be done by connecting cartridges
in parallel (fig. 3) or constructing larger bioreactors.

Fig. 3. A hollow fibre bioreactor system. Eight haemodialysis cartridges are connected in parallel.

2.4. Culture medium

The nutrients for the dense cell culture in the extra capillary compartment are supplied by temperature, pH and PO_2 controlled medium flow via the lumen of the hollow fibres.
A high medium flow is needed to avoid steep concentration gradients both in the axial and radial direction within the capillaries.

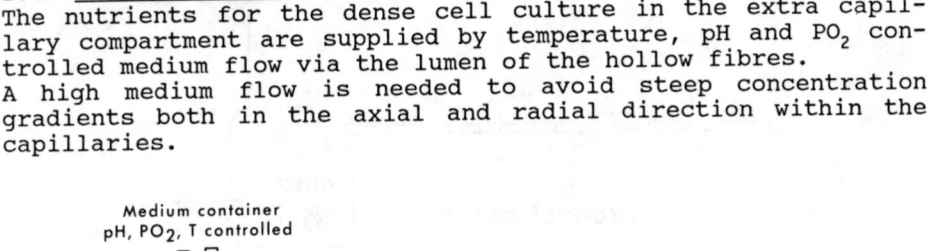

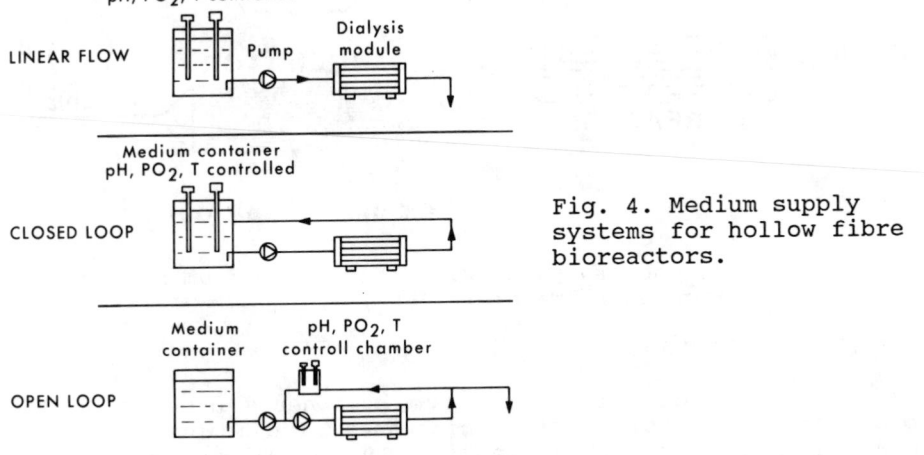

Fig. 4. Medium supply systems for hollow fibre bioreactors.

Fig. 4 shows three systems of medium supply:

· *Linear flow*: culture medium a bioreactor passes only once. The required high medium flux for optimal mass transfer leads to high consumption of medium. The effluent medium will be exhausted for oxygen only. Most of the nutrients will be present at nearly the original concentration and go to the drain.
· *Closed loop*: culture medium passes the fibres many times and is returned to the medium container equipped with sensors and heat- and gas exchangers for adjustment of temperature, pH and PO_2.
The medium in the container is periodically replenished.
· *Open loop*: this system is similar to the closed loop. However, the medium is controlled and adjusted in the loop. Replenishment of the medium is continuously.
A loop system for cell culture is shown in fig. 5. Medium is pumped through a hollow fibre oxygenator to a hollow fibre cell culture bioreactor. After passing the bioreactor the medium is enriched for oxygen and reduced in dioxyde content before it enters the capillaries of the bioreactor again.

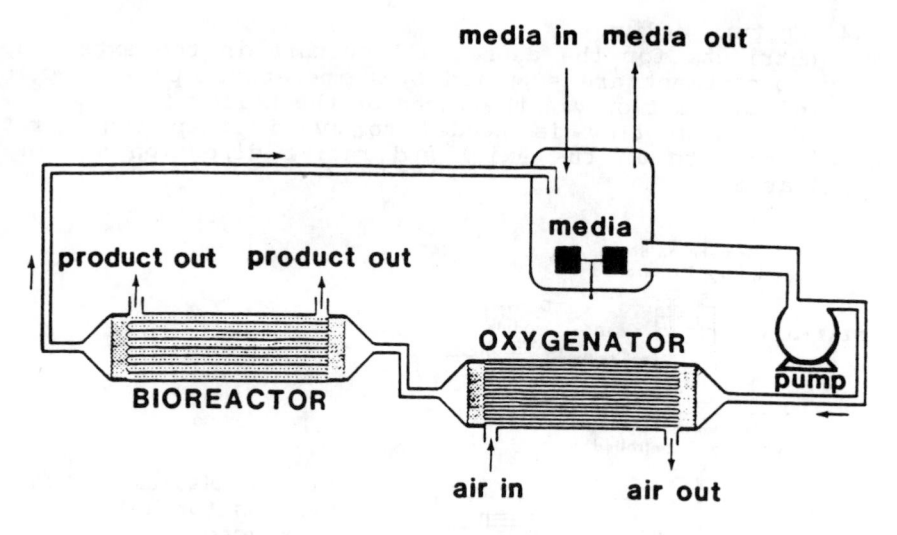

Fig. 5. Bioreactor system with loop for medium supply and oxygenator (Wateler, 1987).

3. SERUM REQUIREMENTS

The requirement of cells for serum components depends on the pore size of the hollow fibre membranes. The membranes with a relatively high cut-off (more than 10,000 Dalton) are permeable for most serum proteins. Moreover, growth factors which are produced by the cells are not retained and are washed out through the membranes. Bioreactors with low cut-off membranes, like dialysis membranes are permeable neither for serum proteins nor for growth factors. Schönherr (1987 a, 1987 b) described a hollow fibre dialysis system with a medium based on Dulbecco's Modified Eagle Medium and Ham's F12, without addition of any protein supplement on which cells could be grown for several months. The results were explained by the assumption that the dense cell cultures became self-supporting for growth factors, which were accumulated in the extracapillary compartment. In contrast, cell cultures in bioreactors with larger pore-size membranes do not retain growth factors and remain serum-dependent.

An advantage of s serum- and protein-free culture system is the absence of undefined components. This facilitates downstream processing to a great extend.

4. CELL CULTURE AND PRODUCTION

Hollow fibre systems have been shown by Endotronics and Schönherr (1987) to be very suitable for the production of mouse monoclonal antibodies. In dialysis cultures products were harvested on a semi-continuous basis. Concentrations of more than 10 g per litre have been attained, which is at the same level

as production *in vivo* in ascitic fluids and two orders of magnitude higher than conventional stirred or airlift cell cultures (see table 3).

Hybridoma	Monoclonal Antibody		
Clone	Concentration (g/l)	Productivity (g/l/24 h)	Total production per culture (g)
A	2,9	0,9	47
B	4,9	1,5	66
C	3,1	0,8	53
D	7,8	2,2	136
E	13,1	2,8	202

Table 3: Monoclonal antibody (MCA) production in hollow fibre bioreactors (Schönherr, 1987 b).

The cells have been grown for several months. Mass transfer over the membranes did not decrease over this period. Dialysis is based on a diffusion process. This process is much less sensitive to clogging than ultrafiltration or microfiltration. The dialysis hollow fibre system has also been used for other types of cells than mouse hybridomas. Human hybridomas and human EBV transformed cell lines, which are even more sensitive to shear forces, have successfully been grown in this system producing antibodies at similar concentrations. The system has also been used for monolayer cell lines like rec.DNA transformed and human tumour cell lines (unpublished results).

5. CONCLUSIONS

· Perfusion systems of dense cell cultures in the extra-capillary compartment of hollow fibre bioreactors were proven to be suitable for the propagation of fragile animal and human cells.
· Dialysis membranes have distinct advantages over ultrafiltration and microfiltration membranes because a completely protein-free medium can be used.
· Products like monoclonal antibodies can be harvested semi-continuously at concentrations of 10 g per litre or more than two orders of magnitude higher than conventional systems over a period of several months.
· The system can be scaled up to an industrial level and is suitable for the synthesis of large scale diagnostic and therapeutic products.

114

REFERENCES

Brock, T.D. (1983). Membrane Filtration. Springer-Verlag Berlin.

Knazek, R.A., Gullino, P.M., Kohler, P.O. and Dedrick, R.L. (1972). Cell Culture on artificial capillaries: An approach to tissue growth in vitro. Science 178, 65-67.

Schönherr, O.T., van Gelder, P.T.J.A., van Hees, P.L., van Os, A.M.J.M. and Roelofs, H.W.M. (1987 a). A hollow fibre dialysis system for the in vitro production of monoclonal antibodies replacing in vivo production in mice. Dev. Biol. Stand. 86, 211-220.

Schönherr, O.T. and van Gelder, P.T.J.A. (1987 b). Culture of Animal Cells in hollow fibre dialysis systems. Animal Cell Biotechnology, volume 3, in press.

Tharakan, J.P. and Chau, P.C. (1986). A radial flow hollow fibre bioreactor for the large-scale culture of mammalian cells. Biotechnology and Bioengineering 28, 329-342.

Vick Roy, T.B., Blanch, H.W. and Wilke, C.R. (1983). Microbial hollow fibre bioreactors. Trends in Biotechnology 1, 135-138.

Wateler, F.R.G. (1987). Membranes in animal cell culture systems. To be published.

Schönherr, O : Culture in hollow fibers

Hope : I have one particular question regarding that last slide that you showed. How many cells were in the system at that time ?

Schönherr : It is difficult to know but approximately 2×10^8 cells/ ml and that is a very dense culture.

Van Meel : The importance of _in situ_ monitoring your production process was stressed earlier today. I do not think that your system allows you to measure actually what is going on in the cellular compartment. Do you think that because of this impossibility, the system is still acceptable by authorities for the production of biologicals for human use ?

Schönherr : That is a very clear question and I have asked it to myself. If we are to discuss that with the authorities, we have to do this on a case by case basis and for each product, we discuss with the authorities.

Merten : Of course you can not monitor the cells directly but do you use some indirect methods such as the consumption of glucose or the consumption of oxygen ?

Schönherr : Yes, these things are very easy to be tested outside in a non sterile way.

Merten : Normally you have only extrapolations with these indirect methods because you never know the physiology of the cell. It is the same situation which prevails in a Roux bottle or in a Roux flask. So you don't have real numbers but only an estimation.

Schönherr : Yes that is true.

Merten :   What is the viability ? Do you have a good viability ?

Schönherr : Yes, we get high viabilities. The distance from membranes
to cells is under 100 μ and that is comparable to the
in vivo situation when you are growing hybridomas in mice.

Drillien : Is it correct that according to the kind of hollow fiber
that you choose, you might do either continuous collection
of your product or collection say, at the end of a long
period of time ?

Schönherr : Yes.

Drouxhet : Interferon production from cells and particularly by
lymphoblastoid cells is well known now. We read a few
months ago in some papers that these cells produce BCGF in
culture medium for some lines and I think it is not
necessary to use hollow fiber systems to be able to use
serum-free medium. We have some media with proteins under
50 μg/ml and we have very good growth and very good
antibody production, in common cell culture, in flask or
in fermentor like Setric's.

Schönherr : Do you preadapt your cells ?

Drouxhet : Yes but within two weeks, growth was satisfactory and when
we tried not to adapt our cells, growth was good also.

Merten :   Did you try to isolate such growth factors ?

Schönherr : No but such a system is very nice of course to isolate
growth factors.

Drouxhet :   These growth factors may be interesting by-products of lymphoblastoid cultures which are now used for human monoclonal antibodies.

Spier :   Have you found the same thing happening for say-recombinant cells- which are not primarily from lymphocyte lineage ?

Schönherr : Yes.

Spier :   Which system do you use for your recombinant cells ?

Schönherr : Many people are using it.

Horaud :   That is a good example of where is the biotechnology at the present moment. Everything is covered by the secret of industry. The secret here is quite open ! These are CHO cells - because of the gene amplification system.

Schönherr : Yes.

Horaud :   So, let us discuss it a little openly ! Did you try CHO cells Ray ? What sorts of cells ?

Spier :   The CHO recombinant cells in suspension are no problem. That is their secret. The question I was asking was basically to follow up the previous question : can you get a serum-free system using a recombinant cell of a non lymphoid lineage which does not have B-cell growth factor ? And the answer to that I think is in some cases you can and in some cases you can't.

PERFUSION CULTURE

WILLIAM R. TOLBERT AND CHRISTOPHER P. PRIOR*

INVITRON CORPORATION
ST. LOUIS, MISSOURI, 63134, U.S.A.

1. INTRODUCTION

In nature growth of microorganisms is generally a batch phenomenon. Individual, microbial cells are completely self-sufficient and proliferate to the extent permitted by nutrients and other conditions in their environment. This is mimicked by batch fermentation methods in which a culture of microorganisms is inoculated into a rich nutrient medium and allowed to grow to its maximum density limited by nutrient depletion and waste product accumulation. From one point of view, microorganisms have evolved specifically to meet the changing conditions that are found in the batch fermentation process. Conversely, cells of higher animals have evolved as integral components of a multicellular organism such that each cell of that organism exists in a highly regulated environment that provides constant levels of nutrients, metabolic products and complex hormonal and biochemical control mechanisms to allow optimum differentiated cellular functions.

In attempting to grow animal cells in vitro, it would seem logical to attempt to approach the in vivo environment as closely as possible. The best method currently available to accomplish this goal involves perfusion of fresh nutrients into the culture and removal of metabolic products. The idea of perfusion associated with mammalian cell culture is not new. Attempts were being made in this direction even early in this century (1). Himmelfarb, et al., reported the first laboratory scale perfusion culture for suspension cells in 1969 (2). This spin filter system was later sold commercially by VirTis Corporation in 500ml and one liter scale versions. In 1972, Knazek et al., published the first paper on the use of hollow fibers for cell culture (3). A number of widely different types of perfusion systems have been utilized and reported in the literature (4,5,6,7). Various approaches have ranged from columns of glass beads or glass helixes through hollow fibers and reeled Teflon tubes to various types of spin filters and rotating cups; all intended to better maintain cellular environment by continuous transfer of nutrients and removal of waste products. Very few of these methods, however, have been applied to large scale production of pharmaceutical products. Partially, this is a problem of scale and partially a problem with the complexity of these systems.

A report in the July, 1987, Biotechnology Journal estimated a 1991 market as large as $23 billion per year in annual sales for products that may be made by animal cell culture (8). That represents double digit millions of liters per year of requirement for animal cell culture capacity. While improvement of mammalian cell expression systems can reduce the total volume required, significant expansion of pharmaceutical-based large scale manufacturing will be necessary to meet this challenge.

*delivered by J. Hope

A. O. A. Miller (ed.), Advanced Research on Animal Cell Technology, 119–137.

Very large batch fermentation systems have been successful at this
scale and beyond for products from microorganisms. However, such systems
may not be optimum for mammalian cell production of new complex protein
biopharmaceuticals due to degradation and inactivation of biological
functions accompanying long exposure to the culture environment. While
perfusion methods with rapid removal of product from the cells can
prevent degradation and maintain biological activity, very few perfusion
systems are available which could meet these capacity requirements. This
paper will review various approaches to perfusion culture and describe
one approach that has been operated at large scale in a "state of the
art" pharmaceutical manufacturing facility.

2. PERFUSION BIOREACTORS

Use of perfusion in large scale mammalian cell culture has a potential
for major benefits in maintaining a constant cellular environment and
providing a means for more closely simulating in vivo conditions. It is
under these conditions that more efficient and more natural production of
biomolecules should be possible. A number of moderate to large scale
perfusion systems have been used commercially and particularly in
pharmaceutical manufacture. These systems generally fall into two major
categories, recirculation or single pass perfusion systems.

The Opticell System currently being sold and utilized by Charles River
Biotechnical Services, Inc. is an example of a small to intermediate
scale recirculation system. Cells are retained within a ceramic monolith
which consists of a ceramic cylinder with a multitude of rectangular
holes penetrating its entire length. Cells reside or are trapped within
these holes while media is perfused from a reservoir through the length
of the device and back to the reservoir. Oxygen and pH levels are
monitored and oxygen added through a hollow fiber oxygenator in series
with the reactor. A number of different cells have been grown utilizing
this system (5,9). Anchorage-dependent cells attach to the inside of the
tubes and anchorage-independent cells such as hybridomas are also somehow
retained within the device. Maximum size monoliths have a surface area
of 18.2 square meters as compared to some large scale microcarrier
systems which can have as much as a 1000 square meters of surface area.
This is then a more intermediate than large scale system.

Very similar systems are used in conjunction with hollow fibers for
growth of both anchorage-dependent and independent cells. In this case,
the cells are retained exterior to the fibers within a generally
cylindrical housing while oxygenated medium recycles rapidly through the
lumin of the fibers. Both Endotronics and BioResponse have used such
systems for cell growth and production of a variety of products. Due to
the difficulty in construction of the hollow fiber cartridges themselves,
these systems are generally limited to a hollow fiber surface area of
approximately one square meter, and while up to ten may be manifolded
together in a single unit, they are still in the intermediate range (10).
Another example of a recirculation system is a fluidized bed device such
as that used by Verax Corporation (5). Generally, such a reactor
consists of a bed of relatively heavy particles that are partially
suspended by a rapid flow of liquid from below. Due to particle weight,
a cleared zone forms above the bed so that the free liquid can be
recycled back to the bottom. In the case of the Verax system, weighted
sponge particles are used to form the bed and cells, such as hybridoma
cells, are trapped within the pores of these sponge particles. A hollow
fiber oxygenator is used in the recycle circuit.

All of these recirculating systems can be converted to perfusion systems by addition of fresh nutrient medium to the recycle circuit and removal of an equal quantity of cell conditioned medium. The recirculation systems themselves provide separation of the cells from the cell conditioned medium allowing this conversion to perfusion to be relatively simple. However, in practice, many of these systems are not set up for continuous perfusion but rather, the recirculating reservoir is periodically changed so that levels of nutrients and waste products have a step function relationship more similar to cyclical batch operation than to perfusion. With this periodic change of the recirculation reservoir, the cells see a continually changing environment from high nutrients, low metabolic products immediately after change and low nutrients, high metabolic products immediately before change. There are several other variations of these systems now coming on the market at a laboratory scale which may eventually be converted to production scale systems.

Several direct perfusion systems have been used for pharmaceutical production. The reactors used by Damon Biotech for their encapsulation technique allow continuous perfusion of nutrients into the vessel and removal of expended medium through an overflow pipe. The Damon system has a mixture of perfusion and batch attributes. While nutrients are continually added and metabolic products are removed, cells and their protein products are retained within the microcapsules throughout the general 20 day production runs (5). At the conclusion of the run, the cells and protein products are harvested batchwise by breaking open the capsules. The process can then be reinitiated with a fresh inoculum of cells in new microcapsules.

Van Wezel introduced a spinning cup separator located on the stirring shaft of the fermentor (11). Cells attached to microcarriers were maintained within the fermentation vessel but outside of the spinning cup. The product containing conditioned medium was aspirated from the cup and replaced by fresh nutrient medium. The system has been used to produce tissue plasminogen activator from Bowes melanoma cells as well as other products. More recently, Reuveney (6) has used a similar device for monoclonal antibody production and reported over 2×10^7 cells/ml with almost 400 micrograms/ml of antibody, significantly higher than batch methods. He noted as disadvantages to this system, its increase complexity and clogging of the rotating filter as early as seven days into a production run. Sato, et al., described use of a settling cone around the shaft of a perfusion reactor to separate cells from the cell conditioned medium (12). This reactor has been used in a size as large as 40 liters for production of alpha interferon from the Namalwa cells (13).

3. INVITRON PERFUSION SYSTEMS

The Invitron Corporation, since mid 1985, has been involved in pharmaceutical production of a number of different products from a variety of cell lines using a perfusion based mammalian cell culture technology. This technology involves three interrelated large scale mammalian cell culture manufacturing systems. They have been designed to closely simulate the natural environment of cells by supplying nutrients and removing waste products as a continuous and carefully regulated process. These systems have been designed to optimize cell density, minimize the requirement for animal serum in the nutrient media and retain flexibility necessary to accommodate the variable growth

characteristics of different mammalian cell lines. In order to provide
as uncomplicated and efficient reactor design as possible, the
environment in which the systems operate has become an integral part of
their control. For example, the reactors containing mammalian cells are
positioned within 37°C incubator rooms. Adjacent are 4°C refrigerated
rooms so that the cell conditioned medium containing product may be
rapidly removed from the high temperature reactor environment to a
refrigerated storage vessel. By storing product at low temperature, most
enzymatic activity is slowed and the opportunity for product degradation
by proteolytic enzymes in the cell conditioned medium is greatly reduced.
This is of significant importance for enhancing the stability and
specific activity of certain human pharmaceutical proteins.

Another very important feature of this production method is the
ability to operate completely without the use of antibiotics.
Antibiotics are frequently used in large scale cell culture to prevent a
single microorganism from contaminating an entire production run. This
is a major economic risk which increases exponentially with the size of
the culture. The production reactors, proprietary aseptic connection
systems and the facility were specifically designed to maintain the
highest level of environmental control in the production process. As a
result, the systems described here are able to operate at high cell
density for many months without antibiotics. The use of antibiotics in
large scale cell culture has several disadvantages. Their use is felt by
some to impact negatively on the safety of the product and can be
detrimental to expression levels of monoclonal antibodies from certain
hybridoma cell lines. Antibiotic use encourages the growth of low levels
of resistant organisms which are difficult to detect and can result in
adulterated, unsafe drug products. Antibiotic use also tends to promote
less than optimum aseptic technique which could lead to other types of
contamination or loss of product integrity.

The first of the three production systems, the perfusion chemostat
system, has been designed for the growth of mammalian cells in
suspension, either as single cells or aggregates of cells (see Figure 1).
A constant environment is maintained in the growth vessel by continually
adding fresh nutrients and removing waste products at a rate proportional
to the number of cells in the vessel. This growth vessel has a
proprietary system used to provide very gentle agitation without cell
damage at the high densities obtained. The filtration unit separates the
cells from the cell conditioned medium in an external recycle circuit and
returns the cells to the growth vessel as fresh medium replaces the
conditioned medium removed. By placing the filtration unit outside of
the reactor, this unit can be replaced periodically when it becomes
clogged and the reactor run continued for as long as several months. A
separate harvest vessel is used to remove excess cells when the desired
density level is attained thereby stabilizing not only the levels of
nutrients and metabolic products, but also the cell number within the
vessel for a constant rate of product production. Dissolved oxygen and
carbon dioxide, as well as pH levels, are continually monitored and
maintained at preset ranges throughout the production run by a
proprietary computer software and process control system.

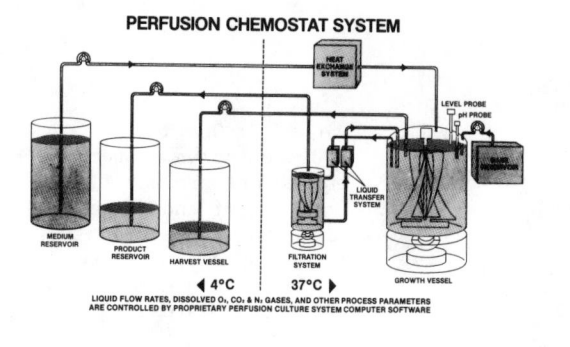

PERFUSION CHEMOSTAT SYSTEM

FIGURE 1: PERFUSION CULTURE SYSTEM: Liquid flow rates, dissolved O_2, CO_2, and N_2 gases, and other processs parameters are controlled by proprietary perfusion culture systems.

The second perfusion system, the microcarrier reactor system, is used for anchorage dependent cells (see Figure 2). This system has added a settling vessel to allow the cells attached to the microcarriers to fall back into the reactor and not be carried to the somewhat higher shear region of the filtration vessel. Filter surface area in the range of 500 to 1000 square meters may be used. Figure 3 shows operation of the 100 liter scale perfusion system within the pharmaceutical manufacturing facility. These reactors are generally operated at cell densities 10 times the maximum batch density, so that each will contain equivalent cells to a 1000 liter conventional vessel.

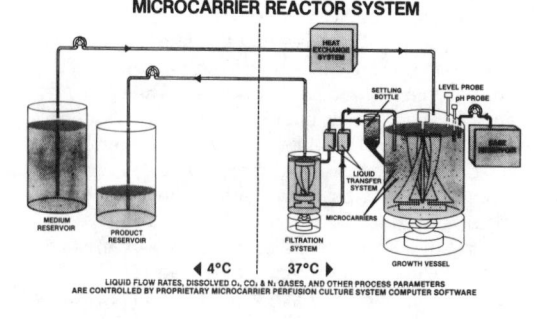

MICROCARRIER REACTOR SYSTEM

FIGURE 2: MICROCARRIER REACTOR SYSTEM: Liquid flow rates, dissolved O_2, CO_2 and N_2 gases, and other process parameters are controlled by proprietary microcarrier perfusion culture system computer software.

FIGURE 3: Operation of a 100L perfusion system has equivalent product capacity of thousands of liters of conventional batch suspension system. A typical perfusion culture run averages 30-90 days.

In the third culture system, the Static Maintenance Reactor (SMR), cells are immobilized at high density in a matrix material so that they may be maintained for long periods of time producing products. Many valuable proteins produced by cultured cells are made more efficiently and abundantly during a non-proliferative phase. In vivo most cells are not actively dividing but are instead expending their energies making and secreting proteins or performing other functions. By simulating the in vivo environment, the SMR enables cells to exist in a state of low or non-proliferation for extended periods of time while producing desired cell derived products. Serum or other expensive medium components used to stimulate proliferation are unnecessary which significantly reduces both upstream and downstream costs. The SMR can be used to maintain both suspension and anchorage-dependent cells at densities approximately a hundred times that obtained in a conventional batch system (see Figure 4). One to two kilograms of cells are grown in either of the previous perfusion systems. They are then concentrated, mixed with a finely divided non-toxic matrix material and pumped as a thick slurry into the SMR. The reactor is filled completely with cells immobilized in the interstices of the matrix material. Nine tenths of the reactor volume is matrix and one tenth evenly distributed cell mass. Porous tubes penetrate the immobilized bed of cells and matrix material to allow perfusion of nutrient medium. A distributed, semi-permeable membrane is also used to allow diffusion of gases. These reactors have been maintained in continuous operation for as long as five months continually producing product (see Figure 5).

STATIC MAINTENANCE REACTOR

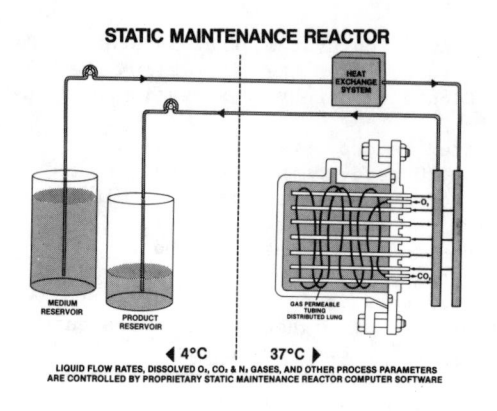

LIQUID FLOW RATES, DISSOLVED O₂, CO₂ & N₂ GASES, AND OTHER PROCESS PARAMETERS
ARE CONTROLLED BY PROPRIETARY STATIC MAINTENANCE REACTOR COMPUTER SOFTWARE

FIGURE 4: STATIC MAINTENANCE REACTOR: Liquid flow rates, dissolved O_2, CO_2, and N_2 gases, and other process paramaters are controlled by proprietary static maintenance reactor computer software.

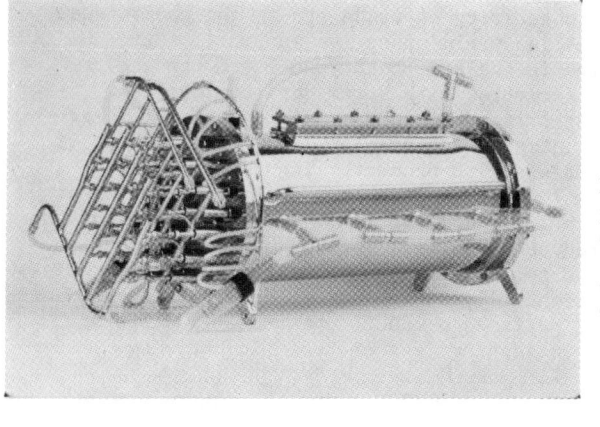

FIGURE 5: The Static Maintenance Reactor daily produces multigram quantitites of medically important proteins. A typical SMR run averages 90 days.

4. PRODUCT INTEGRITY

Perfusion systems for products that are secreted into the medium provide the opportunity for significant improvement in biological activity of over that possible with batch or semi-batch methods. Many processes may occur during the time that secreted proteins are exposed to the cell culture environment which are detrimental to their integrity and biological function. Proteolytic enzymes, for example, may be present from serum supplements or from the cells, particularly if cells are allowed to die during extended production under batch operation. These secreted proteins may also become irreversibly complexed with inhibitor molecules which are contributed by the serum supplement, as with tPA, or which are naturally secreted by the cells. Quite often, cells that secrete active molecules also secrete inhibitors as part of an in vivo feedback control mechanism. Other inactivating processes include aggregate formation and oxidation or other chemical modification of the desired protein structure. In all cases, the rate at which theses inactivating processes occur is related to the temperature of the culture environment. By lowering the temperature, the inactivation process can

126

be greatly reduced if not eliminated. Perfusion systems that separate
the cell conditioned medium containing product from the cells allow the
possibility of rapid temperature reduction and thereby preservation of
product integrity.

We have arranged our computer automated reactor stations in 37°C
incubator rooms immediately adjacent to 4°C cold rooms. Residence time
for product at 37°C is one or two days for a 100 liter perfusion culture
system or microcarrier reactor and only four hours for the Static
Maintenance Reactor. In contrast, batch production methods expose
products accumulating in the reactor to degradative conditions throughout
the length of the run which can extend to weeks in many cases.
Semi-batch systems allow reduced contact time at higher temperatures but
still involve several days at 37°C. Figure 6 shows growth curves of a
recombinant CHO cell line used for production of tPA. The curve labeled
III shows a batch production, the one labeled II a semi-batch, and the I
curve, perfusion production. The shorter product residency time in the
perfusion vessel (hours) compared to semi-batch (days/weeks) and batch
(weeks) improves the specific activity of crude tPA in the conditioned
medium and increases purification yield of active monomeric tPA. The
longer product residency time in the semi-batch and batch operated
vessels resulted in complex formation with serum derived inhibitor
proteins and aggregation of tPA as a consequence of oxidation. These
aggregates possess very low activity and pose an immunogenic threat if
co-purified as trace contaminants with product. Figure 7 shows
corresponding profiles of tPA eluted from the gel filtration column at
the end of the purification process and Figure 8, electrophoresis of the
column fractions on a 10% reduced SDS gel. A major portion of the
protein is irreversibly bound to inhibitor under batch conditions, a less
amount in semi-batch, and essentially no inhibitor complex is shown under
perfusion. In addition, the perfused product contains a significant
portion of single chain tPA indicating less proteolytic activity.

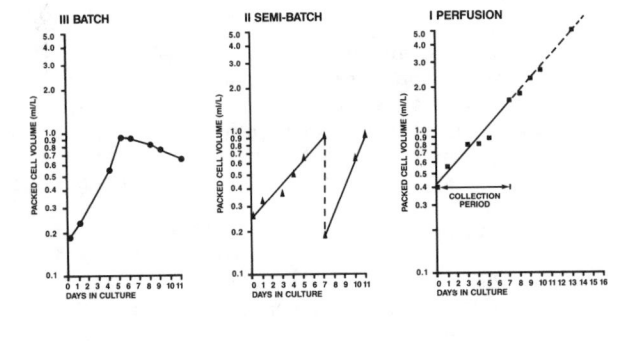

FIGURE 6: Growth
curves for recombi-
nant CHO cells pro-
ducing tPA. Harvests
of cell conditioned
medium containing
7.5% FBS are indi-
cated.

FIGURE 7: Profiles are
purified tPA eluted from
gel filtration column
corresponding to
perfusion (I), semi-
batch (II), and batch
(III) production methods.

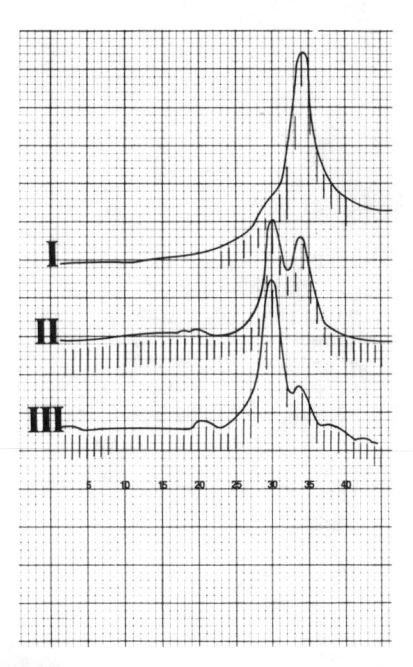

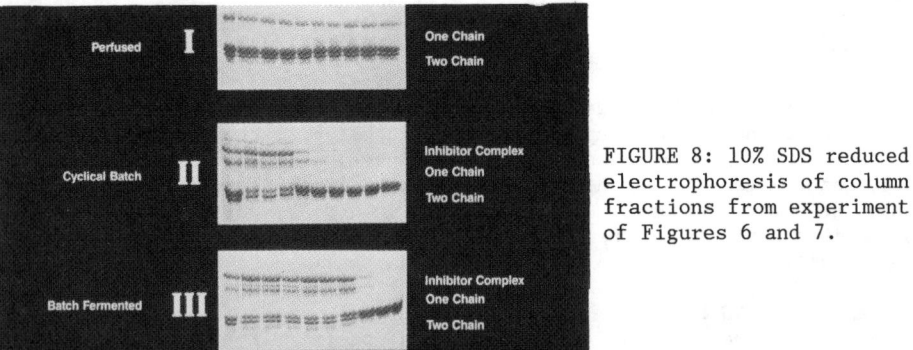

FIGURE 8: 10% SDS reduced
electrophoresis of column
fractions from experiment
of Figures 6 and 7.

In another example, IgM monoclonal antibody was produced and purified.
A reduced silver stained SDS gel (see Figure 9) is shown. The first lane
contains molecular weight standards and the second a batch production
from 5% serum. The third lane is perfusion production also from 5%
serum. The fourth and fifth were perfused with 2% FBS and serum-free,
respectively. The extra band between the heavy and light chains in lane
two is due to enzymatic degradation of the IgM molecule. Immunoblots
proved that all bands were mouse antibody. Perfusion at 5% greatly
minimized this degradation and eliminated it under reduced serum
conditions.

128

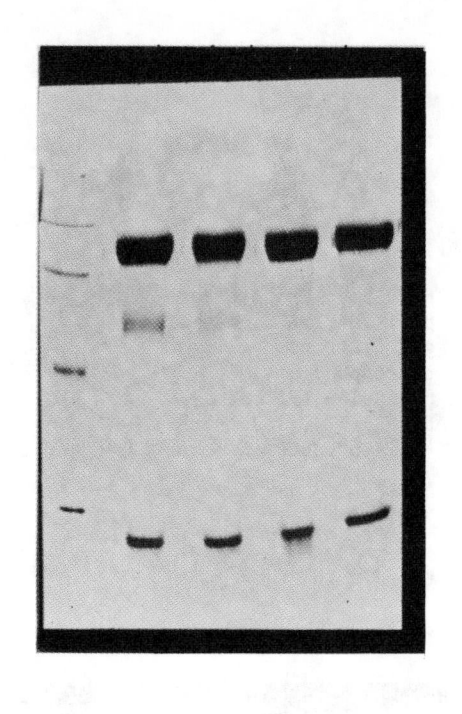

FIGURE 9: 10% SDS reduced electrophoresis of purified IgM. Lane 1 are molecular weight standards, lane 2 is batch produced IgM from hybridoma cells grown in 5% FBS, lanes 3,4,5 are perfusion produced IgM from 5%, 2%, 0% FBS, respectively.

These examples dramatically demonstrate reduced product degradation by perfusion methods when coupled with rapid cooling of the product stream. It is necessary to minimize injection of inactive, complexed or aggregated materials which could be much more immunogenic to the patient than the native molecule. Therefore, the integrity, purity and biological activity of these complex protein molecules is extremely important when they are to be used in human pharmaceutical applications.

5. MANUFACTURING FACILITY

A facility was designed and constructed specifically to implement the previously described cell culture technology for the production of cell-derived products. The plant has an operating capacity sufficient to process in excess of one million liters of cell culture medium per year, which currently makes it the largest plant of its kind in the world. It was designed and constructed to comply with the current Good Manufacturing Practices promulgated by the United States Food and Drug Administration (see Figure 10).

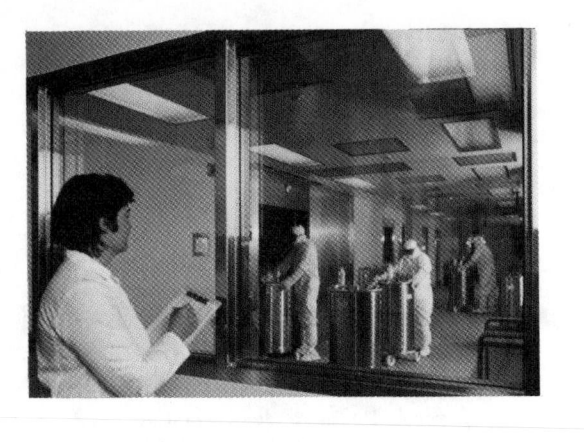

FIGURE 10: Quality
Assurance person observes
the unloading of
autoclave-sterilized
media containers.

In accordance with current practices for parenteral drug facilities,
particular attention has been given to the interior finishes and the
major plant systems. The ceilings, walls and floors in the manufacturing
area are constructed of materials which are easily cleaned and sanitized.
In most cases seamless vinyl, coved at floors and corners, has been used.
All windows, air defusers, and electrical fixtures are flush mounted to
facilitate cleaning. Process gases are supplied to the production area
through stainless steel piping and filtered at point of use. All
utilities are serviced from a mezzanine above the production area to
minimize the need for maintenance personnel entering the contained area.

The heating, ventilation, air conditioning (HVAC) system utilizes
microprocessors to monitor and regulate the flow of air through eighteen
air handlers, four dehumidifiers, and thousands of feet of duct. All air
is supplied to and exhausted from the production area through more than
200 high efficiency particulate air filters (HEPA). Air is supplied to
production rooms at ceiling level and removed at floor level to minimize
turbulence within the room. Air is supplied to the process rooms at a
volume sufficient to provide, 30, 40 or 50 air changes per hour depending
on the requirement of the particular room. The HVAC microprocessors also
centrally monitor temperatures in incubators and refrigerators throughout
the building. An alarm system provides prompt warning of any deviation
from preset temperatures.

Water for the process is provided by a system typical of those
utilized in "state of the art" parenteral drug plants. Incoming city
water is pretreated by carbon filtration, softened by anion and cation
exchange columns and fed to a three hundred gallon per hour Finn-Aqua
distillation unit. Distillate is collected in a 1000 gallon jacketed
tank and circulated through the production area in a welded stainless
steel loop. All tanks, piping and fittings contacting treated water are
constructed of 316L stainless steel. The entire system is maintained at
$80^{\circ}C$ and is designed to provide USP Water for Injection.

Two large (300 cubic foot) autoclaves allow sterilization of materials
and equipment entering and leaving the production area. The vessels are
constructed of stainless steel. Chamber steam is generated at each unit
by passing Water for Injection through a stainless steel heat exchanger.
Both units are microprocessor controlled. Cycles are preprogrammed and
documented in an appropriate manner.

In principle, the flow of raw materials, product, equipment, and personnel is unidirectional through the plant. Material and equipment enter the production area from an autoclave or air lock in the component preparation room and leave the production area through an autoclave or airlock into the equipment cleaning room. Product leaves through a separate airlock and is placed in quarantine for final product test and labeling before shipping.

6. CONCLUSION

This is a very exciting period in the development of biotechnology and its application to human health care. The progression of major new advances in medicine from the research laboratory through manufacture to commercialization will directly or indirectly benefit each of us. The fact that a large number of these biopharmaceuticals will require mammalian cell culture production will dramatically increase the demand for this capability over the next few years. We have described a perfusion-based technology which is well suited for the continual production of these protein-drugs in the absence of antibiotics. The implementation of this technology involves not only the specialized equipment and highly trained and experienced personnel, but also the design and construction of a facility capable of being operated in full compliance with Current Good Manufacturing Practices. Operational experience has demonstrated successful production of a large number of different types of protein products from mammalian cells at a scale and of a quality to be acceptable for pharmaceutical application in humans.

REFERENCES

1. Patterson, M.K., Jr., TCA Manual, 1, 243 (1975)
2. Himmelfarb, P., Thayer, P.S., and Martin, H.E., Science, 164, 555 (1969).
3. Knazek, R.A., Gullino, P.M., Kohler, P.O., and Dedrick, R.L., Science, 178, 65 (1972).
4. Tolbert, W.R. and Feder, J., in Annual Reports on Fermentation Processes, Vol. 6, (G.T. Tsao and M.C. Flickinger, eds.), Academic Press, New York, pp. 35-74 (1983).
5. Feder, J. and Tolbert, W.R., eds. Large-Scale Mammalian Cell Culture, Academic Press, New York, p. 159 (1985).
6. Reuveney, S., Velez, D., Miller, L. and MacMillan, J.D., J. Immunol. Methods, 86, 61-69 (1986).
7. Feder, J., in Advances in Biotechnological Processes, Vol. 7, Upstream Processes - Equipment and Technique, (A. Mizrahi, ed.), Alan R. Liss, Inc. (in press).
8. Ratafia, M., Bio/Technology, 5, 692-693, (1987).
9. Lydersen, J. of Tissue Culture Meth, 4, 147-154 (1983).
10. Hopkinson, J., in "Immobilized Cells and Organells", Vol I, (B. Mattiasson, ed.), CRC Press, Boca Raton, FL, pp. 89-99 (1983).
11. Van Wezel, A.L., in Research Needs in Non-Conventional Bioprocesses, (D.J. Fink, L.M. Curran, B.R. Allen, eds.) Bettelle Press, Columbus, OH, pp. 86-90 (1985).
12. Sato, S., Kawamura, K. and Fujiyoshi, N., J. of Tissue Culture Meth., 4, 167-171 (1983).
13. Fujiyoshi, N. personal communication.

Hope, J. : Perfusion cultures

Spier :   What happens to this business of inserting microcarriers
              in the maintenance reactor. Do you still use that kind of
              system ?

Hope :   When you are using an adherant cell the next convenient
              way to get the kind of inoculum you need to make any sense
              of a reactor of that type is to grow the cells in a
              microcarrier perfusion reactor. So, one would have about
              10-15 L of microcarrier slurry out of that reactor. This
              culture is then put straight into the static maintenance
              reactor. That is obviously not answering to your question
              but I do not think I can answer what you are asking.

Remacle :   I wonder what makes the decision initially to use in one
              case hollow fibers and in your cas these perfusion fermen-
              tors ?

Hope :   Essentially it is a matter of convenience. Ray was saying
              it at the beginning, there is an enormous number of
              systems that are around and one system seems attractive to
              person A and they start working on it, develop it up and
              another system seems attractive to person B.   But one
              advantage of ours - at least over the stirred tank system
              - is that there is no issue at all about whether you can
              grow cells in suspension or hybridomas or recombinant
              cells or whatever, because essentially any cell you can
              either put in suspension or stick on a microcarrier will
              do just as well in this system. I wouldn't like to say
              that you can get the same densities for every cell but I
              would like to stress and everybody should remember this,
              that the reactors are only as good as the cells you put in
              them : the cell stinks, the reactor's output won't be any
              good ; if the cell is great, the reactor will be great and

to some extent the reactor will help or maybe reduce these differences and the type of medium you put through will help, but still, the cell is basically the prime determinant. Only what the reactor does, is provide an environment for the cell to do its job.

Horaud :   My suggestion was to have a sort of talk on large-scale cultivation of animal cells with the speakers of today. Perhaps you can come here Ray, and Otto... My impression is that we have something new in large-scale cultivation of animal cells, namely possibility to have high density cell lines. This seems completely to change the problem of sera in the medium and I would like also to have the opinion of people that we have the chance to have here with us like Dr Sato or Dr Gospodarowicz.

Miller :   In your settling bottle that you showed in the reactor used by In Vitron, I remember that microbeads aggregated to a very large extent so as to produce in fact these spheroids that Prof. Hulser showed us yesterday. Now it seems as if these aggregated cells have their gap junctions permanently closed and although Bill TOLBERT told me it was an advantage to have aggregated microbeads, I am really wondering whether these cells, aggregated in this way in the settling bottle can show metabolic cooperation.

Hope :   What Prof. Miller is alluding to is one of the supposed advantages of the settling bottle. What Dr Tolbert had shown was that as the microcarriers settle down within this bottle, they come into close contact with very little agitation and so, bridging of the microcarriers into large multicarrier clumps can take place. It depends on which angle you are looking at it. If you want to see how many cells you can get into a reactor with a given amount of microcarriers, the ability to form these cross linked

masses helps a lot because you often have - again depending on the cell line and the type of microcarrier - you may have a proportion of your microcarrier which don't actually pick up cells to begin with and the only way for them to become inoculated is to become close enough to another microcarrier for long enough time for the cells to go over. As to your question about closing of gap junctions etc... then indeed there might be problems, but once again, it depends I think on the actual effects of this on the production of whatever product you are looking for. It would depend on the cell type and type of microcarrier and all kinds of things. And also we are not talking about huge great aggregates, we are talking about aggregates of maybe one-to two microcarriers and no more.

What I would be more worried about as a person actually involved in dealing, with one of these thing is that if the aggregates get too big, mass transfer across the aggregates becomes difficult because even though the beads may be solid, in the middle you can get a vast mass of cells and once it starts to get in the 100 microns range, you are starting to have oxygen and nutrients diffusion problems and possibly with the porous microcarriers it might be even more of a problem if you get a large clump of them. But these are the things that are really coming in the heading of bioreactor optimization and process optimization. We only think of them as we come along.

Spier :   Can I ask Gordon Sato to comment on the things that he said about the relationship between high density and growth factor and autocrine functions. Do high density cultures stimulate autocrinal functions ?

Sato :   The topic that was discussed this morning by Dr. Schonherr concerned growth factors produced at high density by the CHO cell. The CHO cell is one of the first cells, well,

one of the first, to have a defined medium made for it.
It was designed by Dick HAM sometimes in the sixties and
its requirements are relatively simple. It just requires
F12 medium and the thing is, as the chemicals got purer,
what HAM discovered is that F12 no longer worked and he
had to add selenium. I think it is an uncommon example
but there are others where cells at high density, especially for maintenance, not growth, are producing enough
factors to sustain viability. I think growth is another
matter to get up to the point where you do have high
density. I don't know how many cell types can be sustained - I don't know, what word to use, at high densities. I would think that this would be rather uncommon
but not terribly uncommon. I know that prostatic cells
can maintain viability at high density without adding
growth factors. I know that hepatoma cells can do this,
so there are large numbers of cells which can do this and
they are mostly long term culture lines. I think that the
comment I would make about these bioreactors - because you
know it is a subject I have not been that interested in -
is that I think that what one should do - I think it was
our last speaker who said that the cells have to be good -
I think what one should do is keep selecting cells for
viability in a bioreactor and that is simply using inocula
from cultures that have gone through a process of
production in a bioreactor. I think the other point is
that more efforts - I don't know how much effort has been
made in industry - but I think that a more efficient
approach, or equally important to making bigger reactor,
is to make cells more productive and I think that we are
learning more and more about gene amplification and almost
to the point where one can make rational approaches in
gene amplification in animal cells. The idea of SHIMKE is
that,if you can disturb DNA synthesis with hydroxyurea at
the time in the synthesis phase when the gene for the

product you are interested in is replicating, then that DNA gets amplified. Not necessarily incorporated into the genome. I think that the other side of getting that particular DNA amplified or produced in large quantities is also to have fairly clever methods for detecting a variant with high amplification.

Spier : I think the bioreactors come into their own when you have optimized the cell. There is a limit to which you can go on the cell and the biotechnologist in combining the discipline of cell biology and chemical engineering has to recognize when he has to go and make a product from a particular given cell basis. Sure, one is always pushing the biology but then you get into diminishing returns. When you are starting to push on the technology, you get into diminishing returns and the thing to do is to get the balance of the two systems.

Schönherr : It is not only of course a question of the cells but also a result of influence of dense culture on the product. You mentioned that already. We didn't do so much on this. From the production of monoclonal antibodies we know that the presence or absence of serum is not so important. You like to have it without serum of course for purification reasons but I can imagine that absence of serum leads to more product inactivation. If I understood well, that is one of your problems with TPA ?

Spier : I'd just like to take upon what Gordon Sato's last comment. This is where we get into the emergence of a new discipline where we are actually combining chemical engineering with biochemistry. I think we are only going through an evolution of the same kind of process by asking people to have a sensitivity and awareness to the biological aspects of what they are doing as well as technical

and physical aspects of the processes. Now, there are two kinds of biotechnology courses. There are the ones which basically teach an awareness of what biotechnology is all about while the second type teaches people how to get to the shopfloor and design the kinds of reactors and downstream processing that allow us to generate products. This latter type of thing that, we are trying to do in Surrey where we have made a biotechnology course which teaches the student biology and chemical engineering from year one. So, we are inculcating the philosophy and understanding that students deal with biological entities at a very early age. But it comes down to this essential problem : how far you can push the biology and how far you can push the technology, what can you expect from either aspect of the system in order to get a truly optimized process and that is the understanding that enables productivity to be improved.

Sato : If you take the example of penicillin where people were trying to build bigger and bigger reactors, real breakthroughs came only when somebody from Eli Lilly went to a downtown market and found a rotten mellon and isolated the strain that produced much more penicillin. It is my view that if bigger schemes are gone to happen, it will be on the biology side rather than on the engineering side. Well that is biased.

Horaud : That's what I said in the morning. Because nobody knows, perhaps in the four, five years to come, we could move back to engineered bacteria. If scientists succeed in making bacteria to secrete and obtain a product that is physiologically active, as long as the product - not like now - is not aggregated not hydrophobic etc... etc... etc, this will be a breakthrough... Nobody knows.
There is a new thing for instance. Scientists have

determined a sequence in bacteria, able to carry the
molecules on the external membrane of the bacteria. This
is a very important breakthrough and I know a group in
Pasteur which develops this system. They are testing it
now by putting different immunogens on the external wall
of a bacteria resembling E.Coli in order to use it as a
veterinary vaccine. So, perhaps 80 % or 90 % of the
equipment that we are buying now with an enormous financial effort will be outdated within ten years. I don't
know. Again it is difficult to predict. I know I am here
with people who like myself are enthousiasts for animal
cells but I cannot be dogmatic and I have to consider the
future.

Sato : My opinion is that the biggest improvement will be made on
the biology side. Some twenty years ago, we developed a
cell line that made growth hormone and that cell made 10 %
of its total protein synthesis, as growth hormone. I
think that is the standard that one can hold up as feasible
and attainable. We didn't get a number for TPA this
morning but I am willing to bet that it is very far off
this 10 % mark.

Spier : When everybody has cell lines that are producing at the
10 % level it is the people who have got the good bioreactor who can actually get the product expressed and put
in the bottle who will win the commercial race.

Sato, G. : <u>Strategy for a rational approach for developing sera</u>
<u>substitutes</u>

Spier : How long do you think it is possible to keep a cell alive
outside the body and maintain , not necessarily
a growing or dividing state ?

Sato : I think actually it is for ever. I mean, there is so much
that we don't know about physiology of any given tissue.
Fats, are a good exemple. We do not know what cells make
up the tissue. There are operational definitions for
preadipocytes, adipoblasts, fully differentiated adipocytes, but we really don't know how many cell types there
are and we don't know how to maintain them in the undifferentiated stem cell state.

Cassiman : Could you comment on the role or the possible role of -
for example - extracellular components or cell conformation in the expression of certain differentiated
functions ?

Sato : Well it is clear that in many cases you need to add
extracellular matrix and in many cases in the past when we
did not have to do this, the cells were producing their
own extracellular matrix. I do not know how specific that
extracellular matrix has to be. When REED who is a
postdoctoral fellow left my laboratory, she said : you've
closed out the classical phase of tissue culture. Now I
am going to start a new phase. I felt very happy about
that but what she was saying was that the extracellular
matrix is very tissue- and cell type-specific and it is
required for the expression of functional characteristics.

---

* This discussion followed the lecture given by G. Sato, of W. Alton Jones
Cell Science Center, Lake Placid, New York; the lecture itself, however,
was not available at the time when "Advanced Research on Animal Cell
Technology" went into production. It was decided by the editor and the
publisher to still include this discussion, as the views expressed herein
justify publication.

She has given pretty good evidence for that in the case of liver.

Cassiman : Do you think that the sensitivity of a cell to a particular growth factor may change as a function of the extracellular matrix ?

Sato : Yes... Yes. Maybe Denis (Gospodarowicz) can say something about that.

Gospodarowicz : For all I know about the extracellular matrix, there is indeed a specificity. This means that if you use an extracellular matrix produced by cells derived from the mesenchyme, it will feed cells derived from mesenchyme from whatever tissue it comes but it will certainly not feed cells derived from the endoderm or the ectoderm. That is the kind of specificity we have in the two types of extracellular matrix we have looked at, one which is produced by ectodermal cells (in fact keratocarcinoma cells) and the other one produced by mesenchymal cells which are of course endothelial cells. So, it looks to me that it is a specificity of the germ layer.
Now, as far as identifying the response of a cell to a given type of growth factor, it is very difficult to even speculate on that because of the experience that EGF and FGF are able to trigger the production of the extracellular matrix by the cell itself. So, I don't think that maintaining them on an extracellular matrix can modify their response to growth factors.

Sato : Well, I think a view that is very different from this would be taken by Nancy Bukharin (Name uncertain). She is not here and I hope I do not overstate her point of view. But she thinks the correct extracellular matrix in the liver keeps these cells from giving a growth response to liver, I mean to growth factors. In her view, I think, the process of regeneration is not only stimulation of parenchymal cells by growth factors but it is also a

process of extracellular matrix which undergoes remodelling in response to a loss of a lot of liver.

Gospodarowicz : If I may interrupt again, I think both of you can be
reconciliated in fact, by the other recent evidence that
growth factors can be found associated with extracellular
matrix - like components. It has been shown in the case of
the pituitary that regeneration involves production of the
extracellular matrix produced by follicular cells but
follicular cells produce an extracellular matrix which is
mostly composed of FGF anyway. So, it depends what the
growth factors will be. In some cases it is simply
associated with the extracellular matrix, in which case
the extracellular matrix will promote regeneration. ..

Sato :   You mean with heparin ?...

Gospodarowicz : Yes, that's correct.

Miller :   If I may jump in the discussion, there is a recent review
of TIBS[3] emphasizing the importance of the overall
morphology of the cell which triggers the metabolism
inside the cell. For instance anchorage-dependent cells,
- this without any growth factors or things like that -,
growing onto microbeads, did not synthesize a given
biological. By making the matrix more hydrophobic, the
anchorage-dependent cells became rounded to different
extents and progressively decreased their multiplication
rate and progressively began to synthesize differentiated
biologicals. It was clear from this review that morpho-

---

(3) Trends in Biochemical Sciences 11 (1986) 437-488.

logy of the cells had an effect on the metabolism, that it modified considerably the cytoskeleton and because the cytoskeleton is related to certain intracellular organelles, the alterations in cell membrane morphology were considered as true transducers from the outside into the inside.

Remacle : Can you comment on the inhibitory factors present in serum ? You cited one example of well defined inhibitory activity of serum upon one cell function, one may conceive that the enhancement of specific functions may depress the functions of other types. Do you know any example of a general inhibitory activity in the serum which will depress proliferation or differentiated functions together ?

Sato : If you take a normal thyroid and put it in a serum-containing medium, by using the old green thumb approach to things, you can eventually isolate epithelial cells which I am pretty sure are of thyroid follicular origin, but they never function. If on the other hand you - in primary cultures - you use a defined medium for thyroid cells, then you get thyroid cells that function, give a growth response to TSH, concentrate iodine, and make thyroglobulin. If you take these cells and give them serum, they die. And I think something is true of David BARUCH's factor which she isolated from mice in defined medium. The cultures don't undergo the crisis nor the sensescence, they don't undergo the selection for heteroploid cells. Cells also die if you give them serum and so they present a great interest to purify these substances. But you see, the interstitial fluid is certainly different from blood or serum, so, possibly these things never get into the interstitial spaces but nevertheless it would be interesting to see what these molecules are.

Spier :   Just to continue the conversation on the inhibitory
             factors, we have been looking at in concentrates from
             hybridoma cells, we found that in serum-free medium they
             can - we think - produce something which inhibits the
             growth of hybridoma cells. But we don't know what it is
             yet.

Sherman :   In response to Dr Remacle's question, retinoids are a good
             example of agents that are inhibitory of proliferation of
             many different kinds of cells but there is a very confu-
             sing caveat and that is that the retinoids are essentially
             only inhibitory in the presence of serum. If you culture
             cells in serum-free medium, then you find that retinoids
             tend to promote cell proliferation which is basically what
             they should be doing in view of what we know about the
             needs for retinoids as vitamins and it gets back to I
             think again what I was mentioning yesterday briefly, that
             the transforming growth factors α and β which can be
             growth promoting for some sorts of cells under certain
             circumstances can be suppressors under others. I wonder
             Gordon whether you might have some comments about that.
             Are growth factors all two-faced ? Or are there growth
             factors which really only promote or in fact might be
             modulating positively or negatively depending on circum-
             stances ?

Sato :   Well, a good example is EGF which inhibits A401 cells.
             You might think that A401 cells are odd cells because they
             have one million EGF receptors but I think this may not be
             true, there may be a lot of normal cell types within the
             body that have lots of EGF receptors and so one could
             easily think of EGF as being growth inhibitory for some
             cells. It is also growth inhibitory for fat cells.

Sherman :   What I am getting at is whether a growth factor in a way
              is a misnomer and whether it shouldn't be called a growth
              regulator.

Sato :   I never liked the term growth factor. My orientation was
              always endocrine physiology and growth factors just come
              about because of the operational use to define substances
              and I think that is wrong. What we are looking at is
              endocrine physiology.

Remacle :   What is your feeling about the possible existence of two
              kinds of growth factors ? The ones for fetal material and
              the others for post natal material, for example for the
              TGF which could correspond to a certain extent to the
              fetal expression of adult growth factors.

Sato :   There could be different regulators and developments when
              hormones are secreted and when cells are responsive to
              them. But I would suspect that these things persist some
              more throughtout life.

Remacle :   Yes, because it seems to be the same with the EGF 1 and 2.

Sherman :   I think you can make a whole list of such related factors.
              Interleukin 1 and FGF is yet another example where there
              is presumably enough homology .to allow two related factors
              to bind to the same receptor and yet enough divergence in
              those proteins to make you think that they should be doing
              something differently but I do not know if there is enough
              evidence that one of these is an embryonic version and
              another one is the adult form.

Cassiman :   Could you explain the effects of these two protease
              inhibitors. By the way, I am not sure I got the name of
              the second protease inhibitor.

Sato :   I think it is called HL 160. The native molecule has really never been well characterized. It has been better characterized in tissue culture. One can only speculate about these. There seems to be a strong correlation between the structure of say the pancreatic trypsin inhibitor and EGF. Evolutionally they could have evolved and have separate functions. I think this is probably not true, that there is something intrinsic about antiprotease activity that has something to do with growth and might have something to do with resistance to granulocyte attack. I think it is an interesting question but I can't give you any answer.

Cassiman :   They are active just in this system ?

Sato :   I don't know of specific tests but we also isolated other protease inhibitors which are very active as growth factors.

Cassiman :   From more than one cell type ?

Sato :   Yes.

GOOD MANUFACTURING PRACTICE IN THE PRODUCTION OF BIOLOGICALS BY
CELL CULTURES

V G EDY

Flow Laboratories Ltd, Irvine KA12 8NB, Scotland

    Good Manufacturing Practice (GMP) is sometimes forgotten
until a production process has been developed, and on occasion
is ignored until a plant is built and ready to run. This is in
part caused by a tendency to view GMP as a nuisance, as
something forced on the scientist or engineer by regulatory
authorities. The aim of this article is to run through the
outlines of a GMP code, emphasising those points most relevant
to cell culture, and bringing out the ways in which the
manufacturer can benefit by GMP compliance.
    There has been a recognition, both by the regulatory
authorities and by manufacturers, that end-product testing
alone cannot provide sufficient assurance of product quality.
Instead, it is necessary to assure complete control of every
stage of the manufacturing process, to be sure that adequate
quality is 'built-into' the product at each step. To provide
guidance to manufacturers, codes of Good Manufacturing Practice
were developed by the regulatory authorities; at first
advisory, these are now mandatory in many situations. There are
a variety of codes, but for the purposes of this article I will
follow in broad outline the regulations laid down by the US
Food and Drug Administration (FDA), as published in the Code
of Federal Regulations (1) (CFR). In fact, there are two main
GMP codes laid down in CFR, part 211 for Finished
Pharmaceuticals, and part 820 for Medical Devices, and
additional codes for Medicated Feeds, Premixes, and Blood and
Blood Components. The various GMPs are all broadly similar in
general intention to the GMP code for Finished Pharmaceuticals.
The various national codes will differ somewhat; the US
Finished Pharmaceutical GMP (21 CFR 211), is chosen for
discussion primarily because the FDA has tended to set the
regulatory pace, and because most has been written about it. A
good general review of the impact of GMP on Biotechnology-
oriented companies is given in (2), although this does not
discuss cell culture.
    The initial parts of the GMP code as laid down in 21 CFR
211 are of general relevance to all manufacturing facilities,
covering formal organisation, personnel, training, buildings
and equipment. A key phrase in considering all these topics
would be 'adequate for the intended purpose'. The structure
and staffing of the organisation must be adequate to run the
intended process. A formal training programme is necessary to
ensure that all staff are adequately qualified for the jobs
they are to perform. Buildings and equipment must be designed,
built and installed and operated so as to be adequate for the

A. O. A. Miller (ed.), Advanced Research on Animal Cell Technology, 147–156.
© 1989 by Kluwer Academic Publishers.

process. They must be easy to clean (and kept clean) and
should be arranged in an orderly manner. It is very important
to arrange units so as to give a clear and logical process
flow, with minimal cross-overs, and with adequate separation of
different processes. Only by starting out with a properly
arranged process flow, can mix-ups and confusion be avoided in
operation.

Proper equipment design is of great importance in
production of materials by cell culture. The processes are
normally aseptic, and clean-in-place/sterilise-in-place
operations have significant advantages. Equipment for
large-scale cell culture is not generally available
'off-the-shelf', and full consideration must be given to ease
of cleaning and sterilisation, and the maintenance of sterility
over prolonged periods. Proper design and specification will
prevent many problems in operation. A preventative maintenance
programme is also part of GMP. Any system or item of equipment
will need occasional repair or overhaul, and to minimise the
likelihood of failure during actual operations, a plan for
routine maintenance, listing what is to be done and when, and
recording what actually was found, is essential.

The most obvious difference in producing something by cell
culture, rather than by classical chemical or extractive means,
is the use of the cell itself. The cell is not strictly
speaking a component of the final product, but it is convenient
to discuss it as if it were. Adherence to GMP requires that a
manufacturer assure himself that all components of his product
are as they should be. It is important that before a component
is released for use in production, it is tested to ensure that
it meets previously laid down specifications for identity,
purity, potency etc; again that it is adequate for the intended
purpose. The strict application of such testing with something
as complex as a living cell is obviously not simple, and it is
a point the regulatory authorities have given much thought to.
There is no list of acceptable and unacceptable cell lines, nor
of the testing required to establish that a cell is, or is not,
acceptable for a particular purpose. In general, the amount of
testing required before a cell line can be considered for use
in a production system will depend upon the specific cell
type, on the nature of the downstream processing of the
product, and to some extent also on the use to which the
product will be put. A cell line used to produce a material
that ends up as an injectable drug will obviously need much
closer study than a cell line used to manufacture a component
of an in vitro diagnostic kit. In broad terms, it will be
necessary to demonstrate, before routine production begins,
that the chosen cell line is stable, or that the limits of
stability are well defined, and that it is free of adventitious
agents, eg. bacteria, fungi, mycoplasma and viruses. Its
karyotype, and a full pedigree, must be established.
Additionally, it may be necessary, if the cell line is
continuous in nature, to determine whether the cells, or cell
components, are tumorigenic in experimental animals (3) and to
design the production and purification processes accordingly.
In routine use, it will be necessary to have formal documented
systems to assure that the cells are as they should be; that no

adventitious agents have crept in, that the correct cell line has been taken from freeze and has not altered somehow in culture or in storage, and that is is used within the number of passages established during development as safe.

Large parts of any GMP code cover documentation; and the tendency of complex systems to generate large amounts of paperwork is often wrongly seen as an onerous requirement imposed by GMP. Whilst it is true that GMP codes do require large amounts of documentation, it is an essential part of any process to maintain full, accurate records. Not only do such records provide proof that the process was properly performed, but also full recording of all variables allows reconstruction of the process in event of failure. Such failure may well not be recognised immediately, and it is not alway possible to predict what will be important in the event of a failure. It is therefore essential that everything is written down. Additionally, complete documentation goes a long way to ensure that the process is performed as it should be , each and every time. All frequently repeated processes are subject to a type of drift, with various changes being deliberately or inadvertently introduced. A complete step-by-step description of the process steps to be followed ensures that deliberate changes are actually introduced, and prevents inadvertent modifications which can cause the process to differ from that originally specified, and from that which the organisation believes is being performed.

All instructions, production protocols, test methods etc, must be fully described in writing, and all processes and tests must be fully and completely recorded at the time they are performed.

A further critical part of a full documentation system, is to maintain complete traceability. It is essential that the product can be followed through from raw material to end user, in a completely detailed fashion. The manufacturer must be able to determine which batches of raw material were processed, when, in what cell culture run, to give what lot of final product which was distributed to which end users.

Documents must be clear and logical, and must be retrievable. Not only is this a requirement of GMP, but if a protocol is confused, so will the operation be, and a document that cannot be found effectively does not exist. To add, if possible, to the emphasis on the importance of proper and complete documentation, the process records are often where regulatory authorities like to start their investigations.

Further sections of the GMPs refer to packaging, labelling, storage, and distribution. There is nothing in these articles that specifically relates to biologicals, and the general principles are, as is normal with GMP, sound common sense. It is necessary to maintain adequate control over the packaging and labelling materials to be used, and to ensure that the product is stored and distributed under the correct conditions, and that distibution is traceable.

Although the emphasis in GMP system is on control throughout the process, testing at all stages, including finished product testing, is still a very important part of GMP. All testing processes must be validated, to ensure that

they actually do have the sensitivity, selectivity and precision that is required, to show that the method actually does test properly for the analyte. Tests should be challenged regularly, to ensure that they are being correctly performed and interpreted.

A major part of the testing effort must be devoted to qualification of the purification method before it is put into routine use. Not only is it necessary to develop and prove methods to purify the product of interest, but also to exclude the possibility that potentially harmful materials (eg viruses or DNA fragments) contaminate the purified product. To this end, it may even be necessary to deliberately challenge the purification, by adding known amounts of representative or 'marker' contaminants, and then to show that they are sufficiently removed by the purification process. Additionally, it is necessary to determine whether the purification process might potentially add harmful material (eg by leakage of ligand from an affinity column) to the final product.

An integral part of any GMP system is showing that equipment and processes all behave as expected. This is the purpose of a validation programme. To accomplish this, the whole production process is broken down into its separate steps, and each process is rigorously tested. For example, in sterilisation of an item of equiment, it is necessary to identify the most difficult-to-sterilise part of the item, and then to demonstrate that the sterilisation cycle used is sufficient to give an adequate assurance of sterility at that point. In an aseptic filling process, it is necessary to validate the level of asepsis achieved. The standard sterility test, where 20 samples are taken at random from a batch of possibly thousands of final containers, actually provides a very low assurance of product sterility. This is a simple consequence of the statistics of sampling; it turns out that with 20 samples from a large batch) the lowest level of contamination that can be detected (with 95% confidence) is 15%, even assuming that the culture method is comletely efficient at detecting the contaminant (4). Validation of the filling process is therefore the only way of assuring that there is a reasonable probability of achieving an acceptable contamination rate. (less than 0.1% probability of contamination.) This is commonly done by filling 3000 containers under normal filling conditions, with bacterial growth medium, and then incubating the lot. The containers are then screened for bacterial growth.

The final part of any GMP programme is ensuring that the whole system is actually working, and to do this it is necessary to have an internal audit programme. Whilst this may seem burdensome, it is far better to detect (and rectify) problems within the organisation, rather than have them pointed out by potential customers or by the regulatory authorities. One useful trick in auditing is to mentally take the part of one of the raw materials, and follow its progress through to final product, asking questions such as 'How does this component get into the process?', 'How can it get contaminated?', 'How can it get lost?'. and so on, checking that the GMP system as operated

provides reasonable answers to the questions.

The end result of setting up and following a GMP programme should be, if the base processes were well conceived, the achievement of a 'state of control'. Indeed, the benefits from being in as near as possible full control of the process, particularly with something as complex as cell culture, are vastly greater than simple compliance with legal requirements. Adherence to a GMP philosophy should result in more consistent production, with fewer failures. Conditions likely to lead to failure should be more rapidly recognised, and when identified, acted upon. Lastly, if a failure does occur, it is more readily traced to causes and these rectified. A GMP approach to the manufacturing of materials by cell culture should lead therefore to greater efficiency and lower costs.

REFERENCES

1. Code of Federal Regulations. Volume 21. General S e r v i c e Administration. Washington DC. (1987) (published annually)

2. Harrison, F.G. Current Good Manufacturing Processes for BioTechnology - Oriented Companies. BioTechnology 3 43-46

3. Petricciani, J.C., P.L. Salk, J. Salk and P.D. Noguchi. Theoretical Considerations and Practical Concerns Regarding the Use of Continuous Cell Lines in the Production of Biologics. In Developements In Biological Standardisation, 50, 15-25, S. Karger, Basel (1982).

4. Tetzlaff, R.F. Regulatory Aspects of Aseptic Processing. Pharmaceutical Tchnology 8 38-44 (1984)

<u>Dr Edy, D.</u> : <u>Good manufacturing practice in the production of</u>
<u>biologicals by cell cultures</u>

Spier : There is one point we might include in this and that is
that if one actually specifies particular conditions, this
is very much more difficult to meet than a <u>range</u> of
conditions within which acceptability can be found.

Edy : You have to specify a range. Never ever when you are
writing a production protocol, specify a value. Never
ever say incubate the cells at 37°. Firstly you don't and
secondly you can't prove that you do. Certainly the cells
are exposed to a temperature between 36° and 38° and what
you should do, and this is common sense and not GMP (Good
Manufacturing Practice) GMP is common sense - I think so -
is you find the edge of failure conditions. You run your
process over a range of conditions and find out what the
limits really are. What you then are able to do is very
rapidly tell whether a drift in temperature or in pH or
whatever matters, or is not really very important or has
actually destroyed your process already. So you never
ever pin yourself down to a unique value of anything.
That is a trick taught by lawyers as well.

Hope : I just like to add to that, Ray, that in tissue culture
particularly in your zeal to conform with GMP, you can
very easily back yourself not just into the corner but
actually under the wellpaper as well and find it is very
very difficult to get out of it and the guidelines are
sufficiently ambiguous - that's a wrong work - leave
enough room, sufficient for interpretation on a case per
case basis if you know what you are doing and your whole
operation has a sort of look of competence and it is quite
obvious everybody is thinking around the right lines, you
can have a sensible degree of control about what you are
doing, I mean sensible limits to what you are doing and
still be in full compliance and yet still have the freedom

to be able to adjust to things as they come. But it is up to you to make sure up front that you don't get backed yourself in a corner.

Spier :   Vic, just a comment, don't do that with the labels.

Miller :   I would like to ask Ray : can you find these guidelines printed in the UK, in the US ?

Edy :   US guidelines are part of the code of federal regulations. The specific part dealing with pharmaceuticals is 21 CFR 211, for making an in vitro diagnostic or 21 CFR 820. The problem with the US guidelines is that they are written in cryptic. They are written in code.   These are thus subject to a wide range of interpretations. The person that does the interpretation is the investigator who comes to inspect you : "I am from the FDA and I come to help you...".

            In Britain there is a code of good manufacturing practice for pharmaceuticals commonly known as the Orange Guide. It is a thinnish book which enunciates very much the same principles as the american code but does so in a way that mere mortals can also understand and I would recommend that unless you are trying to get a product approved by a specific regulatory authority where you have to follow their rules, I would recommend personally to read the UK guidelines.

Cassiman : Is this the same as the Good Laboratory Practice Manual ?

Edy :   Similar code to GLP, yes.

Cassiman : No, because I think for a lot of people here, the Good Laboratory Practice might be more interesting than the Good Manufacturing Practice.

Edy :   Thank you !

Miller :   It is not the same thing I presume ?

Edy :   It is not the same thing.

Spier :   And it is not mandatory ?

Edy :   It depends what you are doing and where you are doing it.

Miller :   Well Good Laboratory Practice is in fact addressed to the people who are handling the products, to save their integrity, to avoid accidents and things like that... Good Manufacturing Practice that's for the product...

From the floor, many voices together : No, no !

Edy :   There is a code of Good Laboratory Practice which is again in 21 CFR but I don't know where. Everything is in 21 CFR... which relates primarily to toxicity testing and safety testing. It is not concerned with laboratory safety that is with not infecting yourself. It is concerned with very similar principles to those I enunciate now : traceability documentation, validation, control of the sample throughout the process. If you are doing a safety study or a preclinical trial, toxicity testing on animals in the US, compliance with GLP is mandatory.

Hope :   Can I make one more comment that GLP - and I speak from my past experience -, the GLP documentation rules are as strict as the GMP. For example, every experiment in your notebook must have a clearly defined aim, the method must be described or at least for further sampling you have to know where all the materials came from and not only do you have to sign it but somebody else has to sign that notebook page after they have inspected it. Otherwise any research that you are doing which ends up going towards a

clinical product could be declared <u>nullum</u> <u>void</u> ; it essentially does not exist. And of course when your are working with patents, situations that becomes essential as well to protect priority, etc... The way it is actually applied I think will vary from company to company. It must not be taken lightly by any means.

Labit Lebouteillier : My question is about environmental acceptabi-
lity. What about identity and stability of genetically engineered cels ? There are the specific productivity, the genomic stability, amplification problems ?

Edy : I don't know the answer to that question. All I know is that you have to be able to show yourself in some way that you are in control of that cell, that you know what it is going to do. There isn't a simple rule. Whether you think that studying the genome is sufficient or studying specific productivity or studying some other parameter of the cell is sufficient. There isn't any easy answer there, no simple, straightforward list of things you have to do. GMP is jobs for life.

Miller : If I may say something, I think that Jim's remark is very important because with the students you have in the laboratory, when they make experiments, they just write that on a sheet of paper and it is very difficult to oblige them to make notes, to keep their notebook in perfect order. I wouldn't say it will be so strict as you are in the industry but I think that's the beginning. It is there that they have to learn how to do that and you can see immediately whether a student can do that or not. It is fundamental to his work if it is a Ph.D. thesis or anything like that.

Edy : Some of the universities are failing in this.

It is no good doing the experiment and then going away and writing it down neatly O.K. It is a lot easier that way because your aims don't look so silly when you see the results. That's again wrong done. You have to write it down as you do it. I think people would be suspicious of a neat pristine notebook without crossings, but they have to write it down. Without it being written down, it doesn't exist.

Spier : I am absolutely certain that there is nobody in any University that complies with that kind of things unless they are on the trail for a patent and they know it and they have been forewarned. But I think that by and large 95 % of all practices are adequate for the purposes for which they are intended.

BIOSENSORS

L.D. Gray Stephens (née Watson)

Cambridge University Biotechnology Centre

INTRODUCTION

Biosensor technology has arisen from an effective synergistic collaboration between micro-electronic engineering and biochemistry, to form an exciting and rapidly developing area of biotechnology. Biosensor devices have potential applications for monitoring environmental pollution, industrial process control, the food industry and throughout the medical field.

Many of the model biosensors which have been investigated to date comprise a biologically sensitive material – an enzyme, multienzyme system, antibody, organelle, bacterial or other cell, or whole sections of mammalian or plant tissues – immobilised in intimate contact with a suitable transducing system to convert the biochemical signal into a quantifiable and processible electrical signal. The general principle of a biosensor is shown in Figure 1.

Figure 1   Schematic view of a generalised biosensor

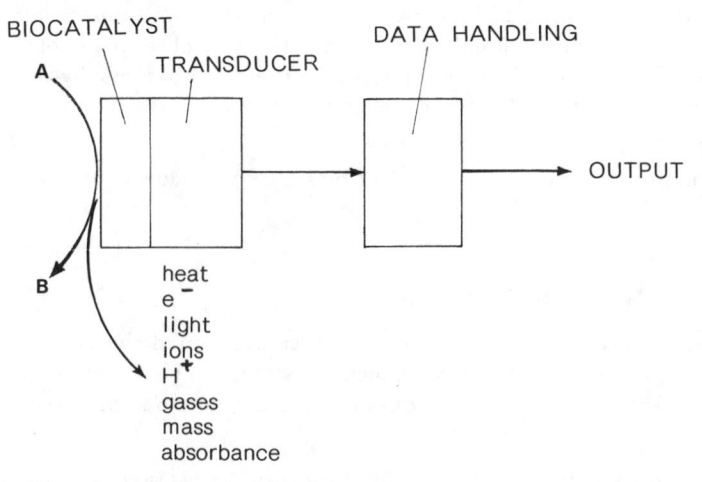

The changes in physico-chemical parameters which occur with the catalysed conversion of substrate A to product B (Figure 1) may be in the form of evolution or uptake of gases (oxygen, carbon dioxide, ammonia), specific ions (ammonium, monovalent cations, cyanide, iodide), heat, absorbance, mass, conductance or electron transfer.

A. O. A. Miller (ed.), Advanced Research on Animal Cell Technology, 157–172.
© 1989 by Kluwer Academic Publishers.

Immobilisation of the biocatalyst to the surface of the device transducer is commonly achieved by chemical cross-linking of the biocatalyst with an inert material and a bifunctional reagent which physically retains the catalyst with an appropriate membrane (1-5). The thickness or permeability of the catalyst layer limits the diffusion rate of substrates through the membrane, and therefore controls the response time for the biosensor to detect the analyte.

Biosensors may be used to monitor the metabolic state of animal cells in culture. As well as monitoring the pH, temperature and dissolved oxygen content (6) of a culture system, other additional parameters such as the concentrations of amino acids, glucose, ionic species (calcium and metal ions) as well as cell bulk, give useful information of the overall metabolic state of the cells in culture. Theoretically, sensing devices could be placed at various positions in a fermentor scheme: they could be placed in-line, either in mid-flow (invasive), or in contact with the outside of the culture vessel (non-invasive), or set a distance away from the vessel (non-contacting). Alternatively, sensors may be: on-line, in a side-loop; off-line, but discharging to waste; or samples may be taken and subsequently analysed. In practice, in situ monitoring (invasive or on-line) means that probes which go inside the reaction vessel or an enclosed flow loop must be rugged, reliable and sterilisable. Off-line sensors, where samples are removed at various times and analysed, are the most convenient sensors to use in practice. A medical glucose sensor has been adapted and experimented for in situ analysis in fermentation systems (7), however, in the fermentation process which may last from 3 to 21 days, there was a build up of end product (peroxide in this case) which poisoned the enzyme. Also, during fermentation, cells will occlude most sensors by sticking to the membrane and growing. Many companies are now concentrating on the development of biosensors for fermentation cultures which are used for off-line monitoring.

Of course the application of biosensors in fermentation cultures is not only limited to following culture medium nutrient concentrations, or other parameters previously described, it has also great potential in monitoring cell products, by-products and contaminants or pyrogens. Biosensor technology impacts at every stage of the fermentation system.

The various principles which are used in biosensor transducers are described below.

POTENTIOMETRIC BIOSENSORS

These sensors operate on the principle of an accumulation of charge density at an electrode surface, resulting in the development of a measurable potential at that electrode. Such measurements are made from non-faradaic electrode processes, with no current flow.

The widest area of potentiometric applications is the utilisation of ion selective electrodes, the change in electrode potential being brought about by the selective association of an ion at the interface of an ion selective membrane.

Potentiometric sensors have been miniaturized as field-effect transistors (FETs).

An example of a multi-function FET device uses the pH sensitive silicon nitrate gate insulator for pH measurement and ionophore-doped polymers as the electroactive films for potassium, calcium and sodium ions (8 & 9).

Development of the FET devices logically progressed to enzyme coupled electrodes (ENFET) and also to immunoassay systems (immunoFET) (Figure 2)

Figure 2   Development of the FET for enzyme and immunosensors

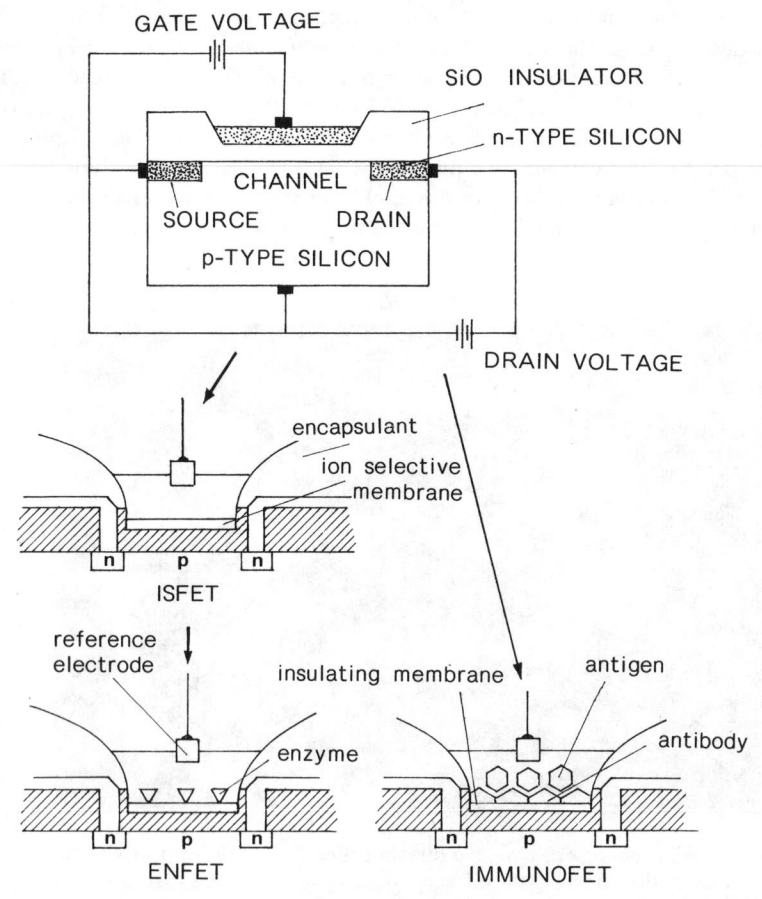

A solid state iodide FET device depends upon the reaction between iodide and hydrogen peroxide. Hydrogen peroxide may be produced, for example, during the oxygen mediated conversion of glucose by membrane entrapped glucose oxidase (10), or from peroxidase labelled antibody systems (11). A pH sensitive FET has been used to measure the strong acid (penicilloic) which is produced by the penicillinase catalysed hydrolysis of penicillin (12).

CONDUCTIMETRIC BIOSENSORS

A measuring principle which is widely applicable to biological systems is the exploitation of solution conductance. Many chemical reactions produce or consume ionic species and thereby alter the overall electrical conductivity of the solution. The monitoring of solution conductance was originally developed as a method for determining chemical reaction rates, but, more recently, it has been applied to enzyme catalysed reactions (13-15).

A microelectronic conductimetric biosensor using urease/urea as a model system, has been developed at the Biotechnology Centre, Cambridge (16 & 17). The device comprises identical pairs or interdigitated multimetal electrodes fabricated on a silicon substrate in a planar configuration (Figure 3). The accompanying instrumentation applies a low distortion sine wave of frequency 1 kHz and of an amplitude of 10 mV about 0 V to the electrode. Amplification was achieved with a standard inverting operational amplifier and the output signal from the amplifier was taken to an RMS-DC converter, so that the response could be monitored on a chart recorder.

Figure 3   The microelectronic enzyme conductimeter: Top view of the silicon chip showing the interdigitated serpentined networks bonded to a 12-pin open-top TO5 package

Jack bean urease was immobilised over the metal network by forming a cross-linked enzyme-albumin membrane with glutaraldehyde (18). Figure 4 shows the output response of the operational amplifier in mV/min as a function of urea concentration in the range of 0.1-10 mM. A linear response for up to 3 min was obtained at all urea concentrations. The immobilised urease conductimetric biosensor also responded to urea present in serum samples, and there was a linear relationship between the urea concentrations determined with the microelectronic device and those obtained from

the major hospital laboratory.

Figure 4   Typical response curves of the microconductimetric sensor to urea
concentration in the range 0.1 - 7.5mM at 30°C in 5mM imidazole-
HCl buffer pH 7.5 containing 10μg.ml   urease

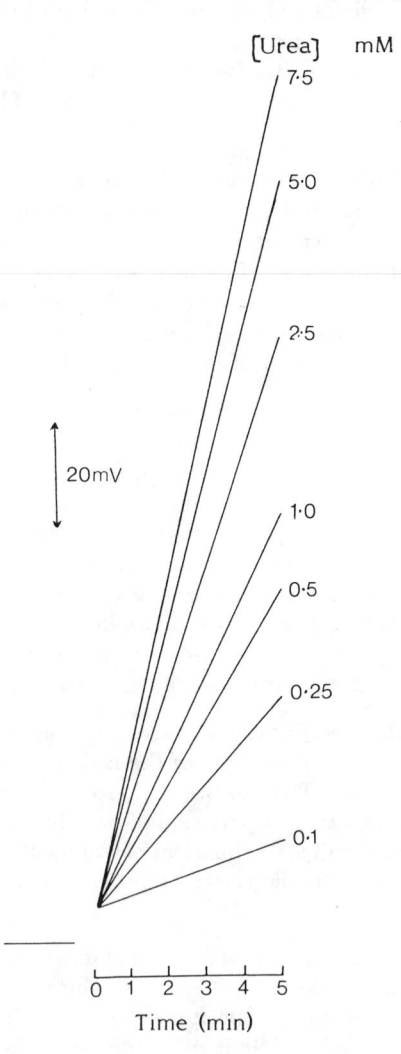

AMPEROMETRIC BIOSENSORS

Amperometric detectors are well established as gas sensors, most notably the Clarke electrode (6) which is used for oxygen measurement.

Enzyme-based amperometric devices involve either a direct or indirect electron exchange between the electrode and the protein and are most appropriate to substrates

employing redox enzymes. Electron exchange can occur, either through the direct electrochemical oxidation or reduction of the active enzyme centre, or the indirect reaction via low molecular weight redox mediators.

An example of a redox enzyme which has been used with an amperometric device is the flavoprotein glucose oxidase, which catalyses the conversion of glucose to gluconolactone. The flavin redox coenzyme part of glucose oxidase, which is reduced in this reaction, can be reoxidised in the presence of dissolved oxygen at a platinum electrode polarised at $+0.6$ V (vs SCE) (19), with the concomitant production of hydrogen peroxide.

Glucose oxidase has been successfully immobilised to a high surface area platinum-ink electrode (20), by entrapment within an electrically grown, conducting polymer matrix such as polypyrrole. Known concentrations of glucose in solution have been assayed by the direct amperometric detection of hydrogen peroxide. The time taken for the device to reach a steady-state current response was in the order of 20 – 30 secs. A double reciprocal plot of glucose concentration against steady-state current revealled a Km of 31 mM which is of the same order of magnitude as that previously published (21).

The alternative, indirect amperometric method uses mediator compounds which include organic dyes (22), hexacyanoferrate (III) (23), dichlorophenol and ferrocene derivatives (24) to accept electrons from glucose oxidase and shuttle them to a suitable electrode. To date ferrocene is reported to provide the most efficient mediation. Variations in oxygen tension in the sample, which may seriously effect the electrode response in the previous method involving hydrogen peroxide formation, are eliminated. By using a mediator compound, such as ferrocene, it serves to move the electro-potential at which the amperometric detection of glucose can be observed, nearer to 0 V. This substantially reduces the redox behaviour of any potentially interfering substances.

Recent work with amperometric biosensors at the Biotechnology Centre in Cambridge has been directed towards the covalent attachment of the mediator, ferrocene, to a polymeric network at the electrode surface. This has been achieved by substituting ferrocene amido alkyl units at the nitrogen atom of pyrrole, followed by electropolymerisation at an electrode surface. Additionally, the physical entrapment of glucose oxidase within the polymer / mediator matrix at the electrode provides an essentially 'reagent-free' amperometric glucose sensor (25).

Dehydrogenase enzymes are an important class of redox enzymes which use the cofactor beta-nicotinamide adenine dinucleotide (NAD^+) and produce the reduced form NADH. An amperometric enzyme electrode has been developed which can electrochemically regenerate the NAD^+ cofactor, so that it may may again be reduced during the enzyme catalysed biotransformations (26). The considerable overpotentials at which NADH oxidation occurs at solid electrodes can be reduced by using the redox mediator hexacyanoferrate (II/III). The mediator was electrodeposited on nickel by anodisation of porous nickel electrodes in aqueous electrolytes containing hexacyanoferrate (II) ions and a cyclic voltamogram of the resultant nickel hexacyanoferrate films revealled two peaks corresponding to the hexacyanoferrate (II/III) redox couple. Steady state current measurements were performed with the modified electrodes

poised at $+0.2\,\mathrm{V}$ (vs Ag/AgCl), in buffered solutions containing NAD^+ and alcohol dehydrogenase, with subsequent addition of ethanol. Results were obtained using both aqueous ethanol and ethanol-doped urine samples.

OPTICAL BIOSENSORS

Optical biosensors are used when a biological system can be coupled to a physically or chemically induced optical change. The optoelectronic sensors cover a multitude of devices of differing complexity and they utilise such diverse principles of measurement as absorbance or transmittance, optical rotation, fluorescence, luminescence, evanescent wave and plasmon resonance.

One of the most simple optical devices utilises the colour change reaction of a pH indicator dye. A fibre optic pH probe based on the use of an indicator dye has been reported, which is suitable for tissue and blood pH measurements in the physiological range 7.0–7.4 with an accuracy of 0.01 pH unit (27).

A related, inexpensive optoelectronic sensor for the detection of serum albumin has also been reported (28). Bromocresol green is covalently attached to a transparent cellophane membrane and sandwiched between a red-light emitting diode and a silicon photodiode/amplifier detector system. At a pH of 3.8, human serum albumin adsorbs to the dye, changes the colour of the dye from yellow to bluish-green and reduces the transmission of red light (630–633 nm) through the transparent membrane. The colour change was proportional to the concentration of albumin present and the effect of albumin on the membrane was completely reversible.

The solid phase optoelectronic sensor may also be exploited to determine substrates such as penicillin, urea and glucose, by co-immobilising to the transparent membrane a pH sensitive dye and appropriate enzymes which generated or consumed protons (29).

Fluorescence or fluorescence quenching is particularly well suited to optical sensing. This method of optical detection can be applied to an immunosensor, by fluorescent labelling of the second added antibody or a labelled substrate and carrying out either a two-site or competitive immuno-assay, respectively.

A considerable effort has been committed to the development of transducers which rely upon the direct detection of antigen becoming attached to an antibody coated surface. The changes in local dielectric properties associated with the binding of antigens or antibodies to a surface can be detected with a variety of optical measurements, which include evanescent wave and plasmon resonance techniques.

In an evanescent wave system, light is propagated through the waveguide by multiple total internal reflections and at each reflection an evanescent light wave is created, which exponentially decays into the medium. The relative refractive indices of the waveguide material and its surrounding medium control the reflection efficiency and the amount of energy coupled into the evanescent. Surface bound molecules alter the refractive index and the local surface dielectric constant of the medium surrounding the waveguide. The two parameters which can be monitored in the evanescent wave

method are the efficiency of longitudinal light transmission along the waveguide, or the diffuse lateral light output by the waveguide.

The extent of penetration of the evanescent wave is of the same order of magnitude as the molecular dimensions of antibody - antigen complexes and therefore it is only surface bound molecules which interact with the light. The sensitivity of this technique is claimed to be in the nanomolar range.

A surface plasmon is a collective motion of electrons in the surface of a metal conductor, excited by the impact of light of appropriate wavelength at a particular angle. For a given wavelength of light a surface plasmon effect is observed as a sharp minimum in light reflectance from the metal surface, known as "Wood's anomaly". The critical angle of minimum light reflectance is very sensitive to the dielectric constant of the medium adjacent to the metal surface and is therefore affected by analytes bound to that surface.

OTHER BIOSENSOR SYSTEMS

Nearly all biological reactions can be followed calorimetrically by determining the heat evolved in the catalytic process. Enzyme activities of an enzyme - immonusensor have been detected by heat of reaction (31,32). However senstivities are low due to the small heat output and there is also a need to shield from relatively large temperature fluctuations of laboratory equipment.

The piezoelectric effect describes the ability of quartz crystals or certain plastics to generate a small electrical charge in response to mechanical stress. An electrical signal can be used to induce a mechanical resonance in the device, which is modified as a function of the attached mass. A number of highly sensitive gas, metal and vapour sensors have been developed using piezoelectric sensors, as well as measuring cell concentrations in a fermentation broth (33,34). Similar crystals have been used as surface acoustic wave (SAW) devices. Surface deposits on such crystals change the velocity of propogation of acoustic waves across the surface. SAW devices have been used as immunosensors for human IgG and Influenza type A virus (35) and can detect nanogram quantities of antigen.

CONCLUSIONS

There are many examples of sensor technology being used in research laboratories, however very few have as yet reach commercialisation. It is envisaged that there will be a progression towards the development of multianalyte sensors as well as their miniturisation.

REFERENCES

1   Zaborsky, O.R. (1973) in Immobilised Enzymes (I. Chibata, ed.), CRC Press, Cleveland, Ohio

2   Chibata, I. (ed) (1978) in Immobilised Enzymes: Research and Development, A Halsted Press Book, John Wiley & Sons Ltd., Tokyo

3   Venkatasubramanian, K. (ed) (1979) in Immobilised Microbial Cells. ACS Symposium Series 106, American Chemical Society

4   Rechnitz, G.A. (1981) Science 214, 287-291

5   Mascini, M. and Guilbault, G.G. (1977) Anal. Chem. 49, 795-798

6   Clark, L.C. (1956) Trans. Am. Soc. Artif. Intern. Organs 2, 41

7   Van Brunt, J. (1987) Bio/Technology 5, 437

8   Sibbald, A., Whalley, P.D. & Covington, A.K. (1984) Anal.Chim. Acta 159, 47

9   Sibbald, A., Covington, A.K. & Cooper, E.A. (1983) Clin. Chem. 29, 405

10   Al-Hitti, I.K., Moody, G.J. & Thomas, J.D.R. (1983) Proceedings Transducers Group, Biological Engineering Society

11   Boitieux, J.-L., Lemay, C., Desmet, G. & Thomas, D. (1981) Clin. Chem. Acta 113, 175

12   Caras, S. & Janata, J. (1980) Anal. Chem. 52, 1935

13   Hanss, M. & Rey, A. (1971) Biochim. Biophys. Acta 227, 630

14   Lawrence, A.J. (1971) Eur. J. Biochem. 18, 221

15   Lawrence, A.J. & Moores, G.R. (1972) Eur. J. Biochem. 24, 538

16   Watson, L.D., Maynard, P., Cullen, D.C., Sethi, R.S., Brettle, J. & Lowe, C.R. (1987) Biosensors in press.

17   Lowe, C.R., Cullen, D.C., Watson, L.D., Sethi, R.S. & Brettle, J. (1987) Patent applied for.

18   Avrameas, S. (1969) Immunochemistry 6, 43

19   Chua, K.S. & Tan, I.K. (1978) Clin. Chem. 24, 150

20   Foulds, N.C. & Lowe, C.R. (1986) J. Chem. Soc. Faraday Trans. 1, 82, 1259

21   Weetall, H.H. (1974) Anal. Chem. 46, 602A

22   Schlapter, P., Hindt, W. & Racine, P. (1974) Clin. Chim. Acta 57, 283

23   Cass, A.E.G., Davis, G., Francis, G.D., Hill, H.A.O., Aston, W.J., Higgins, I.J., Plotkin, E.V., Scott, L.D.L. & Turner, A.P.F. (1984) Anal.Chem. 56, 667

24   Alcksandrovski, Y.A., Bezhikina, L.V. & Rodinov, Y.V. (1981) Biokhimiya 46, 708

25   Foulds, N.C. & Lowe, C.R. (1987) Anal. Chem. submitted for publication

26   Yon-Hin, B.Y. & Lowe, C.R. (1987) Anal. Chem. in press

27   Peterson, J.I., Goldstein, S.R. & Fitzgerald, R.V. (1980) Anal. Chem. 52, 864

28   Goldfinch, M.J. & Lowe, C.R. (1980) Anal.Biochem. 109, 216

29   Goldfinch, M.J. & Lowe, C.R. (1984) Anal. Biochem. 138, 430

30   Cullen, D.C. & Lowe, C.R. (1987) Biosensors in press

31   Aizawa, M. (1983) in Proceedings of th International meeting on chemical sensors 638 Koclansha

32   Mattiason, B., Danielsson, B & Mosbach, K. (1980) Food Process Engineering 2, 59

33   Guilbault, G.G. (1980) Ion-Selective Electrode Rev. 2, 17

34   Ishimori, Y., Karube, I. & Suzuki, S. (1981) Appl. Environ. Microbiol. 42, 632

35   Roederer, J.E. & Bastiaans, G.J. (1983) Anal.Chem. 55, 2333

Dr GRAY-STEPHENS

Schönherr : What about stability of your type of sensors, can you reuse them, are they autoclavable ? and so on. I am asking you a lot of questions and you have lots of different systems of course but just give some answers.

Gray-Stephens : Well, obviously, the actual incorporation of the biological material there does make autoclaving rather difficult. The stability of the glucose oxydase electrode was monitored for at least one month period and it did not fall below eighty percent of the original activity. Now the initial fall in activity may in fact be due to the loose association of the glucose oxydase enzyme actually round the surface of the fairly highly charged support matrix there which falls off. So that might account for that. I carried out the microconductimetric urease sensor myself and that was stable. Again, you can immobilize the urease in the glutaraldehyde-BSA gel and I monitored that for up to two weeks and it was stable.

Spier : If it is stable, where are they ?

Gray-Stephens : Sitting in the fridge in the laboratory.

Spier : Well I mean, why aren't they used ?

Gray-Stephens : Well you see these biosensor devices are actually in the initial research stages. We understand there are an awful lot of further developments that have to actually be made to such devices before they can be introduced into say a fermentor system. You have an awful lot of processes like electrofouling in amperometric systems...

Spier :   So presumably the stability you are talking about is one
             where you take a clean glucose solution and you just look
             at it...

Gray-Stephens : Yes, initially, as a basic research, yes.
             In the case of the urease sensor, we had actually your
             crude human serum sample there, is actually successfully
             and monitored the concentration and accurately to within
             one percent error. To accurately monitor the concentration
             of the urea in the human serum sample is some achievement.

Merten :   There are already sensors available, not directly for the
             fermentation process such as the Yellowstone system.   Do
             you know about it perhaps ?
             It allows one to monitor the glucose concentration. It is
             very much used in food industry.   These probes have a
             quite long stability (some months) and are working quite
             well. Beckman has some system like this and in France
             Setric Génie Industriel has a lactate sensor. There is
             another firm in France, which also has developed a glucose
             sensor. So, there are always some sensors available.

Gray-Stephens : Yes, quite a lot of companies actually in the U.K.,
             they always promise that they are going to bring out
             various sensors and things...

Drouxhet :   In fact, there is a big discrepancy between the number of
             papers dealing with biosensors and the number of patents,
             the very low development of commercially available bio-
             sensors. For example, you mentioned the case of the
             glucose electrode in France. I personally know very well
             this case. In case of the glucose oxydase, some five
             units have until now been sold in France. The problem is
             not one of stability because it is a very stable enzyme
             and once immobilized it is stable for at least one year.
             Many problems occur with these devices when you use
             biocatalysts, but the biggest among them is to know

exactly for what purpose you will develop this type of device.

If you construct a device for fermentation process using media of relatively simple chemical composition, it can work. However if you must use these sensors for more complex media or _in vivo_, then you have to face many troubles and many big problems. So, actually since the discovery of the enzyme electrode in 1962, there are still big problems with these enzyme electrodes and today the tendancy is to construct small disposable devices. The electrochemical site of these sensors for example is waiting for the development of immobilization techniques also to increase the performance. In fact there is a big discrepancy between academic research and the practical applications.

Gray-Stephens : Can I also just add as a point that the microconducti-metric urease sensor has actually been into clinical trial and is used now in the Hammersmith Hospital in London.

Miller : All these sensors are based on proteins and Otto mentioned the problem of autoclaving them. There are other solutions such as adding special antibiotics and you don't need to really autoclave _in situ_.

Anonymous : You can irradiate...

Miller : No, because irradiation may destroy or possibly alter your enzyme. But there is another approach developed in the BATELLE Institute. Due to the presence of carboxyl-, amido- and amino groups, proteins display a characteristic IR spectrum. By recording these spectra and storing them in the memory of a computer, it would be possible afterwards, using a substraction procedure to qualitatively and quantitatively analyze the protein composition of the growth medium without having to rely on enzymatic systems,

by just using optic fibers and so on. Are you aware of
these developments ?

Gray-Stephens : Well, yes.

Miller : What is their future ? Can we expect them soon on the
market ?

Gray-Stephens : It is very difficult to make that type of measuring
principle actually applicable. It is like a problem with
many other of the biosensors devices. It is actually only
specific to the compound, the enzyme or the antibody
species that sits on the surface. There is an awful lot
of problems associated with the specificity, particularly
in optical systems as well. You get surface adsorption
effects... That is the great problem.

Hulser : From one of your slides, I got the impression as if you
could measure one pH unit with about a difference of one
or two volts but Nernst equation tells me that under
optimum conditions, you can measure sixty millivolts for
one pH unit. This is the first question. Second, I have
the feeling that many of these calibrations you showed,
only go into one direction. So, if you have a probe
inserted in a feeding line, where you measure the binding
of antibodies, the system can run only in one direction I
think. When it is loaded, then the experiment is over
isn't it ?

Gray-Stephens : Well, as far as the change in voltage output
properties of the pH indicator, we did incorporate a lot
of amplification as well. We do have associated elec-
tronics with the system so that the actual small magnitude
in change of response that you actually see, can be
amplified up so that you can see the change. So, that can
explain the scale difference.

Hulser :   But still that means you cannot reach more than sixty
            millivolts under optimum conditions even if you have some
            amplification.

Miller :   I think this is a very nasty question but the second one
            is even nastier. Prof. HULZER said... it goes in one
            direction...
            For you it is all right because you use the probe once and
            then throw it in the dustbin but for us, users, first it
            is expensive and...

Gray-Stephens : The whole idea of the microconductimetric biosensor is
            that they are made so incredibly cheap by mass production
            that they can actually be thrown away.

Miller :   You measure the pH say once and then you throw your
            electrode away ?

Merten :   If you know a little bit the group of Klaus MOSSBACH for
            instance, you have maybe heard of the thermistor-based
            enzymic immunosorbent assay. That probe can be used not
            only once but hundred times.   Of course, you need always
            an elution step after the measurement. It is always a
            cycle.

Hulser :   It was only one-direction during the experiment.

Gray-Stephens : It is totally regeneratable and you can actually use
            the urea with it. You actually regenerate and diffuse out
            the ammonium ions.

Merten :   During fermentation, you need a biosensor which is not
            directly in the fermentor and you can use it only once.
            To keep the sensor outside, we have membrane systems, flow
            injection systems.   But one always must run through a
            cycle : you inject the sample with a reagent, you get a
            response, then you have to elute it and equilibrate the

system again. Then starts the next cycle. This is the
only one possibility with an affinity-base biosensor used
in fermentation processes.

Part IV.

Analysis of the Cell's Behaviour

EFFECTS OF RETINOIDS ON GROWTH AND DIFFERENTIATION

MICHAEL I. SHERMAN

DEPARTMENT OF ONCOLOGY AND VIROLOGY, ROCHE RESEARCH CENTER, NUTLEY, NEW JERSEY 07110

1. INTRODUCTION

The retinoids constitute a group of compounds which includes vitamin A (retinol), its metabolites and a multitude of synthetic analogues. Retinol can be oxidized to retinal and these two retinoids are essential for vision and fertility. Apart from eliciting these activities upon cells of the visual and reproductive systems, retinoids appear to have a much more general influence upon cellular behavior: they affect the propensity to differentiate and proliferate. Much of the information to support this view is derived from studies with cultured cells (reviewed in 1). However, it was recognized long ago (2,3) that in animals on diets lacking retinoids, epithelial cells showed evidence of both metaplasia and hyperplasia.

In culture, retinoids tend to promote differentiation, and suppress proliferation, of cells (1), although there are exceptions (reviewed in 4 and 5). In general, retinoic acid is a considerably more potent modulator of gross cellular behavior than is retinol; on the other hand, circulating levels of retinol are two or more logs higher than that of retinoic acid. This raises several questions about the mechanisms by which retinoids influence cellular phenotype in vitro and in vivo. Some of these issues are addressed below, in most instances with mouse embryonal carcinoma cells, the stem cells of teratocarcinomas, as a source of information. We have concentrated our studies on these cells because they resemble cells of the early embryo in several respects, including the ability to differentiate broadly, because they differentiate readily in response to retinoids (as well as several other chemical inducers), and because they are rapidly proliferating, malignant cells whose differentiated progeny tend in most instances to be slower-growing and benign, making them a useful model system for testing the concept of differentiation therapy of cancer.

2. HOW DOES RETINOIC ACID PROMOTE DIFFERENTIATION?

All-trans-retinoic acid can induce differentiation in vitro of cells such as embryonal carcinoma cells (6,7) and promyelocyctic leukemia cells (8) at concentrations in the nanomolar range. Our laboratory has presented several lines of evidence to implicate the cellular binding protein for retinoic acid (CRABP) in the induction of differentiation: (a) all of several tumor-derived embryonal carcinoma cell lines tested possess CRABP (9); (b) various studies have established a good correlation between the ability of acidic retinoids to bind CRABP and to promote differentiation of embryonal carcinoma cells (e.g., 10,11); (c) mutant embryonal carcinoma cells lacking or possessing much reduced levels of CRABP have little or no ability to differentiate in response to retinoic acid (12-14); and (d) differentiation response to retinoic acid can be restored in mutant embryonal carcinoma cells that regain CRABP activity by cell fusion (15) or treatment with sodium

175

A. O. A. Miller (ed.), Advanced Research on Animal Cell Technology, 175–185.
© *1989 by Kluwer Academic Publishers.*

butyrate (16). cDNA for bovine CRABP gene has recently been identified and placed in a eukaryotic expression vector (17,18). In collaboration with Drs. Lena Wei and Chi Nguyen-Huu we are attempting to restore the ability of CRABP⁻ embryonal carcinoma cell mutants to differentiate in response to retinoic acid by transfection of the cells with the expression vector.

How might CRABP mediate retinoic acid induction of differentiation? It appears likely that CRABP is an intracellular transport protein for retinoic acid. Following fractionation of cells exposed to retinoic acid, the retinoid can be found associated with most organelles (e.g., 19). However, control studies suggest that a similar distribution can occur when cells are exposed to [^{3}H]retinoic acid at 4^0C immediately prior to disruption (20). Thus, it is necessary to distinguish between an artifactual organellar distribution of retinoic acid due to its lipophilicity and true organellar interaction. In this regard, studies have demonstrated that retinoic acid can associate in an unsaturable, non-specific manner with isolated nuclei but apparently bind specifically to a finite number of nuclear sites when complexed with CRABP (21-24). It is conceivable, therefore, that CRABP transports retinoic acid into nuclei to interact with, and activate, genes which modulate cellular behavior, including genes which initiate differentiation of embryonal carcinoma cells. Such an activity is unlikely to be strictly analogous to the interaction of steroid receptor holoproteins with the genes they activate (e.g., references 25,26) since the CRABP appears to dissociate from the nucleus once it delivers retinoic acid to its specific sites (23).

No genes have yet been identified as being directly activated by retinoic acid. The observation that retinoic acid induces its own metabolism in cells which possess CRABP but not in CRABP⁻ embryonal carcinoma cells would be consistent with the view that the CRABP holoprotein activates the gene(s) (presumably in the cytochrome P450 family) responsible for the metabolism (27,28). We have obtained preliminary evidence that the mutant cells become competent for induction of retinoic acid metabolism when they regain CRABP activity (29). However, our studies also show that some mutant embryonal carcinoma cells possess CRABP and the ability to induce retinoic acid-metabolizing enzymes but still fail to differentiate upon exposure to retinoic acid (28).

3. HOW DOES RETINOL PROMOTE DIFFERENTIATION?

In initial studies on retinoid induction of differentiation of embryonal carcinoma cells it was reported that retinol was without effect (6). In subsequent investigations, however, it was demonstrated clearly that retinol could promote differentiation of embryonal carcinoma cells, albeit considerably less efficiently than retinoic acid. (30-32). In view of the above discussion of retinoic acid, it is possible that retinol promotes differentiation (a) via metabolic conversion to retinoic acid, (b) by an analogous mechanism, or (c) by unrelated means. There is evidence to suggest that retinol action depends upon metabolism to retinoic acid. Williams and Napoli (32) reported that retinol was about 0.5% as potent as retinoic acid in inducing differentiation of F9 embryonal carcinoma cells using laminin production as an indicator of differentiation; these authors claimed that this potency correlated well with the percent conversion of retinol to retinoic acid by F9 cells. A consistent finding is that embryonal carcinoma cell mutants that had lost responsiveness to retinoic acid also failed to differentiate when exposed to retinol (19).

Other data do not support the view that retinol promotes differentiation of embryonal carcinoma cells via conversion to retinoic acid. Gubler and

Sherman (27) were unable to detect conversion of retinol to retinoic acid by any of several embryonal carcinoma lines. Although the level of sensitivity in the study was only about 1%, this laboratory reported a considerably higher potency of retinol - as much as 10-15% - with some of the cells tested. The latter values were estimated by plasminogen activator assay (30). More recently, we have utilized reduction in cloning efficiency as an indicator of differentiation (since differentiated cells generally fail to proliferate when seeded at clonal density). By this assay, we estimate the potency of retinol as 1.5-3.0% that of retinoic acid (33); equivalent levels of metabolic conversion of retinol to retinoic acid would have been detected in the study of Gubler and Sherman (27). Furthermore, in recent studies we have found that whereas cells selected for lack of responsiveness to retinoic acid also become refractory to retinol, the response of cells to retinoic acid is unaffected by selecting cells which fail to differentiate during long-term exposure to retinol (33).

If retinoic acid is modulated by CRABP, does retinol exert its action in an analogous way, as a ligand of cellular retinol-binding protein (CRBP), a protein which is distinct from, but structurally related to, CRABP (see 34)? Barkai and Sherman (24) have demonstrated that retinol complexed to CRBP binds specifically to embryonal carcinoma nuclei, that the number of binding sites is about the same as that for the retinoic acid-CRABP complex, but, as is the case with liver and testis nuclei (22,35), the two sets of binding sites appear to be different. On the other hand, the CRBP holoprotein complex appears to potentiate the interaction of the retinoic acid-CRABP complex with embryonal carcinoma cell nuclei (24). This observation leaves open the possibility that retinol, together with its binding protein, promotes the interaction of the CRABP holoprotein with nuclei, thus rendering very small amounts of retinoic acid metabolized from retinol more potent than expected. On the other hand, none of these findings eliminate the possibility that retinol, with or without its binding protein, promotes differentiation of embryonal carcinoma cells directly. Data are not currently available to allow a comparison of early changes in gene expression elicited by exposure of embryonal carcinoma cells to the two retinoids. It is perhaps relevant to note that Omori et al. (36) reported that retinol and retinoic acid treatment of rats led to notably different changes in testicular gene expression patterns. Of course, in the latter experiments different cell populations might have been stimulated by the two retinoids.

4. HOW DO RETINOIDS MODULATE CELLULAR PROLIFERATION?

As mentioned above, many investigators have reported that retinoids inhibit proliferation of cells in vitro (reviewed in 1,4). This seems contradictory to the in vivo requirement for retinoids for growth. However, most in vitro studies have been carried out with transformed cells. Furthermore, as Jetten (4) points out, by modulation of culture medium, particularly by modifying or omitting serum, retinoids can promote growth of non-transformed cells and some transformed cells. Despite the antiproliferative effects of retinoids on transformed cells in vitro, there have been relatively few successes in suppressing tumor growth in vivo: retinoids are effective chemopreventive agents in animal models but generally fail to halt the growth, or promote regression, of established malignancies (37). Exceptions which are particularly notable in the context of this discussion are the findings of Strickland and Sawey (38) and Speers (39) that dietary retinoic acid or retinoic acid injected in situ, respectively, could dramatically retard teratocarcinoma growth resulting from inoculation of mice with F9 or PCC4.aza1 embryonal carcinoma cells, respectively. There was evidence from

these studies to suggest that retinoic acid acted by inducing differentiation of the embryonal carcinoma cells to slower-growing, or nondividing, derivatives.

In an expanded series of studies based upon these findings, we have assessed the effects of dietary retinoids upon a number of different embryonal carcinoma lines (40,41). Our observations were not so encouraging as the more limited initial studies: we confirmed that dietary retinoic acid could increase the extent of differentiation of cells in tumors from several embryonal carcinoma lines, including those studied previously; however, a significant reduction in tumor growth rate did not invariably accompany these elevated levels of differentiation. Furthermore, in one embryonal carcinoma line evaluated, Nulli-SCC1, dietary retinoic acid did lead to statistically significant reduction in tumor size, but without any notable effects upon extent of differentiation. Finally, by scoring both growth rate and extent of differentiation in tumors from various embryonal carcinoma lines, it is clear that the two parameters are not necessarily inversely correlated. For example, in animals on a normal retinoid diet (4,000 IU of retinyl palmitate/kg), tumors derived from PCC3-A-1 embryonal carcinoma cells grow twice as fast as those derived from PCC4.aza1R embryonal carcinoma cells (0.18 vs 0.09 mg/day); yet the "differentiation index" (a semi-quantitative histological measure of the extent of differentiation) of the former tumors was more than an order of magnitude greater than that of the latter tumors (31.0 vs 2.6 arbitrary units) (41).

Our in vivo studies lead us to conclude that there is a complex relationship between cell proliferation and differentiation in the dynamics of tumor growth: teratocarcinomas in which cells differentiate readily do not necessarily grow more slowly than those in which cells differentiate less often. We also propose that retinoids can suppress growth of teratocarcinomas by a mechanism which is independent of differentiation. Finally, we have been disappointed by our studies which failed to illustrate that retinoic acid could be successful in differentiation therapy of teratocarcinomas despite its potency in inducing differentiation of embryonal carcinoma cells in vitro. We have, therefore, attempted to learn more about the effects of retinoids upon cell proliferation through a series of in vitro studies.

Retinoic acid treatment of embryonal carcinoma cells in vitro leads to inhibition of proliferation, but differences in cell cycle time are only observed after two or three divisions (42). This effect is not surprising in view of the induction by retinoic acid of differentiation of the cells: differentiated derivatives would be expected to grow more slowly than their undifferentiated parents or to cease cell division altogether, and more than one cell cycle might be required to achieve an overtly differentiated phenotype. It is difficult to determine in such a model system whether retinoids can modulate proliferation of embryonal carcinoma cells by mechanisms unrelated to differentiation. An approach to this problem was to evaluate the effects of retinoic acid on differentiation-defective lines. Our initial results (12) suggested that beyond 24 hr of treatment, proliferation rates of cells from two mutant lines could be slowed by about 50% by 10^{-5} M retinoic acid, a rather high concentration. It is, incidentally, notable that these lines contained little or no CRABP, consistent with the view of Lotan et al. (1) that CRABP is not essential for inhibition of proliferation of some cell types by retinoic acid. In subsequent studies with more moderate concentrations of retinoic acid (10^{-6} M), doubling times of cells from 4 of 5 lines that differentiate in response to retinoic acid were lengthened by 20-45%, whereas there was no

significant effect of retinoic acid on doubling times in cells from three differentiation-defective lines. Our in vitro data to date have not, therefore, led to any firm conclusions about the ability of retinoic acid to inhibit proliferation by a mechanism independent of that involved in cell differentiation.

In an effort to learn more about the dynamics of proliferation vs differentiation of embryonal carcinoma cells in the presence of retinoids, we (33) carried out a series of in vitro studies in which F9 cells were cultured for increasing periods of time in medium supplemented with micromolar amounts of retinoic acid or retinol. Cells were subcultured each time they reached confluence. We routinely took aliquots of the cells to evaluate their ability to grow when seeded at clonal density in the absence of retinoids (as mentioned above, these growth conditions discriminate between undifferentiated and differentiated cells). We observed that after 7 days of culture, the cloning efficiency of retinoic acid-treated cells was reduced by 80% relative to control cultures, whereas retinol was, as expected, less potent, reducing cloning efficiency by only 40%. Increasing periods of exposure to retinoids failed to increase the proportion of differentiated cells, judging by cloning efficiency. To the contrary, by 28 days of continuous exposure to retinoids, the cloning efficiency of cells exposed to retinoic acid or retinol was reduced by only 55 and 20%, respectively. In fact, we maintained F9 cells in the continuous presence of retinol for a full year and failed to evidence levels of differentiation in excess of 50% (33).

The mixture of undifferentiated and differentiated cells in retinoid-treated cultures which persists over long periods of time is analogous to teratocarcinomas which contain such mixtures, yet can often be passaged indefinitely. We were, therefore, interested to determine how some cells remained refractory to retinoid treatment. Based upon our mutant studies, it seemed logical to assume that we were generating genetic variants that were resistant to retinoid treatment. However, if this were the case, we would not have expected to see the persistence for protracted periods of a reasonably high proportion of differentiated cells in the cultures. In a direct test, we (33) maintained F9 cells in 10^{-6} M retinol, 10^{-6} M retinoic acid or no retinoid for 55 days. We then plated cells at clonal density in the absence of retinoids and selected five clones from each culture for expansion and further study. We found that cells derived from the retinol-treated cultures were not significantly more refractory to differentiation following a second round of retinol treatment than were cells not previously exposed to retinol. On the other hand, cells from the original retinoic acid-treated culture were clearly more refractory to subsequent treatment with retinoic acid, although the cells were not fully resistant. Our interpretation of these results is that refractoriness of F9 cells to retinol is epigenetic rather than genetic, whereas at least some genetic change is required for generation of refractoriness to the more potent retinoid, retinoic acid.

5. SUMMARY AND CONCLUSIONS

Retinoids can dramatically influence differentiation and proliferation of embryonal carcinoma cells. Many more questions remain to be asked than have been answered. CRABP appears to mediate retinoic acid-induced differentiation of the cells but the mechanism is yet to be elucidated. Some studies suggest that retinol action in this regard requires conversion to retinoic acid, but other experiments imply that the mechanism of action might be more complicated. In vitro model systems have been constructed which

180

appear to reproduce the behavior of teratocarcinomas in that differentiated cells are continuously generated while proliferation and tumor growth persist. The relative ease with which retinoid-refractory cells (by epigenetic vs genetic mechanisms) can be generated in continuously-treated cultures is consistent with, and could explain, our failure to successfully treat teratocarcinomas by dietary retinoid therapy. Finally, it is assumed by many that the biological activity of retinoids involves direct induction of gene expression. In a recent and comprehensive article, Lippman et al. (43) have reviewed data suggesting that retinoids can modulate activity of almost every protein that has been implicated in regulation of growth and differentiation, including cyclic AMP- and calcium-dependent protein kinases, ornithine decarboxylase, transglutaminase, growth factors, receptors and various oncogenes. It is inevitable that some of these changes will be only secondary consequences of retinoid action. Nevertheless, if we can more clearly elucidate and order the events occurring when cells respond to retinoids, we may gain some profound insights into general mechanisms of the control of differentiation and proliferation.

REFERENCES

1. Lotan R: Effects of vitamin A and its analogs (retinoids) on normal and neoplastic cells. Biochim Biophys Acta 605,33,1980.
2. Wolbach SB, Howe PR: Tissue changes following deprivation of fat soluble A vitamin. J Exp Med 42,753,1925.
3 .Fujumaki Y: Formation of gastric carcinoma in albino rats fed on deficient diets. J Cancer Res 10,469,1926.
4. Jetten AM: Retinoids and their modulation of cell growth. In Growth and Maturation Factors, Vol 3, Guroff G (ed). New York: John Wiley and Sons, 251,1985.
5. Sherman MI: How do retinoids promote differentiation? In Retinoids and Cell Differentiation, Sherman MI (ed). Boca Raton: CRC Press, 161,1986.
6. Strickland S, Mahdavi V: The induction of differentiation in terato-carcinoma cells by retinoic acid. Cell 15,393,1978.
7. Jetten AM, Jetten MER, Sherman MI: Stimulation of differentiation of several murine embryonal carcinoma cell lines by retinoic acid. Exp Cell Res 124,381,1979.
8. Breitman TR, Scolnick SE, Collins SJ: Induction of differentiation of the human promyelocytic cell line (HL-60) by retinoic acid. Proc Natl Acad Sci USA 77,2936,1980.
9. Matthaei KI, McCue PA, Sherman MI: Retinoid binding protein activities in murine embryonal carcinoma cells and their differentiated derivatives. Cancer Res 43,2862,1983.
10. Jetten AM, Jetten MER: Possible role of retinoic acid binding protein in retinoid stimulation of embryonal carcinoma cell differentiation. Nature 278,180,1979.
11. Sherman MI, Paternoster ML, Taketo M: Effects of arotinoids upon murine embryonal carcinoma cells. Cancer Res 43,4283,1983.
12. Schindler J, Matthaei KI, Sherman MI: Isolation and characterization of mouse mutant embryonal carcinoma cells which fail to differentiate in response to retinoic acid. Proc Natl Acad Sci USA 78,1077,1981.
13. McCue PA, Matthaei KI, Taketo M, Sherman MI: Differentiation-defective mutants of mouse embryonal carcinoma cells: response to hexamethylene-bisacetamide and retinoic acid. Dev Biol 96,416,1983.

14. Wang S, Gudas LJ: Selection and characterization of F9 teratocarcinoma stem cell mutants with altered responses to retinoic acid. J Biol Chem 259,5899,1984.

15. McCue PA, Gubler ML, Maffei L, Sherman MI: Complementation analyses of differentiation-defective embryonal carcinoma cells. Dev Biol 103,-399,1984.

16. McCue PA, Gubler ML, Sherman MI, Cohen BN: Sodium butyrate induces histone hyperacetylation and differentiation of murine embryonal carcinoma cells. J Cell Biol 98,602,1984.

17. Shubeita HE, Sambrook JF, McCormick AM: Molecular cloning and analysis of functional cDNA and genomic clones encoding bovine cellular retinoic acid-binding protein. Proc Natl Acad Sci USA 84,5645,1987.

18. Wei LN, Mertz JR, Goodman DS, Nguyen-Huu MC: Cellular retinoic acid- and cellular retinol-binding proteins: cDNA cloning, chromosomal assignment and tissue specific expression. Mol Endocrinol, in press, 1987.

19. Sherman MI, Paternoster ML, Eglitis MA, McCue PA: Studies on the mechanism by which chemical inducers promote differentiation of embryonal carcinoma cells. In Teratocarcinoma Stem Cells, Silver LM, Martin GR, Strickland S (ed). Cold Spring Harbor: Cold Spring Harbor Laboratory, 83,1983.

20. Sherman MI, Coppola GR, unpublished observations.

21. Mehta RG, Cerny WL, Moon RC: Nuclear interactions of retinoic acid binding protein in chemically induced mammary carcinoma. Biochem J 208,731,1982.

22. Cope FO, Knox KL, Hall RC: Retinoid binding to nuclei and microsomes of rat testes interstitial cells. Natr Res 4,289,1984.

23. Takase S, Ong DE, Chytil F: Transfer of retinoic acid from its complex with cellular retinoic acid-binding protein to the nucleus. Arch Biochem Biophys 247,328,1986.

24. Barkai U, Sherman MI: Analyses of the interactions between retinoid-binding proteins and embryonal carcinoma cells. J Cell Biol 104,-671,1987.

25. Mulvihill E, LePenner J-P, Chambon P: Chicken oviduct progesterone receptor: location of specific regions of high affinity binding in cloned DNA fragments of hormone-responsive genes. Cell 28,621,1982.

26. Payvar F, DeFranco D, Firestone GL, Edgar B, Wrange O, Okret S, Gustaffson J-A, Yamamoto KR: Sequence-specific binding of glucocorticoid receptor to MTV DNA sites within and upstream of the transcribed region. Cell 35,381,1983.

27. Gubler ML, Sherman MI: Metabolism of retinoids by embryonal carcinoma cells. J Biol Chem 260,9552,1985.

28. Gubler ML, Levin W, Thomas P, Sherman MI: manuscript in preparation.

29. Gubler ML, Wei L, Nguyen-Huu C, Sherman MI: unpublished observations.

30. Eglitis MA, Sherman MI: Murine embryonal carcinoma cells differentiate in vitro in response to retinol. Exp Cell Res 146,289,1983.

31. Jetten AM, DeLuca LM: Induction of differentiation of embryonal carcinoma cells by retinol: possible mechanisms. Biochem Biophys Res Comm 114,593,1983.

32. Williams JB, Napoli JL: Metabolism of retinoic acid and retinol during differentiation of embryonal carcinoma cells. Proc Natl Acad Sci 82,4658,1985.

33. Schroeder D, Thomas R, Sherman MI: manuscript in preparation.

34. Sundelin J, Busch C, Das K, Das S, Eriksson U, Jönsson KH, Kämpe O, Laurent B, Liljas A, Newcomer M, Nilsson M, Norlinder H, Rask L, Ronne H, Peterson PA: Structure and tissue distribution of some retinoid-

binding proteins. J Invest Dermatol 81,59s,1983.

35. Takase S, Ong DE, Chytil F: Cellular retinol-binding protein allows specific interaction of retinol with the nucleus in vitro. Proc Natl Acad Sci USA 76,2204,1979.

36. Omori M, Chytil F: Mechanism of vitamin A action. J Biol Chem 257,14370,1982.

37. Moon RC, Itri LM: Retinoids and cancer. In The Retinoids, Vol 2, Sporn MB, Roberts AM, Goodman DS (ed). Orlando: Academic Press, 327,1984.

38. Strickland S, Sawey MJ: Studies on the effect of retinoids on the differentiation of teratocarcinoma cells in vitro and in vivo. Dev Biol 78,76,1980.

39. Speers WC: Conversion of malignant murine embryonal carcinomas to benign teratomas by chemical induction of differentiation in vivo. Cancer Res 42,1843,1982.

40. Sherman MI, Thomas RA, Schroeder D, McCue PA: In vivo effects of retinoids on embryonal carcinoma cell growth and differentiation. Abstr First Conf Differentiation Therapy, Sardinia 115,1986.

41. McCue PA, Schroeder D, Thomas RA, Gubler ML, Sherman MI: manuscript in preparation.

42. Rayner MJ, Graham CF: Clonal analysis of the change in growth phenotype during embryonal carcinoma cell differentiation. J Cell Sci 58,331,-1982.

43. Lippman SM, Kessler JF, Meyskens FL Jr: Retinoids as preventive and therapeutic anticancer agents (Part 1). Cancer Treatment Rep 71,391,-1987.

Sherman, M. : <u>Effects of retinoids on growth and differentiation</u>

Miller :   This treatment with butyrate which afterwards allows the
              expression, makes me think of the lactose operon where you
              have a positive and a negative control. You must first
              eliminate the repressor but you also have to form a
              complex between cyclic AMP and the catabolite activating
              protein to push the RNA polymerase in order to allow it to
              transcribe. So it looks as if butyrate could make the RNA
              polymerase able to start at the promoter site.
              Dr Kedinger, do you know the way butyrate acts on
              promoters in eukaryotic cells ?

Kedinger :   I think that is exactly what Dr Sherman said, butyrate
              lowers the interactions between histones and DNA by
              preventing deacetylation.

Sherman :   You get more of higher levels of the histones, so, you are
              neutralizing the lysine charges and therefore the protein-
              DNA interactions loosen.

Baserga :   But it actually depends on the gene Mike ?

Sherman :   Yes it does.

Spier :   How general is this effect of retinoic acid on other tumor
              cells in culture ?

Sherman :   Retinoic acid in general suppresses the proliferation of
              tumor cells in culture but there is as good evidence that
              it always does so by inducing differentiation. These
              cells are particularly susceptible to the differentiating-
              inducing properties of retinoic acid. Promyelocytic
              leukemia cells are very well induced and readily induced
              to differentiate, melanoma cells also. But other cells

like fibrosarcoma cells for instance, are not induced to differentiate with retinoic acid, yet, they will show inhibition of proliferation. So, it is not universal.

Unknown : How did you select actually for your non responsive mutants ?

Sherman : This is very easy. Basically, what you do, you take the cells, you treat them with mutagen and then you just grow the cells at clonal density in the presence of retinoic acid. If the cell differentiates, they won't grow. If they are differentiation-defective then they continue to grow and you get a clone. So, it is very easy.

Roger : In my system - that is thyroid in primary culture - retinoic acid has a wide range of actions and these actions are obtained at very different concentrations. Thus at 10^{-10} molar, retinoic acid inhibits differentiation and stimulates DNA synthesis. At 10^{-8}, it is cytostatic and at 10^{-7} cytotoxic. In other systems, retinoic acid at higher concentrations is also mitogenic. My question is : how can one binding protein explain this wide range of effects ?

Sherman : That's the kind of question that scientists have been trying for sixty years to learn to answer to. I think one has to be very careful in concluding that there is any unitary explanation or any unitary mechanism of action of retinoids. I know that scientists always like to compartimentalise things, they like to simplify ideas and there are many investigators in the field that feel that retinoic acid must be acting only by a single mechanism of action. I have great concern that that might not be the case. For example, these mutants which lack binding protein and do not differentiate in response to retinoic acid can nevertheless be growth-inhibited, can become cytostatic at high concentrations of retinoic acid. So, one must be very cautious before one assumes there is only

one mechanism of action. All I can tell you is that for differentiation, for embryocarcinoma cells, we have evidence that the binding protein is involved but we do not propose or presume to say that this explains the total mechanism of action of retinoids.

Drillien : One of your criteria you are using for measuring differen- tiation, one of the basic criteria, was lack of multipli- cation and inability to form colonies.   Now we heard yesterday a couple of talks that nowadays differentiation is not equal to lack of multiplication.   Differentiated cells can perfectly well multiply even at low cell densi- ties I figure. I would like to hear your comment on this contradiction.

Sherman : I'll comment very quickly. I think it depends on the type of cells you form and whether that cell is terminally differentiated as you will usually see, at the high concentration of retinoic acid that we use, or whether the cell is forced down in an intermediate but not terminal pathway in the differentiation.

Spier : Just adding in to this I think it obviously looks that most of these phenomena go through a period which is reversible when you can go back to the carcinoma state, before there is a commitment to the terminal state which is irreversible.

Phosphoinositide metabolism and control of cell proliferation.

Colin W Taylor, Department of Zoology, University of Cambridge
Downing Street, Cambridge CB2 3EJ. UK.

In most cells initiation of DNA synthesis is inexorably followed by a premitotic phase and then by mitosis, the essential control of cell proliferation is therefore the signal that commits a cell to DNA synthesis. Since cells may be stimulated to divide from various degrees of preparedness these signals differ in their complexity. Nevertheless two general problems emerge that we must address in seeking to understand how mitogens stimulate cells to proliferate. Firstly, DNA synthesis occurs in the nucleus yet mitogens are recognised by their receptors in the plasma membrane: how do occupied receptors communicate, across the cytoplasm, with the nucleus? Secondly, while mitogens are immediately recognised by their receptors, DNA synthesis typically begins many hours later: what happens during this long interval? By the usual criteria of necessity and sufficiency, it is clear that the messengers that link early events at the plasma membrane with the later initiation of DNA synthesis have not been identified. Indeed it is very likely that there are many pathways from cell-surface receptors to the nucleus and these pathways overlap and interact so that no single intracellular messenger is likely to be both necessary and sufficient to commit a cell to DNA synthesis.

Among the early responses of a variety of cells to mitogens

A. O. A. Miller (ed.), Advanced Research on Animal Cell Technology, 187–200.
© 1989 by Kluwer Academic Publishers.

are increased polyphosphoinositide hydrolysis, an increase in the intracellular $[Ca^{2+}]$, stimulation of a Na^+-H^+ antiport leading to an increase in intracellular pH, and increased tyrosine-specific protein kinase activity. Michell (1982) has reviewed early studies which suggested a close association between increased phosphoinositide metabolism and cell proliferation, here I briefly review more recent evidence that has substantially extended our understanding of both the phosphoinositide pathways and their involvement in control of cell proliferation.

Many growth factors share with a large class of hormones and neurotransmitters the ability to stimulate phosphoinositidase C activity and to thereby generate inositol 1,4,5-trisphosphate $((1,4,5)IP_3)$ and 1,2-diacylglycerol (DG). In many transmembrane signalling systems where the receptor and effector enzyme are distinct proteins, including the phosphoinositide system, the two proteins communicate through one of a family of guanine nucleotide-dependent regulatory proteins or G proteins (Taylor and Merritt, 1986). The identities of the G proteins that couple receptors to phosphoinositidase C are unknown, indeed it is not yet certain that all receptors that stimulate phosphoinositidase C do so by directly interacting with a G protein. In Swiss 3T3 cells for example, the receptor for PDGF has an intrinsic tyrosine kinase activity and PDGF also stimulates phosphoinositidase C. However, this response appears to follow a very different time course from the response that follows stimulation by mitogens

such as bombesin, the receptors for which lack tyrosine kinase activity (K D Brown, unpublished observation). It remains to be determined whether these differences reflect fundamentally different mechanisms of receptor-regulation of phosphoinositidase C activity.

Several reports have suggested that pertussis toxin distinguishes the mitogenic pathways activated by those receptors with tyrosine kinase activity from those that appear only to stimulate phosphoninositidase C: only the latter appear to be inhibited by the toxin (Letterio et al., 1986; Chambard et al., 1987). Our results are not consistent with this attractively simple conclusion since we have found that in Swiss 3T3 cells responses to a wide range of mitogens are similarly attenuated by pertussis toxin treatment. Although in Chinese hamster lung fibroblasts pertussis toxin blocked thrombin-stimulation of phosphoinositidase C (Paris and Pouyssegur, 1986), in Swiss 3T3 cells the effects of pertussis toxin on mitogen-stimulated phosphoinositidase C activity were either non-specific (unpublished observation) or in the case of bombesin there was no effect (Lopez-Rivas et al., 1987). These results suggest that while pertussis toxin may, in some cells, selectively attenuate the mitogenic response only to those receptors believed to couple directly to phosphoinositidase C; there is presently no convincing evidence to suggest that the effect is directly attributable to involvement of a pertussis toxin-sensitive G protein in coupling receptors to phosphoinositidase C.

The products of the ras proto-oncogenes are membrane-

associated proteins which bind GTP and exhibit GTPase activity, characteristics that they share with the family of G proteins which couple receptors to their effector enzymes. These properties, the frequent occurence of ras oncogenes in human tumours, and the observations that microinjection of p21^{ras} proteins into NIH 3T3 fibroblasts leads to rapid initiation of DNA synthesis (Bar-Sagi and Feramisco, 1986) and that serum-stimulation of DNA synthesis can be blocked by microinjection of a monoclonal antibody to p21^{ras} (Mulcahy et al., 1985), has fueled speculation that p21^{ras} may couple receptors for mitogens to their intracellular effectors. More specifically, the increased phosphoinositide metabolism that accompanies transfection with mutant ras genes suggests that the effector enzyme may be phosphoinositidase C (Fleischman et al., 1986). More direct evidence for that conclusion is provided by studies of NIH 3T3 cells that have the normal N-ras gene under control of a glucocorticoid-driven promoter. In these cells, overexpression of normal p21^{N-ras} correlates with increased DNA synthesis and inositol phosphate formation in response to bombesin or bradykinin (Wakelam et al., 1986). The simplest interpretation of these results is that p21^{N-ras} couples each of these receptors to phosphoinositidase C and that this coupling is an important control of DNA synthesis. A problem with these experiments is that DNA sythesis or inositol phosphate formation can only be measured many hours after overexpression p21^{N-ras}. There are many other changes in cells that overexpress p21^{ras} including activation of

phospholipase A_2, a decrease in cellular cyclic AMP, an increase in transcription of a gene related to that for PDGF, and increased secretion of TGF-α ; it is therefore very difficult to directly ascribe the increased receptor coupling to phosphoinositidase C to the overexpression of p21$\underline{ras}$ or to relate that change to the increased DNA synthesis. In other fibroblast lines overexpression of p21$\underline{ras}$ enhanced inositol phosphate formation in response to bradykinin but attenuated the response to PDGF (Parrier et al., 1986). It remains to be seen whether this reflects an important difference between regulation of phosphoinositidase C by receptors with and those without tyrosine kinase activity.

While many mitogens trigger an increase in intracellular Ca^{2+} activity, the pattern of the response differs for different mitogens. Most striking are the different time courses and dependencies on extracellular Ca^{2+}. In Swiss 3T3 cells, bombesin, vasopressin and prostaglandin $F_{2\alpha}$ each elicit a rapid increase in intracellular $[Ca^{2+}]$ that is relatively independent of extracellular Ca^{2+} (Hesketh et al., 1985). By contrast, the responses to EGF and perhaps to PDGF are more dependent on extracellular Ca^{2+} and, at least for PDGF, the response occurs only after a lag of 2-3 minutes (Hesketh et al., 1985; Lopez-Rivas et al., 1987; K D Brown, unpublished observations).

The mechanisms whereby mitogens such as EGF stimulate Ca^{2+} influx at the plasma membrane are unknown though at least for EGF it is clear that the increase in cytosolic $[Ca^{2+}]$ occurs without detectable formation of inositol phosphates.

Mobilization of intracellular Ca^{2+} pools is much better understood. It has become widely accepted that $(1,4,5)IP_3$ is the intracellular messenger that mediates the effects of cell-surface receptors on intracellular Ca^{2+} pools (Berridge and Irvine, 1984). This generalisation has been extended to Swiss 3T3 cells where $(1,4,5)IP_3$ stimulates Ca^{2+} release from an intracellular pool and each of the mitogens that elicits mobilization of intracellular Ca^{2+} also stimulates formation of $(1,4,5)IP_3$ (Berridge et al., 1985). Specific receptors on the endoplasmic reticulum bind $(1,4,5)IP_3$ (Spat et al., 1986) and that leads to opening of a Ca^{2+} channel through which Ca^{2+} leaks to the cytoplasm. In both platelets and sea urchin eggs the effects of $(1,4,5)IP_3$ on Ca^{2+} release are greater as the pH is increased (Clapper and Lee, 1985; Brass and Joseph, 1985). The increase in cytosolic pH that follows activation of the Na^+-H^+ antiport by protein kinase C may thereby increase the sensitivity of the endoplasmic reticulum to $(1,4,5)IP_3$. Both limbs of the phosphoinositide pathway may therefore contribute to mobilization of intracellular Ca^{2+}.

Another early signal associated with the action of mitogens is the increase in intracellular pH that follows activation of the Na^+-H^+ antiport: as for the Ca^{2+} signal, there appear to be several ways of activating it. Protein kinase C can activate the Na^+-H^+ antiport by increasing its affinity for intracellular H^+ (Moolenaar et al., 1984). In addition EGF in fibroblasts and mitogenic lectins in lymphocytes can activate Na^+-H^+ exchange without detectable activation of protein

kinase C. That different mitogens can trigger the same responses, increases in both intracellular [Ca^{2+}] and pH, but by very different pathways further underscores the difficulty of attempting to establish which signals are necessary and sufficient for initiation of DNA synthesis.

An outstanding feature of the phosphoinositide system is the complex metabolism of each of the initial intracellular messengers. These pathways may either inactivate their signalling properties or generate additional messengers.

In addition to its role as an activator of protein kinase C, DG is also the substrate of a kinase which phosphorylates it to phosphatidic acid. This reaction serves not only to recycle DG to phosphatidylinositol, but phosphatidic acid may itself be a regulator of cellular activity since it has been reported to stimulate both phosphoinositidase C and DNA synthesis (Moolenaar et al., 1986). A second pathway of DG metabolism may also be important in generating further active messengers since the sequential removal of the two fatty acids of DG liberates free arachidonic acid from which prostaglandins and other eicosanoids are synthesised.

The metabolism of (1,4,5)IP$_3$ is even more complex. One pathway begins with a 5-phosphatase that dephosphorylates (1,4,5)IP$_3$ to (1,4)IP$_2$ and a second pathway begins with a 3-kinase that phosphorylates (1,4,5)IP$_3$ to (1,3,4,5)IP$_4$. Either of these pathways inactivates the ability of (1,4,5)IP$_3$ to mobilize intracellular Ca^{2+}, but it is not yet certain whether either product or the subsequent products of their complex dephosphorylations are themselves intracellular messengers.

Already there are hints that $(1,3,4,5)IP_4$ may play a role in regulation of of Ca^{2+} entry at the plasma membrane (Irvine and Moor, 1986) and that $(1,4)IP_2$ may regulate the activity of DNA polymerase (D Busbee, personal communication). Future studies will surely determine whether additional messengers are to be found among the many inositol phosphates formed from $(1,4,5)IP_3$.

The lipid substrate of the phosphoinositide pathway, phosphatidylinositol 4,5-bisphosphate (PIP_2), is a minor component of the plasma membrane and must be constantly replenished by successive phosphorylations of phosphatidylinositol if the cell response is to be sustained. The major function of PIP_2 appears to be as substrate for phosphoinositidase C and since its turnover is very rapid, the availability of PIP_2 is clearly a potential site at which the phosphoinositide signalling pathway may be regulated. Unfortunately, the kinetic details of the enzymes involved in phosphoinositide metabolism are poorly defined, it is not therefore known whether phosphoinositidase C is always saturated with substrate or whether an increase in substrate availability could lead to increased formation of DG and $(1,4,5)IP_3$.

An interesting relationship between two early reponses to mitogens, increased tyrosine kinase activity and phosphoinositide hydrolysis, was suggested when certain oncogene products with tyrosine kinase activity (pp60^{v-src}, pp68^{v-ros}, middle t/pp60^{c-src}, pp120^{v-abl}) were found to have

associated phosphoinositide kinase activity (Macara et al., 1984; Sugimoto et al., 1984). More recently it has been found that for most of these proteins, the phosphoinositide and tyrosine kinase activities are properties of distinct though perhaps associated proteins. Furthermore while specific antibodies can completely precipitate the proteins with tyrosine kinase activity, the same antibodies scarcely deplete cells of their phosphoinositide kinase activity (Macdonald et al., 1985). Although these results suggest an interaction between two very important mitogenic pathways, we cannot yet be certain of the importance of the interaction in either control of normal cell proliferation or during transformation. A hint that the interaction may be important is provided by Swiss 3T3 cells where insulin, the receptor for which has tyrosine kinase activity, and bombesin synergistically stimulate both DNA synthesis and inositol phosphate formation (Heslop et al., 1986).

Although phosphoinositide hydrolysis is a common response to mitogens, and the intracellular messengers generated may be important regulators of the ionic events associated with mitogen action, we are still far from understanding the relationships between these early signals and the initiation of DNA synthesis.

References.

Bar-Sagi, D and Feramisco, J R (1986) Science 233, 1061-1068

Berridge, M J and Irvine, R F (1984) Nature 312, 315-321

Berridge, M J, Brown, K D, Irvine, R F and Heslop, J P (1985)

J. Cell Sci. Suppl 3, 187-198

Brass, L F and Joseph S K (1985) J. Biol. Chem. 260, 15172-15179

Chambard, J C, Paris, S, L'Allemain, G and Pouyssegur, J (1987) Nature 326, 800-803

Clapper, D L and Lee, H C (1985) J. Biol. Chem. 260, 13947-13954

Fleischman, L F, Chahwala, S B and Cantley, L (1986) Science 231, 407-410

Hesketh, T R, Moore, J P, Morris J D H, Taylor, M V, Rogers, J, Smith, G A and Metcalfe, J C (1985) Nature 313,481-484

Heslop, J P, Blakeley, D M, Brown, K D, Irvine, R F and Berridge, M J (1986) Cell 47, 703-709

Irvine, R F and Moor, R M (1986) Biochem. J. 240, 917-920

Letterio, J, Coughlin, S and Williams, L (1986) Science 234, 1117-1119

Lopez-Rivas, A, Mendoza, S A, Nanberg, E, Sinnett-Smith, J W and Rozengurt, E (1987) Proc. Natl. Acad. Sci. 84, 5768-5772

Macara, I G, Marinetti, G V and Baldozzi, P C (1984) Proc. Natl. Acad. Sci. 81, 2778-2732

Macdonald, M L, Kuenzel, E A, Glomset, J A and Krebs, E G (1985) Proc. Natl. Acad. Sci. 82, 3993-3997

Michell, R H (1982) Cell Calcium 3, 429-440

Moolenaar, W H, Tertoolen, L G J and de Laat, S W (1984) Nature 312, 371-374

Moolenaar, W H, Kruiger, W, Tilly, B C, Verlaan, I, Bierman, A J and de Laat S W (1986) Nature 323, 171-173

Mulcahy, L S, Smith, M R and Stacey, D W (1985) Nature
313, 241-243

Paris, S and Pouyssegur, J (1986) EMBO J. 5, 55-60

Parrier, G, Hoebel, R and Racker, E (1986) Proc. Natl. Acad.
Sci. 84, 2648-2652

Spat, A, Bradford, P G, McKinney, J S, Rubin, R P and Putney,
J W Jr. (1986) Nature 319, 514-516

Sugimoto, Y, Whitman, M, Cantley, L C and Erickson, R L (1984)
Proc. Natl. Acad. Sci. 81, 2117-2121

Taylor, C W and Merritt, J E (1986) Trends Pharmacol. Sci. 7,
238-242

Wakelam, M J O, Davies S A, Houslay, M D, McKay, I, Marshall,
C J and Hall, A (1986) Nature 323, 173-176

Taylor, C. : <u>The phosphoinositide pathway</u>

Spier :   Thank you very much Dr Taylor for that sort of almost mind
           boggling insight into the complexes of the intermediate
           metabolism and energy and information transfer within the
           cells. Please let us have a go and see if we can make
           this relevant to the production of TPA or something... Do
           you have any ideas how you could make it relevant to the
           production of tissue plasminogen activator ?

Taylor :   Well, I thought of the implications of this... in clinical
           biochemistry where certainly this pathway is fundamentally
           important. Lithium is one of the most widely used anti-
           manic depressive treatment. I certainly thaught of the
           implications of this pathway. I can't establish a link
           with bulk culture itself. We do not yet know the links
           between the phosphoinositide pathway and initiation of DNA
           synthesis. We do have some links between events at the
           plasma membrane and transcription of certain oncogenes.
           Some of the early oncogenes like c-fos and c-myc. Their
           transcription certainly can be modulated by products of
           the phosphoinositide pathway. The relevance of trans-
           cription of the oncogenes to initiation of DNA synthesis,
           the links between transcription of these various oncogenes
           and DNA synthesis are still elusive. So, no, I don't
           think we can go from this pathway directly to defined
           applications.

Baserga : You have a problem there besides spatial and temporal
           requirements, it is a real problem of gating if you want
           to call it that way. A lot of these things happen even
           when cells are not stimulated to proliferate. In fact,
           they are induced even when cells are inhibited to proli-
           ferate by differentiating agents. So, there is a third

problem here. These things start and something comes in
at a later point and the cells do not take it to DNA
synthesis. I do not say proliferation.

Taylor :   One thing bears on what you are saying. I am sure you are
aware that αl adrenergic agonists in liver play their part
in regulating glycogen metabolism, in the short term,
under acute conditions of stimulation. The same agonists
under conditions of prolonged stimulation can lead to DNA
synthesis. So, the same pathways and the same receptors
may under different temporal sequences of stimulation lead
to very very different responses.

Baserga :   But in the A431 cells that Rodrigo BRAVO uses, those are
cells for instance that are inhibited by EGF, not simu-
lated, and yet EGF does everything that it does to cells
stimulated to proliferate.

Taylor :   Right.

Miller :   I think that the applications of the metabolisms to
biotechnology is straightforward. You have here two
classes of operations. One which acts immediately, the
one we just have seen and with Mike Sherman we have seen a
delayed action. Because the genes cannot be turned on all
the time, they must possibly be on for a certain time,
accumulate in the cytoplasm the stuff you want, and then
certain signals make these accumulated substances in the
cytoplasm to be secreted and produced immediately. I
think that what we have just learned, happens on two such
levels. One level is on the long term and the other one
on the short term. There are examples in the litterature
like that of ferritin synthesis.
Ferritin is a protein which stores iron especially because
high intracellular concentrations of iron are damaging the
DNA. Now, if you add iron to animal cells, the time taken
to transcribe ferritin messenger RNA will be so long as to

lead to cell intoxication and cell death. Cells must have ferritin messenger RNA molecules stored in the cytoplasm, ready to make the ferritin which immediately takes the toxic iron.

I think that we really ought to consider two types of mechanisms in the cell. An immediate one which responds quickly and a more long-term one. One can of course always speculate but I think this phosphoinositide pathway belongs to the first category. This in in order to partially answer the question of Ray : how can you use this knowledge to the good of biotechnology ?

Baserga : There are several things that we have to remember. One thing is that there is a lot of redundancy of the genome and this is becoming increasingly apparent. What you do with ras, you can do it with at least four other onco-genes. So, there is redundancy and you can short circuit the cell cycle. You can bypass half of the cell cycle and get the cell into DNA synthesis. That is one thing you have to remember for biotechnology. The other thing you have to remember and which is important is the connection between the early microseconds events and cellular DNA synthesis. I have been trying to simulate cells as some of you know for the past thirty years and no matter what you do, it takes twelve, fifteen, sixteen hours - you can microinject the SV40 T antigen straight into the nucleus, it still takes you 16-20 hours. To anybody who has been trained in biochemistry, this means only one thing and that is that there must be a stoichiometric component. Enzymes go much faster than that. The cell must be counting something.

SOMATIC CELL GENETIC ANALYSIS OF GROWTH CONTROL.

J. Szpirer[1], M.Q. Islam[2], G. Levan[2] & C. Szpirer[1]

[1] Départment de Biologie Moléculaire, Université Libre de Bruxelles, Belgium
[2] Department of Genetics, University of Goteborg, Sweden

These last years, a growing number of altered genes have been implicated in neoplastic development. These genes, in their normal guise, are supposed to be involved in the control of cell growth. Two seemingly different groups of such growth control genes have been identified by different approches, on the one hand studies of highly oncogenic retroviruses and transfection experiments and on the other hand somatic cell hybridization experiments and studies of hereditary cancers.

The former studies allowed the discovery of oncogenes which are capable of transforming certain populations of relatively normal cells into in vivo tumorigenic cells and in vitro transformed cells. The oncogenes are activated versions of normal proto-oncogenes which are supposed to be involved in important stages of the cell cycle; the link of some of them with growth factors, growth factor receptors or signal transductors is already clear (for review see ref.1).

The somatic cell hybridization experiments, which preceeded the discovery of oncogenes, suggested that the normal genome contains tumour suppressor genes, i. e. genes that are capable of suppressing the tumorigenic phenotype. These tumour suppressor genes are supposed to be involved in the control of proto-oncogene expression or to code for products that antagonize proto-oncogene functions; for this reason these genes are sometimes called anti-oncogenes (for review see ref.2).

The aim of this paper is to present and discuss results obtained using the somatic cell hybridization approach. A first and extensive series of experiments in this field has been performed by Henry Harris, George Klein and co-workers (3). These authors showed that the somatic cell hybrids formed between normal and tumour cells exhibited a normal phenotype as long as they retained a relatively complete chromosome complement of the normal diploid parental cell. These results were confirmed and extended by others and especially by Stanbridge and co-workers (4).

The observation that the tumorigenic phenotype can be suppressed by the introduction of normal genetic information into a tumorigenic cell seems contradictory with the reports that oncogenes act in a dominant fashion. However, it is observed that the transfected oncogene is generally amplified in the transformed recipient cell (5). Moreover, in naturally occuring malignant tumours, it has been found that the presence of a mutated oncogene is often associated with the absence of the normal allele (6). In such a context where the dosage of different interacting genes may be largely unequal, it appears not genetically rigourous to use the terms dominance and recessivness. Finally, it is now well established that the transforming activity of certain oncogenes strongly varies between different cell types or different cell stages (for review see ref.7). In conclusion, this paradox is probably more apparent than real

A. O. A. Miller (ed.), Advanced Research on Animal Cell Technology, 201–213.
© 1989 by Kluwer Academic Publishers.

and a better understanding of the role played respectively by the proto-oncogenes, oncogenes and tumour supressor genes will certainly reconcile these seemingly discordant observations.

What is the evidence that suppression of the tumorigenic or transformed phenotype is genetic? As mentioned above, hybrids between normal and tumour cells exhibit a normal phenotype as long as they retain a relatively complete chromosome complement of the normal diploid parental cell. Sometimes, malignant or transformed segregants appear amongst the hybrids. Detailed cytogenetic analysis of the non-tumorigenic hybrids and the resulting tumorigenic segregants revealed that the malignant phenotype reappeared when certain chromosomes donated by the normal parental cell were eliminated from the hybrids (8-10). The reversibility of the suppression of the tumorigenic phenotype strongly suggests that genes on specific chromosomes provide trans-acting control of growth properties and that this regulatory control is removed from the hybrid cell upon loss of a relevant chromosome which carries the tumour suppressor gene. Such an elimination of a tumour suppressor gene is supposed to account for the development of hereditary human cancers such as retinoblastoma (11) Wilms tumour(12) and adenocarcinoma in familial adenomatous polyposis (13), where the loss of a specific locus is critical to the transformation process. The identification of tumour suppressor genes is the final aim of the somatic cell genetic approach.

In an attempt to identify genes responsible for suppressing transformation in hepatoma cells, we have derived hybrids from the fusion of mouse hepatoma cells (of the line BWTG3) with normal diploid cells of an adult rat. The normal diploid cells chosen were of two different tissular origins: hepatocytes (14) and fibroblasts (15). The isolated hybrids were tested for in vitro growth properties associated with cell transformation, namely saturation density and anchorage independence.

1- The fusion of the mouse hepatoma cells with normal diploid rat hepatocytes yielded proliferating hybrids at a very low rate (i.e. 0.001 hybrid per 100 treated cells). However, 45 hybrid clones could be isolated (LB hybrids); all were transformed. Their chromosome composition was determined. It appeared that most of them (31 out 45) had doubled the number of mouse chromosomes (2S hybrids), i.e. the chromosome complement derived from the hepatoma parental cells. The 14 hybrids that did not doubled the number of mouse chromosomes (1S hybrids) were subjected to detailed cytogenetic analysis (16, 17). The rat chromosome composition of the 1S hepatoma-hepatocyte hybrids is given in table 1.

This table shows that amongst these hybrids, only 2 clones retained rat chromosome number 8. On the other hand, the 2S hybrids were tested for the presence of rat chromosome 8; amongst the 31 hybrids of this type, 13 retained rat chromosome 8. This chromosome is thus preferentially lost in the hepatoma-hepatocyte hybrids that did not doubled the mouse hepatoma genome.

From these results, we conclude that rat chromosome 8 confers specifically a selective growth disadvantage to the hepatoma x hepatocyte hybrids: the non mitotic state of the normal hepatocyte appears to counteract proliferation of the hybrids and the loss of chromosome 8 would be required to allow continuous growth of these hybrids. Alternatively, doubling the hepatoma genome seems to counterbalance the negative effect of rat chromosome 8. In other words, we suppose that rat chromosome 8 carries a growth control gene.

Table 1: Chromosome composition of the 1S LB hybrids.

Hybrids	Rat Chromosomes																					
	1	2	3	4	5	6	7	8	9	10	11	12	13	14	15	16	17	18	19	20	X	Y
LB150	(-)	+	+	+	-	-	+	-	+	+	+	+	+	-	-	+	(-)	+	+	(-)	+	-
LB161	-	+	+	+	+	+	+	-	+	+	-	(-)	+	+	+	+	+	+	+	(+)	+	-
LB210I	-	-	-	-	-	-	-	-	-	-	-	-	+	+	-	+	-	+	-	-	+	-
LB251	+	+	(-)	+	-	(-)	+	-	-	+	-	+	+	-	-	-	+	-	+	-	+	-
LB260	-	+	+	+	-	-	+	-	-	(+)	+	-	-	+	-	+	+	-	-	-	+	-
LB330	-	+	(-)	+	-	+	-	-	-	+	-	+	-	-	-	+	(+)	-	-	-	+	-
LB360B	-	-	+	+	+	-	+	+	+	+	+	+	+	+	+	+	+	+	+	+	+	-
LB510	+	+	+	+	(-)	-	+	-	-	-	-	+	+	+	+	+	+	-	(-)	+	+	-
LB600	+	+	+	+	+	(+)	+	-	(-)	+	+	+	+	+	+	+	-	+	+	(-)	+	-
LB630	(-)	-	+	+	(+)	+	+	-	+	-	+	+	+	(-)	+	+	-	+	+	(-)	+	-
LB780	-	+	+	+	+	-	+	-	-	+	+	-	+	-	-	-	+	+	-	(+)	+	-
LB810	-	+	+	+	-	+	+	+	-	+	+	+	+	+	+	+	+	-	+	(+)	+	-
LB860	-	+	+	+	-	-	+	-	+	-	+	+	+	-	+	+	+	+	-	(+)	+	-
LB1040	-	-	+	+	(-)	+	+	-	-	+	+	+	-	-	+	+	-	+	-	+	+	-

2- The other fusion was performed using the same mouse hepatoma cells, but here they were fused with normal diploid rat fibroblasts. In this case, the hybrids(called BS hybrids) arose at a much higher rate (i.e. 0.1%) than in the case of the hepatoma x hepatocyte hybrids; 1S and 2S hybrids were obtained. The 1S hybrids were tested for their ability to clone in agar and it was found that all but one 1S hybrids were suppressed for the transformed phenotype. By selection of rare colonies in agar, transformed segregants were obtained from some of these hybrids. Chromosome analysis was performed for the transformed segregants and the suppressed hybrids from which they derived. It was observed that rat chromosome 8 was retained in the BS hybrids (transformed as well as suppressed), but rat chromosome 5 was eliminated from the transformed segregants. Interestingly enough, the transformant clones derived from the BS181 hybrid, instead of showing a complete loss of chromosome 5, showed only a partial deletion in the central region of this chromosome. The results of this analysis are given in Tables 2 and 3.

We conclude that rat chromosome 5 also carries a growth control gene which is located in the region of chromosome 5 covered by the deletion in the BS181 transformed subclones.

Table 2: Chromosome composition of some BS hybrids and their transformed derivatives (clones h and a)

Hybrids	1	2	3	4	5	6	7	8	9	10	11	12	13	14	15	16	17	18	19	20	X	Y
											Rat Chromosomes											
BS181	+	+	+	+	+	+	+	+	+	+	+	+	+	+	+	+	+	+	+	+	+	+
BS181h1	+	+	+	+	*	+	+	+	+	+	+	+	+	+	+	+	+	+	+	+	+	+
BS1B1a2	+	+	+	+	*	+	+	+	+	+	+	+	+	+	+	+	+	+	+	+	+	+
BS181a3	+	+	+	+	*	+	+	+	+	+	+	+	+	+	+	+	+	+	+	+	+	+
BS181a4	+	+	+	+	*	+	+	+	+	+	+	+	+	+	+	+	+	+	+	+	+	+
BS181a5	+	+	+	+	*	+	+	+	+	+	+	+	+	+	+	+	+	+	+	+	+	+
BS140	+	+	+	+	+	+	+	+	+	+	+	+	+	+	+	+	+	+	+	+	+	-
BS140h1	+	+	+	+	-	+	+	+	+	+	+	+	+	+	+	+	+	+	+	+	+	-
BS140h2	+	+	+	+	-	+	+	+	+	+	+	+	+	+	+	+	+	+	+	+	+	-
BS140h3	+	+	+	+	-	+	+	+	+	+	+	+	+	+	+	+	+	+	+	+	+	-
BS140a1	+	+	+	+	-	+	+	+	+	+	+	+	+	+	+	+	+	+	+	+	+	-
BS140a2	+	+	+	+	-	+	+	+	+	+	+	+	+	+	+	+	+	+	+	+	+	-
BS140a3	+	+	+	+	-	+	+	+	+	+	+	+	+	+	+	+	+	+	+	+	+	-
BS100	+	+	+	+	-	+	+	+	+	+	+	+	+	+	+	+	+	+	+	+	+	-

* means that the chromosome 5 present in these cells is deleted

Table 3: Phenotype of the BS hybrids is correlated with the presence or absence of a region of rat chromosome 5

Hybrids	Rat Chromosome 5 nb/cell	Cloning efficiency (%) on plastic	in agar
BS181	0.9	14	$5 \cdot 10^{-5}$
BS181h1	1del	35	3
BS1B1a2	1del	52	1.5
BS181a3	1del	45	2.5
BS181a4	1del	10	1
BS181a5	1del	22	5
BS140	1.3	10	$2 \cdot 10^{-4}$
BS140h1	0	15	0.1
BS140h2	0	13	1.5
BS140h3	0	25	0.15
BS140a20	0	30	5
BS100	0	29	0.25

1del means that one copy of a deleted chromosome 5 is present

The fusions of the same tumour cell with a normal cell either of the same differentiation or of a different differentiation revealed the existence of two different growth control genes: one which is located on rat chromosome 8 and the other which is located on rat chromosome 5. The two genes are involved in the suppression of the transformed phenotype but probably in different contexts. Since it is generally observed that progressive multiplication of cancer cells is associated with a block of normal differentiation completion, we believe that the two transformation suppressor genes revealed by this analysis do play a role in cell differentiation but at different steps or in different lines. The suppressor gene located on chromosome 8 would be involved more specially in hepatic differentiation.

Isolating tumour or growth suppressor genes is important both in the contribution of understanding oncogenesis and in the opportunity for production of anti-tumour propteins which could have therapeutic potentialities

REFERENCES

1. Bishop JM: Trends in oncogenes. Trends Genet. 10, 245–249, 1985
2. Harris H: The genetic analysis of malignancy. J. Cell Sci. Suppl. 4, 431–444, 1986.
3. Klein G, Bregula U, Wiener F & Harris H: The analysis of malignancy by cell fusion: I. Hybrids between tumour cells and L cell derivatives. J. Cell Sci. 3, 659–672, 1971.
4. Stanbridge E, Channing J Der, Doersen CJ, Nishimi RY, Peehl DM, Weissman BE & Wilkinson JE: Human cell hybrids: analysis of transformation and tumorigenicity. Science 215, 252–259, 1982
5. Bos JJ, Toksoz D, Christopher CJ, Verlaan-de Vries M, Veeneman GH, van der Eb AJ, van Boom JH, Janssen JWG & Steenvoorden CM: Amino-acid substitutions at codon 13 of the N-ras oncogene in human acute myeloid leukaemia. Nature 315, 726–730, 1985.
6. Spira J, Wiener F, Babonits M, Gamble J, Miller J & Klein G: The role of chromosome 15 in murine leukemogenesis. I. Contrasting behavior of the tumor vs. normal parent-derived chromosome no.15 in somatic hybrids of varying tumorigenicity. Int. J. Cancer 28, 785–798, 1981.
7. Klein G & Klein E: Conditioned tumorigenicity of activated oncogenes. Cancer Res. 46, 3211–3224, 1986.
8. Jonasson J, Povey S & Harris H: The analysis of malignancy by cell fusion. VII. Cytogenetic analysis of hybrids between malignant and diploid cells and of tumours derived from them. J. Cell Sci. 24, 217–234, 1977.
9. Evans EP, Byrtenshaw MD, Brown BB, Hennion R & Harris H: The analysis of malignancy by cell fusion: IX. Re-examination and clarification of the cytogenetic problem. J. Cell Sci. 56, 113–130 1982.
10. Srivatsan ES, Benedict WF & Stanbridge EJ: Implication of chromosome 11 in the suppression of neoplastic expression in human cell hybrids. Cancer Res. 46, 6174–6179, 1986.
11. Murphree AL, Benedict WF: Retinoblastoma: clues to human oncogenesis Science 221, 1028–1033, 1984.
12. Orkin SH, Goldman DS & Sallan SE: Development of homozygosity for chromosome 11p markers in Wilms' tumour. Nature 309, 172–174, 1984.

13. Solomon E, Voss R, Hall V, Bodmer WF, Jass JR, Jeffreys AJ, Lucibello FC, Patel I & Rider SH: Chromosome 5 allele loss in human colorectal carcinomas. Nature 328, 614–616, 1987
14. Szpirer J, Szpirer C & Wanson JC: Control of serum protein production in hepatocyte hybridomas: Immortalization and expression of normal hepatocyte genes. Proc. Natl. Acad. Sci. USA 77, 6616–6620, 1980.
15. Szpirer C & Szpirer J: Suppression of the transformed phenotype of hepatoma cells after hybridization with normal diploid fibroblasts. Exp. Cell Res. 125, 305–312, 1980.
16. Szpirer J, Levan G, Thorn M & Szpirer C: Gene mapping in the rat by mouse–rat cell hybridization: Synteny of the albumin and alpha-fetoprotein genes and assignment to chromosome 14, Cytogen. Cell Genet. 38, 142–149, 1984.
17. Szpirer J, Islam MQ, Cooke N, Szpirer C & Levan G: Assignment of three rat genes coding for plasma proteins: transferrin, the third component of complement and beta–fibrinogen to chromosome 8, 9 and 2. Cytogenet. Cell Genet. (in press) 1987.

Szpirer, J. : <u>Somatic cell genetic analysis of growth control</u>

Spier :   If you are saying that the genes that suppress the trans-
            formed state are producing materials which we can then use
            for - if you like - tumor suppression, you said in chro-
            mosome 8 - I think it was ferritin ...

Szpirer :   No, the transferrin gene is on chromosome 8 and we use it
            as a marker to look whether chromosome 8 is present or not
            in the 25 hybrids. Transferrin could be implicated in
            some human cancers there are chromosomes where the rear-
            rangement and the breakpoints are just near transferrin
            but I don't know if here in this case, transferrin has
            something to do with what we observe.

Spier :   Have you any thoughts about chromosome 5 ?

Szpirer :   Chromosome 5 is a homolog, by banding homology with
            chromosome 4 which was found by HARRIS and coworkers to
            contain tumor suppressor genes and mouse chromosome 4
            which is homologous to rat chromosome 5 is homologous to
            human chromosome 1 which was found by STANBRIDGE to carry
            a tumor suppressor gene. So, we believe that the one
            suppressor gene we suppose to exist on chromosome 5 in the
            rat is the same as the number 4 of HARRIS and the 1 of
            STANBRIDGE but we have a deletion and we shall be able to
            characterize it first.

Sherman :   Just to follow up on your last comment that your deletion
            might allow you to characterize it further, what is the
            size of your deletion ?

Szpirer : Well, it's quite big but we will try to add markers in the deletion and then to know where to look for. It will be easier if we can associate the supposed tumor suppressor gene with a linked gene.

Sherman : Yes, I take your point ... but by the same token it is only fair to note that HARRIS and STANBRIDGE and others have known the existence of these suppressor genes in these chromosomes for a long time and there is an awful lot of walking to do along the chromosome before you identify the gene that is presumably relevant to the observed phenotype.

Szpirer : I think so yes, but you can see a difference of phenotype when you don't have the specific chromosome, so, we have to take this observation into account to try to find these genes.

Sherman : May I ask a second question ? One thing that I was not certain of is, how do you select these hybrids ? Were they selected in soft agar or in monolayer cultures ?

Szpirer : The mouse hepatoma lines we have used is deficient for GPRT and so, we select the hybrids in HAT medium. Afterwards, we plate the hybrids in agar to look whether they are transformed or not and for some of them we could pick up some very rare segregants.

Sherman : My question then is whether it is in fact possible that the different "suppressor genes" that you are looking at, are in fact acting at different levels, one perhaps at the level of immobilization that you might expect in hepato-cyte times hepatocyte case, and the other at the level more of transformation since the fibroblast partner presumably would continue to grow.

Szpirer :   The only thing I can say is that when you get hybrids between hepatoma and hepatocytes, they are immortalized but they are also transformed so you get the two steps together in this case.

Baserga :   If I remember correctly the experiment of STANBRIDGE, the tumor suppressor gene is tumor-specific : one gene will suppress the Wilms tumor and another gene will suppress the retinoblastoma. So you will have to have a tumor suppressor gene for each tumor.

Szpirer :   Probably not for each tumor but for groups of tumors.

Van Meel :   In normal somatic cell genetic analysis, you start with defined specificity, let's say an enzyme activity and then you try to assign this activity to a certain site on the genome. Now you are talking about something... let's say undefined such as cell proliferation and transformation. So, I have a bit of a problem to understand how you can align such a broad spectrum of activities to one specific site in the genome map.

Szpirer :   Well, probably, we can do so because we look at a special activity. We see that when this gene is present, something changes in the phenotype of this particular kind of hybrid. So, we suppose that only one gene is involved since only one chromosome is lost but maybe there is more than one important gene on this chromosome.

Van Meel :   But how can you be sure that, when chromosome 8 is present, there is really activity from that chromosome ?

Szpirer :   Well, as regards chromosome 8, it is difficult to say but as regards chromosome 5, when it is present, the hybrid behaves as a normal cell and when it is absent, the hybrid is transformed.

So, I suppose the chromosome is important and if I go back to the experiments of STANBRIDGE, he made microcell hybrids. With only one chromosome 1 being present in the transformed cell, the transformed phenotype was corrected. So I suppose that is a good reason.

Sherman :   I just wanted to actually address the question that Renato posed which you partially answered and for example, in the case of the retinoblastoma, BENEDICT and his colleagues have provided evidence that that suppressed gene is probably also implicated in osteosarcoma because those individuals who have congenital retinoblastoma which is treated, then have a very high incidence thereafter of osteosarcoma and to a lesser extent of fibrosarcoma. My question is : does that mean that if there are forty or more oncogenes, that there are going to be forty or more suppressor genes. Why are people picking up with the chromosomes studies what seems always to be the same class of cancers with the hybrid studies ?

Szpirer :   Also for the oncogenes, people are working with 3T3. In the beginning they always found the same oncogenes. Probably for the tumor suppressor gene which is the same. It is because people are often using for instance fibroblasts as tests because it is easy to get normal diploid fibroblasts in culture. So perhaps this is the reason why most often now, the people found the same kind of suppressor gene and to answer your question about the number of suppressor genes, I don't know but maybe some of these suppressor genes will be the normal version of an oncogen. You can imagine that there is a competion between the normal gene and the product of the oncogene.

Sherman :   I think in some of the recent studies on colorectal cancer by FORRESTER et al from PERRUCHO's lab and VAN DER EB in another lab, the implication there is, that, in terms of the existence altered ras protein or ras gene, you may or

may not see the presence of the wild type ras protein.

I am not sure from these studies whether we can clearly feel confident that the presence of the normal oncogene must be noticed or that you must have elimination of the normal oncogene in order to see the malignant phenotype, or that the presence of the normal oncogene will be protective. In fact, the expression level as I remember, the levels of expression of the mutant versus the normal ras, were not greatly affected.

Szpirer :   Does it not appear to be the case that you have a competition between the normal and the altered product, but in other kinds of oncogenes ?

Baserga :   There is another experiment by Bruce HOWARD. Are you familiar with it ? He took the DNA from human diploid fibroblasts and transfected into HeLa cells. He gets very slowly growing HeLa cells.

HeLa as you know, if you spill them on this table, they will grow. But when he transfected them, all of a sudden, they became very quiet. But there is a very important observation there because it implies an epigenetic mechanism. The system works if you keep the human diploid fibroblast, quiescent, for several days before you prepare its DNA. Then they have this inhibitory action. If you keep them quiescent only for 24 hours, they don't have any inhibitory action. The genes are still there, so it almost implies an epigenetic mechanism.

Spier :   I think that during embryogenesis, two of the oncogenes, I think ras and myc are active and then they obviously get turned off so there are natural mechanisms that we need to have in order to suppress these activities and I don't think you need more than three different suppressor genes to antagonize the protein kinases or the ras type oncogenes or perhaps the myc and fos type oncogenes. Do yo want to comment on that ?

Szpirer :   I have no comments.

Spier :   You would agree then that perhaps we don't need forty anti
oncogenes in order to get suppression, we only need three.

Sherman :   I think first of all, I am not so sure that ras and myc
are turned off in non-embryonic or in adult cells. I
think you can find them active in the cellular virgin
protooncogen. They are active in many cell types but I
think also that this gets us back again to the whole
question Alain raised this morning about the immediate-
early effects and the subsequent effects and it maybe that
oncogenes or suppressor genes can act to counteract these
very early events that you are talking about but there may
be later events involved in the progression of malignancies, that we know very little about now or suppressor
genes may be just as effective. I would not presume that
there could be as many as forty or more critical steps in
the events that lead ultimately to the tumor progression
and the refractory tumors that could be service targets
for suppressor gene action.

Szpirer :   There may be only three functions, but the same functions
are made by different suppressor genes which are expressed
at different times or on different cells.

Miller :   How many suppressor anti-oncogenes are known in human
today ? I know the Wilm's, the retinoblastoma, the
colorectal carcinoma and there is another one, the exotic
bilateral acoustic neurofibromatosis, the BANF.

Szpirer :   The only thing I know for sure is that only the retinoblastoma tumor suppressor oncogene has been isolated and
that it is the only one to be really well characterized.

Sherman :   There are also two others - I don't believe yet published - there is evidence for suppressor genes in hereditary congenital lung and breast cancers.

Miller :   That is a way of looking, assessing, making an analysis of the possible pathways, looking at these oncogenes, you can see whether the gene products belong to the same or different metabolic pathways and thus make a linkage analysis.

Szpirer :   But this requires isolation of the gene product and for the moment only one suppressor gene has been isolated.

PROMOTER ORGANIZATION OF EUKARYOTIC PROTEIN-CODING GENES

C. KEDINGER*, H. BOEUF, D. ZAJCHOWSKI, P. JALINOT, C. HAUSS, B. DEVAUX, G. ALBRECHT and P. JANSEN-DURR

LGME du CNRS, U.184 de l'INSERM, 11, rue Humann, 67085 STRASBOURG Cédex, France.

1. INTRODUCTION

The identification and analysis of the control elements of eukaryotic genes has largely benefited from the development of site directed mutagenesis technology. Classical genetics is not readily applicable to higher eukaryotes essentially because of their genome complexity. New methods for the isolation and cloning of genes and, above all, the chemical synthesis of DNA, allow precise in vitro mutagenesis by deleting, inserting or altering defined parts of a given gene. The biological activity of these mutants can then be tested in vitro, by using appropriate cell-free transcription systems (1, 2), or in vivo, by putting the manipulated gene back into a cellular environment and measuring its expression after transient (3) or stable (4) cell transformation.

In parallel to the delineation of the nucleotide sequence elements required for the efficient transcription of a particular gene, the protein factors which contribute to the promoter activity by interacting with these elements can be identified. The specific binding of these factors to their recognition sequence, as visualized by DNase I footprinting (5, 6) or electrophoretic band-shifting experiments (7), constitutes a convenient functional assay to follow their purification from crude cell extracts.

We present here some experimental approaches which we have followed to dissect the promoter structure of a human adenovirus gene, the early EIIa gene (EIIaE) of adenovirus type 2 or 5 (Ad2 or Ad5). The expression of this gene which codes for an essential DNA-binding protein involved in viral DNA replication is controlled by a promoter lacking the canonical TATA box sequence. Furthermore, although it exhibits detectable constitutive activity, this promoter is strongly stimulated by the products of the viral immediate early gene EIa. Thus the EIIaE promoter constitutes an attractive model system for the study not only of an atypical eukaryotic promoter, but also of the mechanisms of gene induction. In the following we will however restrict ourselves to the mapping of the sequences required for efficient constitutive activity of this promoter, illustrating the technical strategies which have been applied to reach this goal.

2. PROCEDURES

2.1. Recombinant plasmids

All constructions have been previously described : pEII (8), Mp8EII5', Mp8EII3', pMT5 and LS WT (9) and will only be briefly recalled in the corresponding figure legends.

The 5'-deletion series derived from pEII (8) was obtained by exonuclease Bal31 digestion from the SmaI site at position −250 with respect to the EIIaE major capsite (EIIaE1, +1). A XbaI linker sequence was inserted at each deletion end-point and the XbaI(deletion end-point)-HindIII(position +719) fragments of the resulting plasmids were recloned

A. O. A. Miller (ed.), Advanced Research on Animal Cell Technology, 215–229.
© 1989 by Kluwer Academic Publishers.

between the XbaI (−250) and HindIII (+719) sites of pEII, in place of the wild type sequence. By this procedure a series of unidirectional 5'-deletion mutants was generated (see Fig. 1A), which we named according to the position of each mutant's deletion end-point.

The linker-scanning (LS) derivatives were constructed by the method of McKnight and Kingsbury (10) and the procedure specifically used here is outlined in Fig. 2 (see also ref. 9).

2.2. Cells and extracts

HeLa cells were grown in Eagle minimum essential suspension culture medium supplemented with 7% newborn calf serum, collected by centrifugation and frozen in liquid nitrogen in a buffer containing 50 mM Tris-HCl, pH 7.9, 10 mM α-thioglycerol, 30% glycerol (final concentration). Whole cell extracts (WCE) were prepared by the procedure of Natarajan et al. (11) and nuclear extracts (NE) were obtained essentially as described by Dignam et al. (12). Transcription factors were partially purified by chromatography of the WCE on heparin sepharose and DEAE-cellulose columns as previously described (13). The material eluted from the latter column with 0.15 and 0.25 M KCl were the DE0.15 and DE0.25 fractions, respectively.

2.3. In vivo expression assay

HeLa cells in monolayers were transfected by calcium phosphate coprecipitation (14) with 2 to 5 μg of a given plasmid, and the final DNA concentration per 10-cm Petri dish was adjusted to 15 μg with M13 RF DNA. Cytoplasmic RNA was extracted 36 h after transfection and specific transcripts were analyzed by the quantitative S1 nuclease assay (15), using the single-stranded probe shown in Fig. 1A.

2.4. DNase I protection

The standard DNase I footprinting assay (5) was used with slight modifications for experiments with NE (16) or DE fractions (17). Unlabelled carrier DNA was poly(dI-dC) (50 ng per 20 μl binding reaction). Usually DNase digestion was carried out in the presence of 20 ng DNase I (1600 Kunitz units/mg) for naked DNA, 40 to 100 ng for the DE fraction-containing samples and 200 to 600 ng for those including the NE.

2.5. Gel electrophoresis DNA binding assay

The electrophoretic band-shift assay of Fried and Crothers (7) was used under the conditions described (18). Briefly, specific protein fractions were incubated with a 5'-end-labelled probe in the presence of carrier poly(dI-dC). After 25 min at 25°C the mixture was electrophoresed on a 4.5% non-denaturing polyacrylamide gel. The gel was then fixed and dried before autoradiography.

For the methylation interference mapping the end-labelled DNA probes were methylated with dimethyl sulfate (19) before incubation with the protein fractions. After electrophoresis as above, the gel was directly autoradiographed and the DNA fragments corresponding to the retarded and unbound probe were electroeluted, treated with piperidine and analysed on 8% sequencing gels (19).

2.6. Genomic footprinting

HeLa cells, at about 60% confluency, were infected with adenovirus at 100 pfu/cell. Six hours post-infection, the cells were collected and disrupted in the presence of 0.3% NP40 (20). The nuclei were recovered by centrifugation through a sucrose cushion and digested with DNase I in the presence of 3 mM $CaCl_2$. DNA was then extracted, restricted with EcoRI,

fractionated on a 8% sequencing gel, and electroblotted to a nylon membrane (6). The blotted DNA was then hybridized with a highly labelled (6) single-stranded EIIaE probe and exposed for autoradiography.

3. RESULTS
3.1. External-deletion mapping of the EIIaE promoter
Progressive deletion of sequences located upstream of the transcriptional startsite of a gene allows the positioning of the 5' limit of the promoter sequences essential for maximal transcription. From the functional analysis of the series of deletion mutants shown in Fig. 1 it appeared that the EIIaE promoter sequences extend next to position -88, since pEII-88 retains full activity (see Fig. 1B and C). In fact careful analysis of a

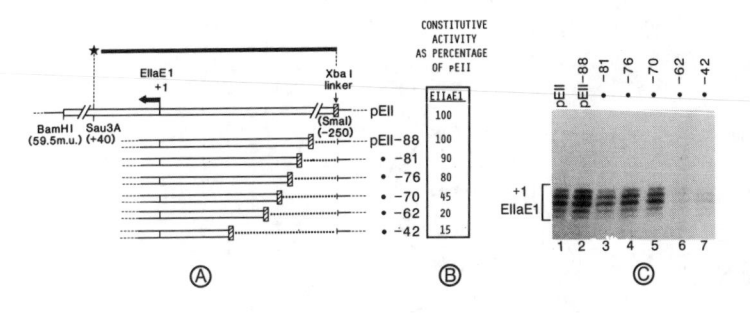

FIGURE 1. Construction and analysis of 5'-external deletion mutants of the EIIaE promoter. A) Structure of the pEII series derived from the parental, undeleted recombinant shown on the top. This recombinant (pEII) contains the entire EIIa transcription unit (open box) between 76.0 map units (m.u.) (SmaI site at -250) and 59.5 m.u. (BamHI site at +6800). The hatched box indicates the position of the XbaI linker which was inserted at the SmaI site (pEII) and at each deletion end-point (pEII derivatives). The dotted lines represent the deleted portions of the EIIaE promoter. The horizontal arrow indicates the position of the major capsite (EIIaE, +1) and the direction of transcription. The probe, 5' end-labelled at the Sau3A site (+40), used for the S1 nuclease assays is shown on the top. B) Constitutive transcription from the pEII series was quantitated from _in vivo_ transfection experiments similar to that shown in panel C, and expressed relative to pEII-specific transcription. C) S1 nuclease analysis of cytoplasmic RNA from cells transfected with the pEII series. The bracket indicates the probe fragments protected by the EIIaE-specific transcripts.

more extensive set of mutants revealed the existence of an additional but weaker promoter element (region C, in Fig. 7), located between positions -148 and -110 (8). Furthermore, it was found that sequences around -100 with weak inhibitory effects on EIIaE promoter activity were masking the stimulatory effect of the weak element located further upstream in the wild-type promoter.

3.2. Linker-scanning mutagenesis of the EIIaE promoter
External deletions do not provide conclusive information about the position and relative importance of additional promoter elements which could be located downstream of a strong element. To define more precisely the sequence components involved in the EIIaE promoter function, we have therefore constructed (see ref. 9 and Fig. 2) a series of mutants with

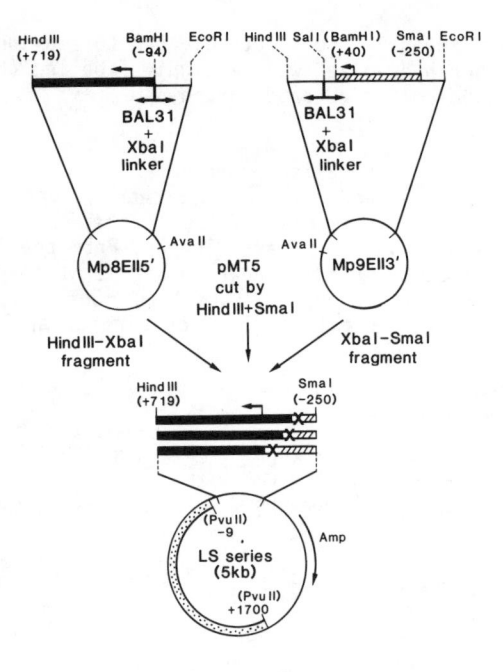

FIGURE 2. Construction of the LS series of EIIaE promoter mutants. The parental EIIaE fragments (boxes) inserted into the polylinkers of the M13mp8 and mp9 vectors are indicated on the top. Recombinants Mp8EII5' and Mp9EII3' are families of 5' and 3' deletion mutants produced by Bal31 exonuclease digestion from the BamHI and SalI restriction sites, respectively, with an XbaI linker sequence inserted at the deletion endpoints. Appropriate sized HindIII-XbaI fragments excised from the 5' deletion family (shaded box) and XbaI-SmaI fragments (hatched box) from the 3' one were recombined at their XbaI ends (X), between the HindIII and the SmaI sites of the pMT5 vector. This vector, which provides splice and poly-adenylation sites to the EIIaE sequences, contains the rabbit β-globin PvuII fragment extending from −9 to about +1700 with respect to the globin cap site (stippled area on the lower diagram), inserted between the HindIII and PvuII sites of pBR322. The EcoRI to HindIII polylinker sequence of M13mp12 has also been inserted between the EcoRI and HindIII sites into this vector. The resulting constructs constitute the LS series of EIIaE promoter mutants. The arrow pointing to the left indicates the direction of transcription from the EIIaE major mRNA start site. The AvaII restriction site and the β-lactamase gene (Amp) are included as landmarks. Restriction sites in parentheses have been lost during construction.

clustered base substitutions in the region between −98 and +1 (Fig. 3A). In these mutants the wild-type sequence was substituted by the synthetic XbaI linker, CTCTAGAG, at fixed positions spanning the promoter region. Analysis of the EIIaE1-specific transcription from these recombinants (tabulated values in Fig. 3A and see Fig. 3B as an example of the functional test) clearly revealed three essential elements in this region : the sequences between −33 and −19 (pseudo-TATA box, T1 in Fig. 7), between −48 and −39

(upstream region A in Fig. 7), and between -91 and -62 (upstream region B in Fig. 7).

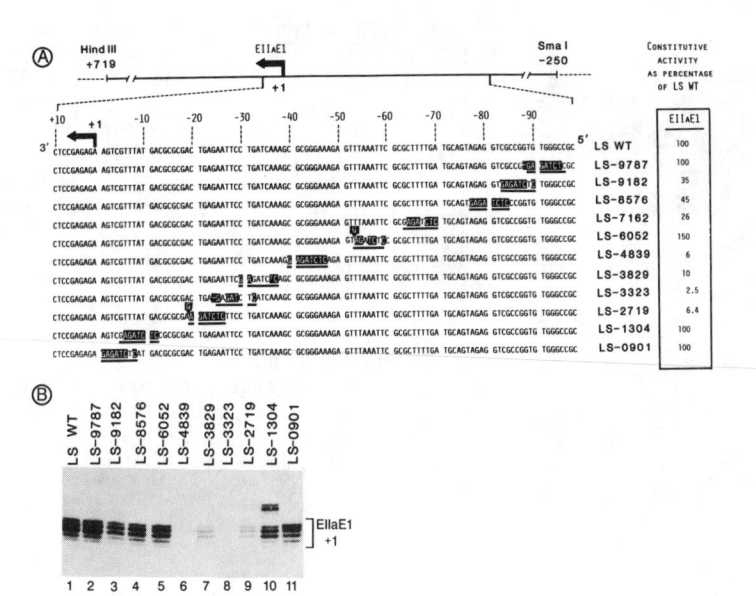

FIGURE 3. Analysis of the LS series of EIIaE promoter mutants. A) The SmaI-HindIII fragment of the EIIaE transcription unit is depicted, with the nucleotide sequence of the LS series given between -98 and +10. The linker substitution nucleotides are underlined and those altering the parental sequence are shaded. Constitutive transcription from the EIIaE1 site of the transfected LS series is given on the right, relative to the LS WT transcriptional activity. B) Representative S1 nuclease analysis of cytoplasmic RNA from cells transfected with the LS series.

3.3. In vitro DNA-binding experiments

   3.3.1. DNase I footprinting. Sequence elements involved in promoter function generally correspond to binding sites for essential transcription factors. To directly visualize the interaction of specific proteins with the control elements identified on the EIIaE promoter region by the mutational analysis, DNase I protection experiments have been performed. As shown in Fig. 4 (lanes 2-4) for the EIIaE non-coding strand, NE proteins strongly bind to sequences comprised within elements B (-82/-68) and C (-146/-128), and more weakly interact with the T1 region (-36/-19). No protection was detected over element A, but the presence of cognate proteins on this region was suggested by the DNase I hypersensitive sites at positions -50 and -38. A strong hypersensitive site at -84 and a slight protection between -105 and -88 were also observed. We suspect that these latter interactions may reflect the negative effect on EIIaE transcription of the sequences located between elements B and C (see above).

   A partial separation of the factors present in the NE preparation has been achieved by DEAE-cellulose chromatography. The DE0.15 fraction generated the same protection pattern as the NE on the EIIaE upstream regions, but failed to interact with the T1 element (Fig. 4, lane 5). On the other hand, fraction DE0.25 (lane 6) only protected the T1 region, indicating

that the pseudo-TATA box binding factor has been selectively eluted in this fraction.

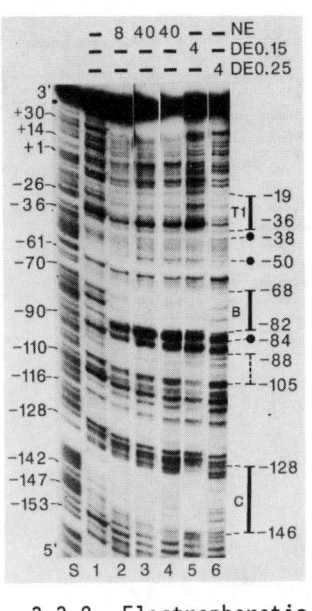

FIGURE 4. In vitro DNase I footprinting on the EIIaE promoter region. The protection against DNase I digestion provided by the NE (lanes 2-4), DEO.15 (lane 5) and DEO.25 (lane 6) fractions is shown on the non-coding strand of the SmaI(-250)-PvuII(+62) EIIaE fragment (5' end-labelled at the PvuII site). The amounts (µg) of protein used in each assay are indicated on the top. The DNase I digestion pattern of the naked probe is given in lane 1. The correspond G + A sequence ladder (S) is also shown, with numberings relative to the EIIaE1 capsite (+1). The major effects of the nucleo-protein interactions are schematized on the right : strongly and weakly protected sequences are marked by solid and dotted bars, respectively, and DNase hypersensitive sites by heavy dots, with corresponding coordinates. T1, B and C refer to the promoter elements (see Fig. 7) spanning the protected sequences.

3.3.2. Electrophoretic mobility shift assay. An alternative method of visualizing specific binding of proteins to the promoter sequence, in the presence of cell extracts, is based on the change in mobility of a promoter DNA fragment, consecutive to the binding of a factor to its recognition site. To avoid random protein-DNA interactions, the binding assay is carried out in the presence of appropriate amounts of non-specific competitor DNA which traps non-specific DNA-binding proteins. A typical result is shown in Fig. 5A, in the case of the EIIaE promoter. Three sets of retarded bands, corresponding to complexes Cx1, Cx2 and Cx3, were found consistently, whether using WCE (lane 2) or NE (lane 3).

The precise localization of these three binding sites has been determined by methylation interference studies (Fig. 5B). The specific protein-DNA complexes were formed on probe DNA which had been pretreated with dimethyl-sulfate. The complexes were resolved by electrophoresis and the methylated DNA present in each band was extracted and processed to produce the G ladders. The G residues which participate to the specific binding of a factor appeared as weaker bands when compared to the pattern of unbound probe DNA (Fig. 5B, lanes 1 and 4). Thus Cx1 (lane 2) involves the sequence between -80 and -63, Cx2 (lane 5), the sequence between -128 and -113 and Cx3 (lane 3), the sequence between -142 and -128, on the non-coding strand of the EIIaE promoter. These complexes therefore correspond to factors bound to the B element (Cx1) and to the C region (Cx2 and Cx3). Proteins interacting with the sequences further downstream (elements T1 and A) have not been visualized under these conditions.

3.4. In vivo genomic footprinting
To examine whether the protein binding pattern observed in solution, under the in vitro incubation conditions, does actually mimick the in vivo pattern, the high resolution genomic DNA footprinting assay of Church and

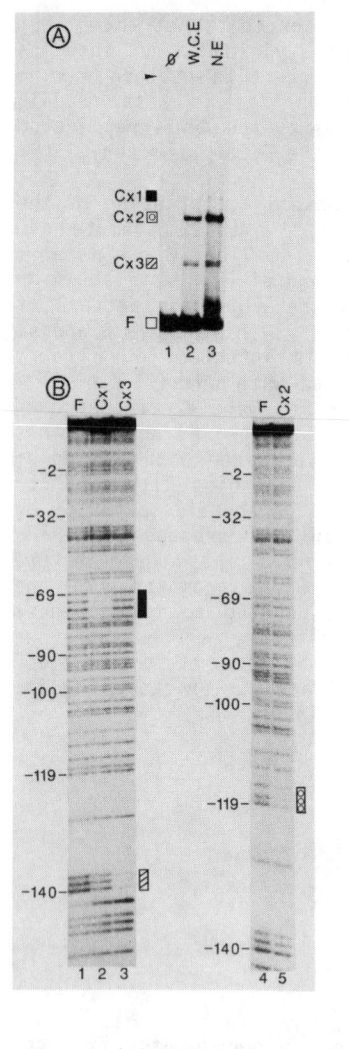

FIGURE 5. Electrophoretic band-shift analysis of crude cell extracts. A) Gel retardation pattern generated by the WCE and NE preparations after incubation with a Sau3A(+40)-BamHI(-169) probe (see ref. 18), 5' end-labelled at the Sau3A site. The positions of unbound probe (F) and retarded complexes (Cx1, Cx2 and Cx3) are indicated with respective symbols. Probe alone (φ) was migrated in lane 1. The arrow-head points to the top of the gel. B) Mapping of the Cx1, 2 and 3 binding sites (non-coding strand) by methylation interference (see Procedures). Numbering on the left is given with respect to the EIIaE1 capsite (+1). The boxes (symbolized as in panel A) span the G residues whose methylation prevents the formation of the respective complexes.

Gilbert (6) may be used. In the present case HeLa cells were infected with adenovirus type 5 (Ad5). Intact nuclei, prepared 6 hours post-infection, were treated with limited amounts of DNase I before the total nuclear DNA was purified and digested to completion with a given restriction enzyme (EcoRI). Samples were fractionated by electrophoresis on denaturing poly-acrylamide gels and the DNA fragments transferred and linked to a nylon membrane. Finally, the blots were hybridized with a specific probe to visualize the EIIaE promoter fragments (Fig. 6). Compared to the naked DNA digestion profile, the non-coding strand of the viral genome showed a clear protection of the EIIaE B promoter element and to the upstream portion of the C element. A weaker protection over region A was also observed, with a DNase I hypersensitive site at -27, within the T1 element. Since at 6 hours post-infection, a time at which viral DNA has not replicated, the EIIaE promoter is fully active, it is likely that the protection pattern shown in Fig. 6 essentially reflects that of the transcribed viral genome. If the

same analysis was repeated at 20 hours post-infection (data not shown, but see ref. 21), a time where viral DNA replication has yielded a net increase of viral DNA molecules of about 200-fold, the protection pattern was no longer observed. This loss of the specific footprints is most readily explained by the dilution of the parental templates by the newly replicated molecules which lack detectable proteins bound to the EIIaE promoter.

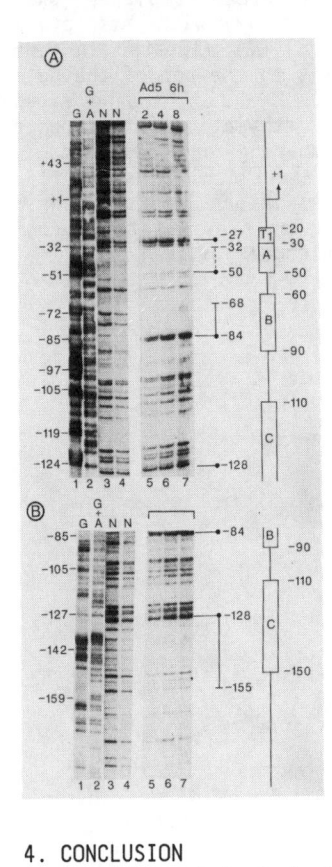

FIGURE 6. Genomic DNase I footprinting of the non-coding strand of the EIIaE promoter in early infected cells. The G and G + A ladders of the non-coding strand of EIIaE are shown in lanes 1 and 2, and the digestion pattern of naked viral DNA (N) is given in lanes 3 and 4. Nuclei from HeLa cells infected for 6 hours with Ad5 were digested with DNase I for 2, 4 or 8 min and DNA samples were electrophoresed for 6 (panel A) or 3.5 (panel B) hours (lanes 5-7). The blot was hybridized with a single-stranded EIIaE probe extending from -280 to -140 and labelled to high specific activity by primer elongation in the presence of ^{32}P-labelled deoxynucleotides. The scheme on the right represents the EIIaE promoter with the lettered boxes corresponding to the elements essential for constitutive expression (see Fig. 7). The major effects of the nucleo-protein interactions are summarized alongside the autoradiogram as in Fig. 4.

4. CONCLUSION

A series of technical approaches can be used to decrypt the promoter control signals present in eukaryotic genes. In the present report, a combination of these complementary techniques has led to a rather elaborate picture of the adenovirus EIIaE promoter (summarized in Fig. 7), showing a close correlation between the promoter sequences required for efficient transcription and the binding sites for particular proteins. It can be safely inferred therefore that these DNA-binding proteins correspond in fact to the specific factors which participate to the formation of active transcription complexes on this promoter.

Ultimately, the purification of the various transcription factors will allow the reconstitution, on the promoter DNA, of the minimal multi-protein complex which, in the presence of RNA polymerase, will initiate specific transcription. The elucidation of the molecular processes involved in this crucial step of gene expression will require the fine characterization of these factors and, in particular, of their interactions between themselves and with the DNA double-helix.

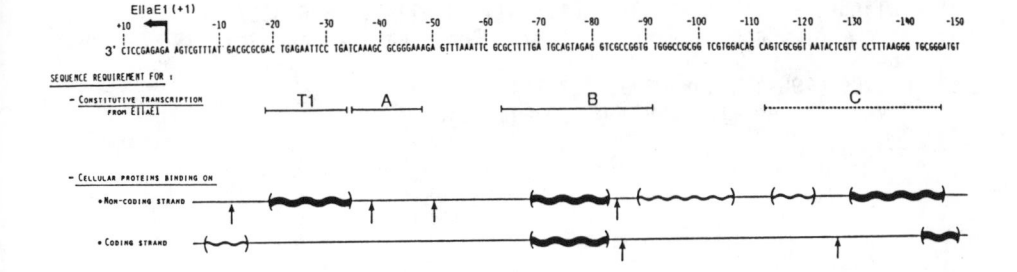

FIGURE 7. EIIaE promoter organization as deduced from in vivo and in vitro studies. The nucleotide sequence (non-coding strand) of the EIIaE promoter region is represented with the major startsite (EIIaE1) indicated. The sequences required for maximal EIIaE constitutive in vivo promoter activity are marked by bars positioned under the corresponding nucleotides and denoted T1, A, B and C. Element C is dotted because it corresponds to the weakest of these promoter domains. The nucleoprotein interactions, as deduced from the results of both in vitro and in vivo protection experiments are schematically represented by the wavy symbols, with the heavy and thin drawings depicting strong and weaker interactions, respectively. DNase I hypersensitive sites are noted by vertical arrows.

ACKNOWLEDGEMENTS
We thank J.M. Egly and M. Chipoulet for generous gift of materials, the cell culture group for technical assistance, C. Kister, C. Werlé, E. Hugonot and B. Boulay and the entire secretarial staff for preparing this manuscript. This work was supported by the CNRS, the INSERM, the Association pour le Développement de la Recherche sur le Cancer, and the Ligue Française contre le Cancer.

REFERENCES

1. Weil, P.A. et al. (1979) J. Biol. Chem. 254, 6163-6173.
2. Manley, J.L. et al. (1980) Proc. Natl. Acad. Sci. USA 77, 3855-3859.
3. Graham, F.L. and van der Eb, A.J. (1973) Virology 52, 456-457.
4. Scangos, G. and Ruddle, F.H. (1981) Gene 14, 1-10.
5. Galas, D. and Schmitz, A. (1978) Nucleic Acids Res. 5, 3175-3170.
6. Church, G.M. and Gilbert, W. (1984) Proc. Natl. Acad. Sci. USA 81, 1991-1995.
7. Fried, M.G. and Crothers, D.M. (1981) Nucleic Acids Res. 9, 6505-6525.
8. Boeuf, H. et al. (1986) Cancer Cells - DNA tumor viruses, Cold Spring Harbor Laboratory 4, 203-215.
9. Zajchowski, D.A. et al. (1985) EMBO J. 4, 1293-1300.
10. McKnight, S.L. and Kingsbury, R. (1982) Science 217, 316-324.
11. Natarajan, V. et al. (1984) Proc. Natl. Acad. Sci. USA 84, 6290-6294.
12. Dignam, J.L. et al. (1983) Nucleic Acids Res. 11, 1475-1489.
13. Moncollin, V. et al. (1986) EMBO J. 5, 2577-2584.
14. Banerji, J. et al. (1981) Cell 27, 299-308.
15. Mathis, D.J. et al. (1981) Proc. Natl. Acad. Sci. USA 78, 7383-7387.
16. Wildeman, A.G. et al. (1986) Mol. Cell. Biol. 6, 2098-2105.
17. Miyamoto, N.G. et al. (1985) EMBO J. 4, 3563-3570.

18. Jalinot, P. et al. (1987) Mol. Cell. Biol. 7, 3806–3817.
19. Maxam, A.M. and Gilbert, W. (1977) Proc. Natl. Acad. Sci. USA 74, 560–564.
20. Wu, C. (1984) Nature 309, 229–234.
21. Devaux, B. et al. (1987) Mol. Cell. Biol.

Kedinger, Cl. : Promoter organization

Sherman :   Would you care to comment on the importance of that minus
             30 sequence that was not a TATA sequence. What are the
             implications of that in the EIIA ?

Kedinger :   I think the implication is probably not so dramatic. We
             have been struck by the differences compared with the
             consensus sequence. If that is the case, then certainly
             different factors will interact with that sequence.
             To check that, we did a competition experiment. We used a
             synthetic oligonucleotide encoding the TATA sequence and
             added that oligonucleotide into a test tube where we were
             measuring specific transcription from the EII gene and to
             our surprise, specific transcription was not out sugges-
             ting probably that indeed the same factor is recognizing
             both the consensus and the non-consensus sequence. Then,
             you are forced to conclude that in fact, what is important
             in that sequence element is not so much the sequence
             itself but its surroundings. From other people's works in
             the States and in several different laboratories it turns
             out that interaction between transcription factors loca-
             lized next to the TATA box are crucial for transcription
             complex formation. So, I think it must have a meaning of
             course, but we don't know the meaning. However, we can
             explain the differences.

Sherman :   How much can you alter the nucleotide sequence and still
             get adequate transcription. I mean.. you can't really
             have it both ways. If you need your specificity, your
             deletion analysis says that you need that region there and
             now you are saying that maybe it is not so important like
             the sequence, so...

Kedinger :   Yes, you are perfectly right. The sequence is important
            as well. You cannot - as you see - maybe you remember the
            mutant which was destroying the region round
            minus 30 was only 2 % of wild type activity. So, the
            sequence is indeed very important. What I meant was that
            the sequence alone does not do it. You have interactions
            between transcription factors and sequence elements
            certainly.

Miller :   Do you have answers for ribosomal and transfer RNA genes ?

Kedinger : In the case of ribosomal RNA genes, sequences have been
            described, localized in the spacers which separate each
            ribosomal transcription unit and in fact these sequences
            have been named enhancers as well, but it is likely now
            that they correspond essentially to sequences which
            attract factors essential for ribosomal DNA transcription.
            They attract because there are several of them, one next
            to each other. So the factors will bind to those regions
            first and then it is thaught that these factors are
            transferred to the specific promoter which is nextdoor.
            In fact, if you compare the sequences of these so-called
            enhancers-ribosomal enhancers, you find some homologies
            between the sequences localized in the spacer with the
            sequences localized in the promoter region that is plus
            and minus 150.

            So, you see, it is a trap for factors. Once the factors
            are on the DNA, it is just a matter of bending the DNA and
            transferring the factor on the specific transcription
            site.

Miller :   Regulation of these genes cannot follow the same pattern
            as messenger genes otherwise you have havoc in your
            expression.

Kedinger :   That's right.

Miller :   Do you have similarities between the ribosomal RNA promoters from various cells such as HeLa, CHO... If yes, that would indicate structural analogies among all the RNA polymerases I.

Kedinger :   As far as I know, and that's what I said, there was no consensus sequence found within the ribosomal genes.

Baserga :   Transcription of ribosomal RNA is species-specific. You take a ribosomal promoter of mouse and you put it in human cells, it doesn't work.

Kedinger :   That's right and you find the same with extracts. Cellular extracts are very specific, they only transcribe in vitro the genes from the same organism. That is true. But what is interesting is that genes have been cloned now which code for subunits of RNA polymerase I for example or which code for different ribosomal proteins and in front of these genes, sequences have been found which are present in all genes related to the ribosomal machinery. In other words, RNA polymerase I or the ribosomal proteins and the other proteins related to the translation machinery are targetted by the same consensus sequence in their gene, not in the gene they transcribe.

Drillien :   If you consider a given initiation complex do you have all the factors that independently bind to DNA sequences or do you have also protein-protein interactions ?

Kedinger :   Oh yes, yes absolutely.

Drillien :   If so, do they happen before, like a preformed complex which would reach a given sequence or do they have to relate to proteins already bound to the DNA ?

Kedinger : That is an interesting comment. Indeed not all factors bind to DNA. Those which I have shown to you bind to specific sequences but in addition to those factors, there are factors binding to those already bound to DNA. So, you clearly heave protein-protein interactions and in fact we think that the EIA-mediated induction of these early genes transcription is related not only to an increased binding of factors but also to an increased protein-protein interaction because EIF for example does not bind to DNA itself. It binds to the factors. So, I am sure and I agree with you that factor-factor interactions are certainly as important as protein-DNA interactions.

Drillien : Does that mean that when a given factor reaches the DNA sequence, it gets bound, changes its configuration and opens now a recognition site for another protein factor ?

Kedinger : Eventually yes. In addition to that, some factors bind to the polymerase and if you don't have these factors you will have specific transcription going on but at a very low rate.

Spier : Obviously we are getting close to lunch and it is time to have a go at speculation. The ribosome itself is a translation thing which consists of about seventy proteins. How many proteins do you think you are going to see involved in the transcription operation ?

Kedinger : Certainly not as many as seventy. You said seventy ?

Spier : Yes.

Kedinger : Up to now, I would say half a dozen factors have been identified. Not yet purified but identified which are probably different. But as it turns out, by comparing the factors involved in different genes, not only ribosomal

genes but several messenger RNA, the factors involved in
these transcriptions, some of them are common, some of
them are different. So I think the number will certainly
be more than ten perhaps but not more than fifteen.

Spier :   The transactivating factors that you get in retrovirus
systems, are they incorporated into your thinking ?

Kedinger : Yes, yes, each of these transactivating fractions are
different but maybe the mechanism in which they are
involved is similar, so... I think it is too early to
answer that question. We have to purify all these factors
individually. What we need is to reproduce a wonderful
work which has been done for prokaryotic systems and
especially for the lambda system where we have such a
beautiful work being done. You crystallize things, see
exactly how it is hooked to the DNA and then can mutage-
nize specific sites in order to change the specificity for
example. That is exactly what we need to do for eukaryotic genes.

Part V.

Modification of the Cell Genome

TAILORING OF AN ANTI-HUMAN PLACENTAL ALKALINE PHOSPHATASE
IMMUNOGLOBULIN USING GENETIC ENGINEERING.

A.VAN DE VOORDE, V.FEYS, P.DE WAELE, P.CASNEUF, W.FIERS.

Lab.Molec.Biol.,State Univ. Ghent, Ledeganckstraat 35, 9000
Ghent, Belgium.

1.HUMAN PLACENTAL ALKALINE PHOSPHATASE.
    Human placental alkaline phosphatase (hPLAP;
E.C.3.1.3.1.) is a member of a group of enzymes named
according to the organ in which they predominate. Other
members of this group include an intestinal isoenzyme (IAP)
and a tissue non- specific isoenzyme found in liver, bone and
kidney (LAP).
    hPLAP normally occurs in the microvilli of the
syncytiotrophoblast (1) and can be detected in sera of
pregnant women in rising concentrations starting from the 13th
week of pregnancy (2, 3). The enzyme acts as a dimer with a
M.W. of 130 kDa and has a subunit M.W. of 67 kDa. It is a
metalloenzyme containing 4 Zn^{2+} atoms per molecule and it
appears to be glycosylated (for reviews see 4, 5). Its partial
amino acid sequence has been determined by amino acid
sequencing (6). Its complete amino acid sequence was deduced
from the nucleotide sequence of the cloned cDNA (7).
    hPLAP is a highly polymorphic protein for which more
then 20 allelic variants were described (8), but only 3
alleles make up 6 homozygotic and heterozygotic phenotypes
that occur in 98% of all phenotypes found, while the remaining
2% are made up of rare variants (9, 10).
    The renewed interest in hPLAP stems from the fact that
apart from its eutopic expression in the serum of pregnant
women it was also found in sera and tumor tissues of various
cancer patients (11-19). The enzyme found in tumor tissues
displays the same pattern of phenotypic variation, although
historically 3 major ectopic variants were described : the
"Regan" variant, later identified as the normal placental
enzyme (20), the "Nagao" variant, shown to closely resemble
the rare D-variant (20, 21) and the "Kasahara" variant,
thought to be a heterodimer composed of hPLAP and the fetal
form of hIAP (22). hPLAP or its Nagao variant (recently
renamed PLAP-like AP) has also been detected in very low
amounts in various types of normal tissues, including thymus,
cervix, endometrium, Fallopian tube and testis (23-26). The
hPLAP-like AP can regularly be found in the serum of smokers
(27, 28).
    Previously, the detection of hPLAP was severely hampered
by the close resemblance of the biochemical and biophysical
properties of hPLAP and hIAP. Usually, the distinction
between the 3 isoenzymes (PLAP, IAP, LAP) was based on

A. O. A. Miller (ed.), Advanced Research on Animal Cell Technology, 233–249.
© 1989 by Kluwer Academic Publishers.

characteristics such as heat-inactivation (29), sensitivity to uncompetitive inhibitors (4, 30), differential migration in acrylamide, agarose or starch gels (4) and differentiation of these isoenzymes was also attempted by immunological means (31, 32). The latter method is seriously hampered by the cross-reactivity between the intestinal and the placental isoenzymes, due to the fact that the two enzymes share a number of tryptic peptides (33) and on the level of the nucleotide sequence almost 90% homology in the translated regions is found between the two enzymes (34). Until recently, the cross-reactivity between hPLAP and hIAP could only be eliminated by intensive adsorption of the polyclonal sera used (35). With the advent of the hybridoma technology, monoclonal antibodies were prepared that not only differentiate completely between hPLAP and hIAP but also enable the partial discrimination of hPLAP and the PLAP-like AP variant (2, 21, 28, 36, 37). Monoclonal antibodies directed against hPLAP thus allowed the accurate and sensitive detection of the enzyme. Assays, developed in various laboratories, used the fact that anti-hPLAP monclonal antibodies do not interfere with the activity of the enzyme and as such the antigens own enzymatic activity could be used for quantification with sensitivities as low as 50 micrograms per liter using p-nitrophenylphosphate and 0,5 micrograms per liter using 4-methyl-umbelliferylphosphate (2, 32, 38).

In order to evaluate hPLAP as a tumor associated protein a multicenter study was set up under supervision of Prof. M. De Broe (Dept. Nephrol., Univ. Hospital Antwerp, Belgium) in which 17 institutions in 8 countries were engaged. In this study the monoclonal antibody 327 was used (39) to determine the amounts of hPLAP found in the sera of 506 cancer patients. The results showed that elevated hPLAP levels (>0,1 U/l) occurred in 90% of all seminomas, 73% of testicular cancers, 48% of ovarian cancers, 20% of lung cancers, 15% of gastro-intestinal cancers and in 10% of breast cancers. A comparison of tumor marker distribution in several types of cancer allowed a rating according to sensitivity (40). It was shown that hPLAP is a first choice marker for testicular cancers and a strongly advised marker for ovarian cancers, mainly due to its low frequency of false positives which usually can be ascribed to interference from smoking (27, 38, 41, 42). hPLAP was also detected in butanol extracted, homogenised tumor tissues and in histochemical paraffin sections of tumor tissues. Here a serious discrepancy was found between the number of seropositive patients as compared to tissue positive patients (17, 19). In all cases substancially more tissue-positivity was detected than seropositivity (for breast cancer : 5,2% seropos. versus 43% tissue pos.;for ovarian cancer: 25 to 54% seropos. versus 68 to 94% tissue pos.;pooled results from ref. 17, 19, 25, 38).

In view of these findings, monoclonal antibodies directed against hPLAP have opened new therapeutic possibilities, such as the possible use of tumor-directed monoclonal antibodies for radioimmunolocalization and radioimmunotherapy, and as such have spurred numerous studies. At first these were performed on nude mice xenografted with human tumors, but now

they are executed increasingly and often already routinely
on cancer patients , using antibodies directed against CEA
(43), HMFG2 (44), NDOG2 (45), melanoma antigens (46), AFP (47)
or hCG (48). Monoclonal and polyclonal anti-hPLAP antibodies
were used in a nude mice system (49) and in the detection by
radiomimmunolocalization of tumors of the testis, ovary and
cervix (50).

2.CRITERIA FOR THE USE OF ANTI-HPLAP ANTIBODIES AS TUMOR-DIRECTED VEHICLES.

In order to be useful as reagents in drug or toxin
targeting, in immunolocalization or in immunotherapy
anti-tumor monoclonals should comply with several criteria:
their antigen recognition pattern must be carefully checked
for cross-reactivities, even with low-affinity epitopes and on
all possible normal tissues (if possible also including fetal
ones), the surface localization and the accessibility of the
epitope on tumor tissue must be controlled, the turnover rate
and the absence of immunomodulation of the antigen should be
measured and antibodies should be selected for high Kaff
values.

In order to validate the potential application of
anti-hPLAP monoclonal antibodies, we investigated several of
these parameters using the anti-hPLAP antibodies from the
hybridoma cell line E6. This hybridoma was produced at our
instistute by fusion of splenocytes derived from A/J mice
immunised with partially purified hPLAP of all major
phenotypes and SP2/0-Ag14 myeloma cells (Flow). The resulting
E6 antibodies were of the IgG2b,kappa subtype and show an
absolute specificity for hPLAP (2). They were shown to be very
useful reagents for the immunohistopathological determination
of hPLAP on normal and tumor-derived tissues (17, 19, 25, 26).

Binding studies using purified hPLAP or hPLAP expressed
on the surface of HeLa or KB cells showed that E6 bound to its
antigen with a relatively high Kaff of $2x10^{10}$ M (fig.1).
^{125}I-labeled E6 was used in saturation binding studies on
HeLa cells, indicating that on this established tumor cell
line the antigen was present at a density of 300,000 hPLAP
molecules per cell (fig.2) and staining patterns using
fluorescein-labeled antibodies indicated a rather patchy
antigen distribution.

When E6 antibodies were bound to hPLAP-expressing tumor
cell lines the resulting antibody-antigen complexes were
internalized at the relatively low speed of 10 to 30 molecules
per min per cell at 37 C (fig.3).

On the basis of these results we conclude that the E6
antibodies complied with the criteria of specificity and
affinity as well as with the availabilty of the antigen on
tumor cells. However, the slow internalization kinetics of the
bound antibodies are unfavorable to their use as
drug-targeting vehicles where large amounts of drugs are
needed to kill a cell (52). This feature should pose no
problems for their use in toxin-targeting, localization or
immunotherapy.

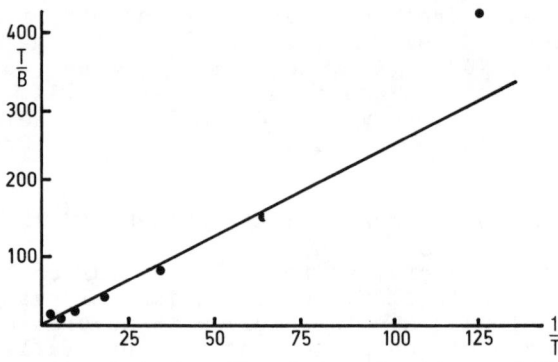

Fig.1 Langmuir plot for the determination of Kaff of E6 with purified hPLAP.

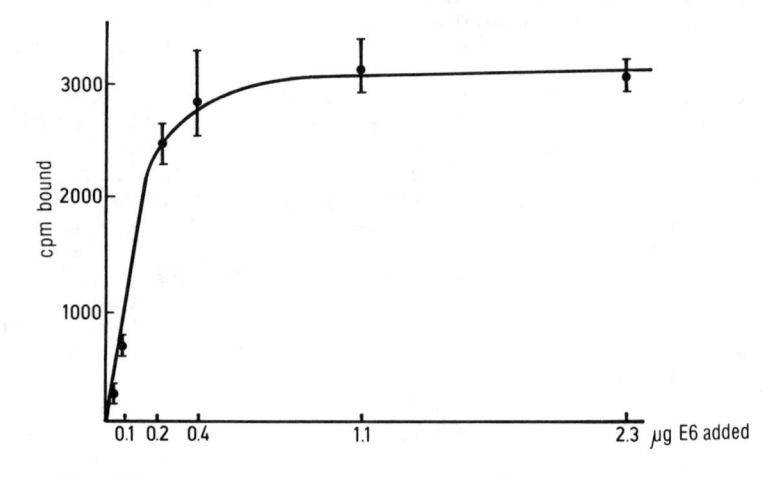

Fig.2 Saturation binding of ^{125}I-labeled E6 on hPLAP-expressing KB cells.

3. cDNA CLONING OF THE INFORMATION OF THE E6 ANTIBODY.

The use of mouse-derived monoclonal antibodies in the treatment of human cancer patients is hampered by problems of aspecificity as a consequence of Fc-mediated interactions (53), the elicitation of immune interactions (54), difficulties in penetrating tumor tissue, neseccitating the use of F(ab')$_2$ or F(ab') molecules (55) or of problems arising from immunomodulation (56).

Therefore several workers explored the possibility of cloning the information for the H and L chains into plasmids, hoping that coexpression of these plasmids in a suitable host would result in the production of functional

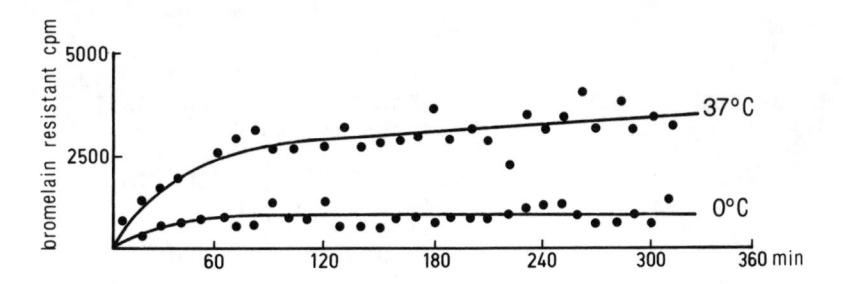

<u>Fig.3</u> Internalization of ^{125}I-labeled E6, measured as the
difference between the amount of bromelain-resistant
cpm (51) at 37 C (internalization) and 0 C (no
internalization).

antibodies. Coexpression in E.coli has resulted in no
detectable antibody activity (57), but upon expression in
yeast cells (58), in COS or CHO cells (59) or upon
cotransfection in mouse myeloma cells (60), functionally
active antibodies could be recovered.

We decided to clone the H and L chain information for
the synthesis of anti–hPLAP antibodies with the aim of
allowing the production of genetically engineered $F(ab')_2$ and
$F(ab')$ molecules, of mouse–human chimeric antibodies, of
bispecific monovalent antibodies or of antibodies tailored for
optimal use in radioimmunolocalization and therapy.

We therefore prepared mRNA from E6 hybridoma cells,
purified it by oligo dT–cellulose chromatography and sucrose
gradient centrifugation, transcribed it into cDNA (61) and
cloned it into the PstI site of pBR322, resulting in the
plasmids pBRE6H (heavy chain) and pBRE6L (light chain). DNA
sequencing (62) confirmed the identity of the H chain as an
IgG2b and of the L chain as a kappa subtype. The DNA sequence
allowed the deduction of the amino acid sequence (fig.4).

L CHAIN

```
-20         -10          -1/1                           30
  MSVPTQVLGLLLLWLTDARC DIQMTQSPASLSVSVGESVTITCRASENIY
  Signal peptide        Mature protein
           40          50          60          70          80
  SNLAWYQQKQGKSPQLLVYVATKLVDGVPSRFSGSGSGTQYSLKINSLQS

           90         100         110         120         130
  EDFGSYYCQHFWDTPFTFGSGTKLDMKRADAAPTVSIFPPSSEQLTSGGA

          140         150         160         170         180
  SVVCFLNNFYPKDINVKWKIDGSERQNGVLNSWTDQDSKDSTYSMSSTLT

          190         200         210
  LTKDEYDRHNSYTCEATHKTSTSPIVKSFNRNDC
```

H CHAIN

```
 -19         -10          -1/1          10          20          30
  MEWIWIFLFILSGTAGVQS QVQLQQSGAELAEPRASVKLSCKASGYTLT
  Signal peptide        Mature protein
           40          50          60          70          80
  SYGISWVKQRTGQGLEWIGEIYPGSGNSYFNEKFKGKATLTVDKSSSTAY

           90         100         110         120         130
  LHLSSLTSEDSAVYFCAGPRQVGLLPFGYWGQGTLVTASAAKTTPPSVYP

          140         150         160         170         180
  LAPGCGDTTGSSVTLGCLVKGYFPESVTVTWNSGSLSSSVHTFPALLQSG

          190         200         210         220         230
  LYTMSSSVTVPSSTWPSQTVTCSVAHPASSTTVDKKLEPSGPTSTINPCP

          240         250         260         270         280
  PCKECHKCPAPNLEGGPSVFIFPPNIKDVLMISLTPKVTCVVVDVSEDDP

          290         300         310         320         330
  DVQISWFVNNVEVLTAQTQTHREDYNSTIRVVSALPIQHQDWMSGKEFKC

          340         350         360         370         380
  KVNNKDLPAPIERTISKIKGIVRAPQVYILSPPPEQLSRKDVSLTCLAVG

          390         400         410         420         430
  FSPEDISVEWTSNGHTEENYKDTAPVLDSDGSYFIYSKLNMKTSKWEKTD

          440         450
  SFSCNVRHEGLKNYYLKKTISRSPGK
```

Fig.4 Amino acid sequence as derived from the nucleotide
        sequence of the H and L chain of the E6 antibody,
        represented in the one-letter code.

The deduced amino acid sequence of the mature H chain consists of 456 amino acids (calculated M.W. 50,6 kDa), preceeded by a signal sequence of 19 amino acids. The sequence shows a strong conservation with published mouse gamma 2b constant regions (97% amino acid homology) and also with gamma 2a constant regions (76% amino acid homology). The non-homology with the published gamma 2b sequence is located in the hinge region (1 conservative amino acid change), in the CH2 domain (4 conservative amino acid changes) and in the CH3 domain (5 conservative amino acid changes and the replacement of an uncharged with a charged residue). We assume that the amino acid changes we found are more likely to be strain-dependent allelic variantions rather than the result of point mutations with a dominant CH2 and CH3 domain localization (63). The Cys residues, normally all involved in disulfide bridges, are absolutely conserved, as is the Tryp at position 36 in the variable region (64).

The deduced amino acid sequence of the L chain shows a mature protein of 214 amino acids (calculated M.W. of 23,5 kDa), preceeded by a 20 amino acids long signal peptide. Here the homology with published mouse kappa amino acid sequences amounts to 100% for the constant region of the protein. In the variable region of the sequence we find a small aberration from the previously published sequence, as we see a Met at position 106 in the J2 region where normally Ile is found (65). We ascribe this to a point mutation or a structural difference between Balb/c and A/J germline sequences.

In general the signal peptides of the H and L chain obey the general rules for signal peptides: their length is close to the main weight of lengths of a pool of prokaryotic and eukaryotic signal sequences, the -5 to -2 positions contain Gly and Ala (H chain) and Leu and Ala (L chain), while the positions upstream from -5 are very hydrophobic (66). In view of the role of the L chain in promoting H-chain transport (67) it may be of importance to note the highly hydrophilic stretch in the L chain signal peptide between -4 and -1 .

4.CONSTRUCTION OF EUKARYOTIC EXPRESSION VECTORS CONTAINING THE H OR L CHAIN INFORMATION.

The relevant sequences containing H or L chain information were excised from the plasmids pBRE6H or pBRE6L and recloned in the multi-insertion site of the shuttle vector pSV23P. This vector has an SV40 origin of replication, an early promotor before the insertion site, splice signals from the SV40 small t antigen and the SV40 polyadenylation and termination signals (68). The resulting plasmids, pSV23PE6H and pSV23PE6L, are shown in fig.5.

In order to diminish Fc-related immunogenicity of recombinant antibodies and to enhance the potentiation of tumor-directed ADCC and C'-dependent lysis upon use in humans we constructed a mouse-human chimeric H chain. For this purpose, we isolated the human constant gamma3 gene from the human HaeIII-AluI Charon 4A library (69) using a CH1-specific probe (gift from Dr. T.H.Rabbitts). The isolated

gene was subsequently recloned as a HindIII-EcoRI fragment in plasmid pAT153. From this, a HinfI-SalI fragment containing the COOH-terminal part of the CH1 domain, the hinge region and the CH2 and CH3 domains from the human genomic clone was ligated to the HinIII-HinfI fragment of pSV23PE6H and to the appropiately opened vector pSV23M (identical with pSV23P but for the abscence of splice signals). The resulting plasmid pSV23MHY3 (fig.6) thus contains the information for the synthesis of an IgG chimeric H chain consisting of the anti-hPLAP-directed V_H (mouse), the chimeric CH1 domain (predominantly mouse, terminal part human) and the all-human hinge, CH2 and CH3 domains of the gamma 3 subclass. On the basis of the respective amino acid sequences and the exact fusion position of the sequence we can predict that this chimeric protein should not differ too much from the normal as in the newly formed CH1 domain 66 amino acids out of 97 remain of mouse origin, and of the remaining 37 of human origin 21 are identical with the original mouse sequence (overall homology of 83,5%). On the level of protein folding no aberrations are expected because the relative positions of the Cys residues and hence of the disulfide bridges remain unaltered.

With the aim to produce recombinant $F(ab')_2$ fragments of the chimeric protein we inserted a multistop linker (MSL; sequence: 5'-CTTAGTTAACTAAG) at a blunted, unique MstII site in the CH2 domain. The resulting expression vector pSV25MHY3F2 should now contain a truncated CH2 domain consisting of 13 amino acids, terminating with (CH2 amino acid number 27:Pro)-Asp-Leu-Val-Thr. Moreover, as the MstII site was located in front of the normal glycosylation site, the newly synthesized shortened H chain should not be glycosylated. For similar purposes the whole-mouse H chain plasmid pSV23PE6H was opened and blunted at the unique XhoI site and the same MSL linker was ligated onto it. Recloning of the resulting HindIII-HindIII fragment containing the H chain fragment into the expression vector pSV25P produced pSV25PE6HF2, a plasmid that should allow the synthesis of a truncated mouse H chain terminating with (amino acid number 243:Leu)-Asp-Leu-Val-Asn.

5.EXPRESSION OF RECOMBINANT IMMUNOGLOBULINS IN NON-LYMPHOID CELLS.

The expression of genes in eukaryotic cell systems can be effected either in a transient system or in a permanent expression system. We opted for the use of COS cells (70) as an efficient and fast transient system. These cells contain the SV40 large T antigen in an integrated state. Transfection with SV40-based shuttle vectors in these cells allows not only efficient transcription but also limited plasmid DNA replication, so that an additional gene dosage effect augments the amount of product expressed.

In our expression experiments we transfected each of the plasmids pSV23PE6H, pSV23MHY3, pSV25MHY3F2 or pSV25PE6HF2 together with the L chain plasmid pSV23PE6L using the DEAE-dextran method (71) at 0,5 microgram plasmid DNA per 3.5×10^4 cells. After 96 h the cell supernatant was collected

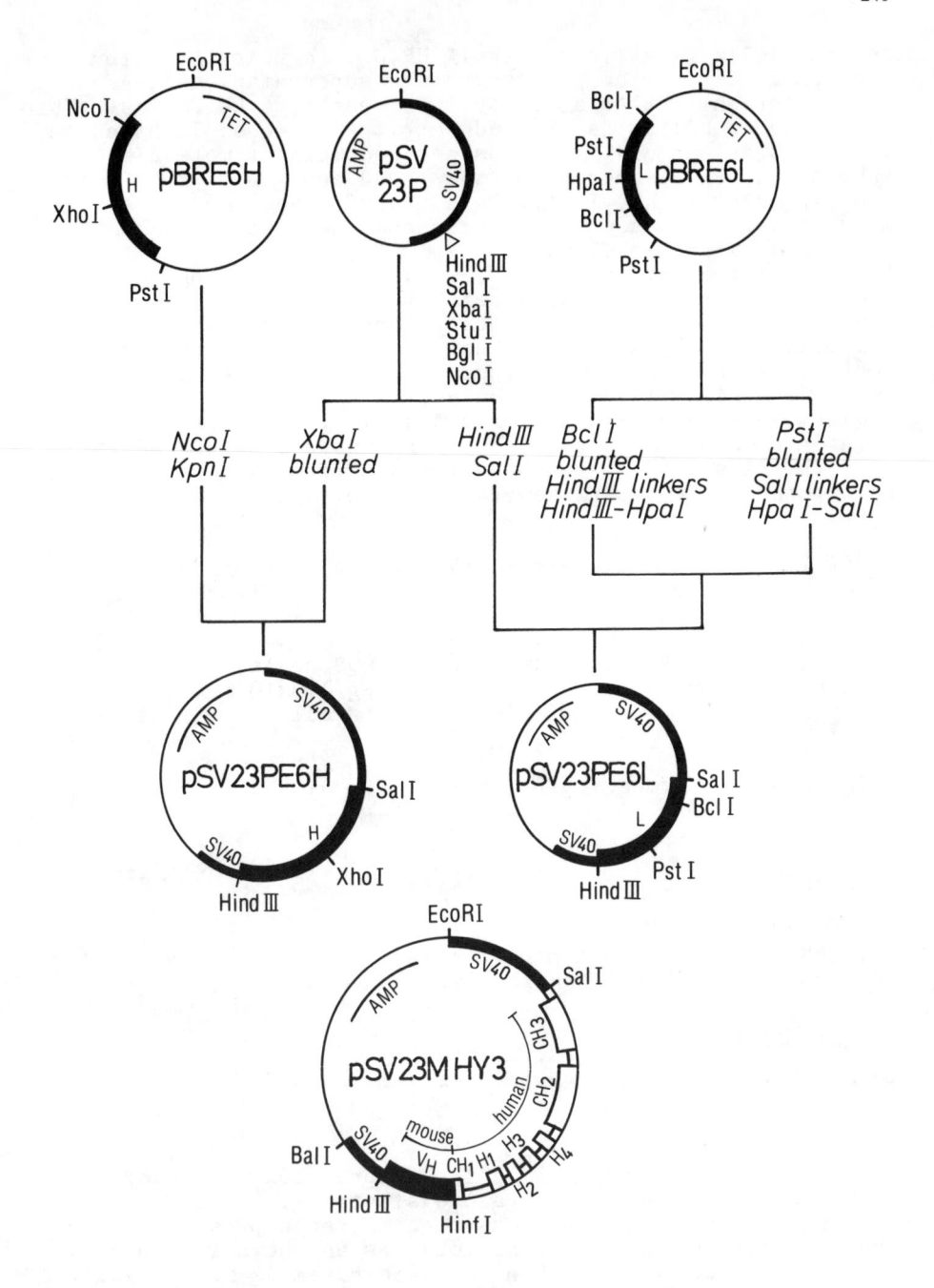

Fig.5   Schematic overview of the construction of the
         expression plasmids containing the mouse H chain
         (pSV23PE6H) and L chain (pSV23PE6L) and the structural
         arrangement of the mouse-human chimaeric H chain
         plasmid pSV23MHY3.

and the cells were lysed with 1% NP40 in hypotonic buffer. For each pair of plasmids we assayed the supernatant and the lysate for anti-hPLAP activity in a sensitive ELISA test. This consisted of solid-phase-bound rabbit anti-mouse light chain, used to trap our newly synthesized recombinant molecules. We than added hPLAP and assayed the bound antigen on the basis of its own enzymatic activity. This activity was used as a measure for the presence of immunocompetent, completely assembled immunoglobulins.

For permanent expression, we used transfection into CHO-DHFR cells (DHFR = dihydrofolate reductase). The same plasmid pairs were used as above, supplemented with the plasmid pAdD26SVp(A)-3 which contains the DHFR gene under control of the Adenovirus promotor (72). The presence of the latter plasmid allows quick selection on the basis of the acquired DHFR characteristics and amplification of cointegrated genes is possible using increasing concentrations of methotrexate (73). Transfection was executed with the calciumphosphate technique (74) using 21

TABLE I. Expression levels in the supernatant after transfection into COS or CHO cells without methotrexate amplification.

Plasmid combination	Product expected to be formed	COS cells, ng/3.5×10^4 cells	CHO cells, ng/5×10^5 cells
pSV23PE6H pSV23PE6L	mouse Ig2b,kappa	10	80
pSV23MHY3 pSV23PE6L	mouse-human chimera	5	50
pSV25MHY3F2 pSV23PE6L	mouse-human chimera, F(ab')$_2$ molecules	5	12
pSV25PE6HF2 pSV23PE6L	mouse Ig2b,kappa, F(ab')$_2$ molecules	0,2	3

micrograms of DNA per 5×10^5 cells. Expression levels of DHFR-resistent clones was again evaluated with the ELISA test.

Under these circumstances we obtained expression of fully functional,recombinant immunoglobulins as shown in table I. Transfection of separate L or H chain plasmids never resulted in a positive signal in the ELISA assay. Western blot analysis of concentrated supernatant or lysate samples of transfected cells showed the presence of bands at the expected M.W. for

complete mouse immunoglobulins, mouse-human chimeric immunoglobulins and for mouse-human chimeric $F(ab')_2$ molecules (Fig.6).

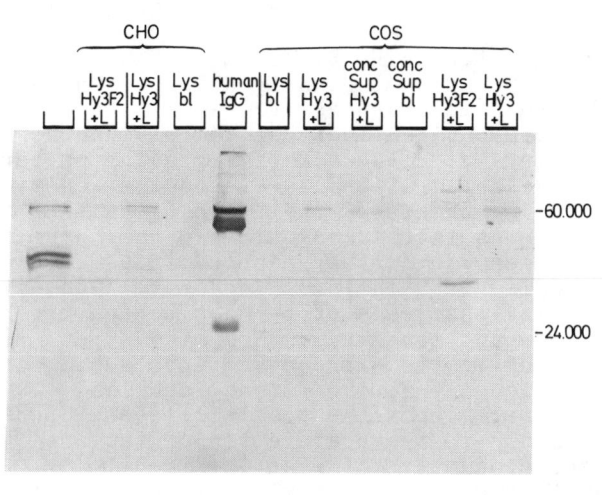

Fig.6   Western blot analysis of the supernatant (Sup) or the lysates (Lys) of COS or CHO cells, transfected with pSV23MHY3 + pSV23PE6L, pSV25MHY3F2 + pSV23PE6L or with pSV25M as a control (bl). Samples of lysates or concentrated supernatant were loaded on a 7% polyacrylamide gel under reducing conditions. Bands were detected with Nordic rabbit anti-human IgG3 or Promega goat anti-rabbit IgG-Fc immunoglobulin conjugated with alkaline phosphatase.

The fact that functionally reassociated $F(ab')_2$ molecules can be recovered upon transfection of L chain plasmids with truncated H chain plasmids indicates that the presence of the CH2 and the CH3 domains is not required for the correct reassociation. It also seems to suggest a natural affinity between the V and the CH1 regions of the H and L chains, allowing the correct formation of all intra and inter S-S bridges and gaining additional stability by hinge region interactions. In this respect it may be noted that the recovery of functional $F(ab')^2$ molecules is higher upon transfection with the mouse-human chimeric plasmids than in the corresponding whole mouse $F(ab')^2$ constructs. If the hinge region were to play an important role in structural stability this would be logical since the hinge region of the human gamma3 subclass is extended over 87 amino acid residues, grouped in 4 subregions, containing stable poly-Pro helical regions cemented by 11 S-S bridges. In contrast, the mouse gamma2b hinge region only has a short Cys-Pro-Pro-Cys region with a mere 4 S-S bridges (75, 76).

244

6.CONCLUSION.
     Human placental alkaline phosphatase is an enzyme that
can serve as a very useful marker in the detection of ovarian
cancer and of seminoma. The monoclonal antibody E6,
developed at our institute, has an absolute specificity for
the enzyme and allows its very sensitive detection in sera,
tumor tissue extracts and histological sections of tumor
tissues.
     The monoclonal antibody E6 binds to its antigen with a
high Kaff of $2x10^{10}$ M and as the antigen is present at high
epitope density with a low turnover value it should be a good
candidate for radioimmunolocalization, radioimmunotherapy and
immunotoxin targeting. With the aim of improving on certain
parameters for these applications we cloned the genetic
information for the synthesis of the H and L chain of this
antibody as cDNA clones into eukaryotic expression vectors.
The cloned information was further adapted in a construction
of a mouse-human chimeric H chain. In order to reduce the M.W.
of the antibodies we inserted a stop-codon behind the hinge
region of both mouse and mouse-human H chains. Upon
transfection into eukaryotic expression vectors all constructs
sofar yielded functionally active immunoglobulins.
     Our results clearly show the possibilities of genetic
engineering technology in the preparation of tailored
antibodies and the production of $F(ab')_2$-like
antibodies.Production of immunoglobulins provided with an
engineered toxin, an immunomodulator or even tailed for
optimal labeling with radioactive tracers is now within our
possibilities.

REFERENCES
  1. Carlson RW, Wada HG, Sussman HH: The plasma membrane of
     human placenta. Isolation of microvillus membrane and
     characterization of protein and glycosylation subunits.
     J.Biol.Chem.251, 4139-4146 (1976).
  2. De Groote G, De Waele P, Van de Voorde A, De Broe ME,
     Fiers W: Use of monoclonal antibodies to detect human
     placental alkaline phosphatase. Clin.Chem. 29, 115-119
     (1983).
  3. McLaughlin PJ, Gee H, Johnson PM: Placental-type alkaline
     phosphatase in pregnancy and malignancy plasma: specific
     estimation using a monoclonal antibody in a solid phase
     immunoassay Clin.Chim Acta 130, 199-209 (1983).
  4. Fishman WH: perspectives on alkaline phosphatase
     isoenzymes. Am.J.Med. 56, 617-650 (1974).
  5. Moss DW: Alkaline phosphatase isoenzymes. Clin.Chem. 28,
     2007-2016 (1982).
  6. Ezra E, Blacker R, Udenfriend S: Purification and partial
     sequencing of human placental alkaline phosphatase.
     Biochem.Biophys.Res.Comm. 116, 1076-1083 (1983).
  7. Millan JL: Molecular cloning and sequence analysis of
     human placental alkaline phosphatase. J.Biol.Chem. 261,
     3112-3115 (1986).
  8. Gogolin KJ, Slaughter CA, Harris H: Electrophoresis of
     enzyme-monoclonal antibody complexes. Studies of human

placental alkaline phosphatase polymorphism.
Proc.Natl.Acad.Sci.USA 78, 5061-5065 (1981).

9. Robson EB, Harris H: Genetics of the alkaline phosphatase polymorfism of the human placenta. Nature 207, 1257-1259 (1965).

10. Donald LY, Robson EB: Rare variants of placental alkaline phosphatases Am.Hum.Genet. (London) 37, 303 (1974).

11. Fishman WH, Inglis MI, Stolbach II, Krant MJ: A serum alkaline phosphatase isoenzyme of human neoplastic origin. Cancer Res. 28, 150-154 (1968).

12. Jacoby B, Bagshawe KD: Placental-type alkaline phosphatase from human tumor tissue. Clin.Chim.Acta 35, 473-481 (1971).

13. Higashino K, Hashinostume M, Kang K-Y, Takahashi Y, Yammamura Y: Studies on a variant alkaline phosphatase in sera of patients with hepatocellular carcinoma. Clin.Chim.Acta 40, 67-81 (1972).

14. Cadeau Y, Blackstein HE, Malkin A: Increased incidence of placental-like AP in human breast cancer. Cancer Res. 34, 729-732 (1974).

15. Keller JA, Bush RS, Malkin A: Placenta-like alkaline phosphatase in gynaecological cancers. Cancer Res. 36, 269-271 (1976).

16. Wada HG, Shindelman JE, Ortmeyer AE, Sussman HH: Demonstration of placental alkaline phosphatase in human breast cancer. Int.J.Cancer 23, 781-787 (1979).

17. Van de Voorde A, De Groote G, De Waele P, De Broe ME, Pollet D, De Boever J, Vandekerckhove D, Fiers W: Screening of sera and tumor extracts of cancer patients using a monoclonal antibody directed against human placental alkaline phosphatase. Eur.J.Cancer Clin.Oncol. 21, 65-71 (1985).

18. Eerdekens MW, Nouwen EJ, Pollet DE, Briers TW, De Broe ME: Placental alkaline phosphatase and CA125 in sera of patients with benign and malignant diseases. Clin.Chem. 31, 687-690 (1985).

19. Nouwen EJ, Pollet DE, Schelstraete JB, Eerdekens MW, Hansch C, Van de Voorde A, De Broe ME: Human placental alkaline phosphatase in benign and malignant ovarian neoplasia. Cancer Res. 45, 892-902 (1985).

20. Inglis MR, Kirley S, Stolbach II, Fishman WH: Phenotypes of the Regan isoenzyme and identity between the placental D-variant and the Nagao enzyme. Cancer Res. 33, 1657-1661 (1973).

21. Millan JL, Stigbrand T: Antigenic determinants of human placental alkaline and testicular placental-like alkaline phosphatases as mapped by monoclonal antibodies. Eur.J.Biochem. 136, 1-7 (1983).

22. Hada T, Amuro Y, Higashino K: Studies of intestinal-like alkaline phosphatases. Stigbrand T and Fishman WH(ed) A.R.Liss Inc.New York, 235-242 (1984).

23 Chang CH, Angelis D, Fishman WH: Presence of the rare D-variant heat-stable placental type alkaline phosphatase in normal testis Cancer Res. 40, 1506-1510 (1980).

24. Goldstein DJ, Rogers CE, Harris H: A search for trace expression of placental-like alkaline phosphatase in

246

non-malignant human tissues: demonstration of its
occurrence in lung, cervix, testis and thymus.
Clin.Chim.Acta 125, 63-75 (1982).
25. Van de Voorde A, Serreyn R, De Boever J, De Waele P,
Vandekerckhove D, Fiers W: The occurrence of human
placental alkaline phosphatase in extracts of normal,
benign and malignant tissues of the female genital tract.
Tumour Biology 6, 545-553 (1985).
26. Nouwen EJ, Pollet DE, Eerdekens MW, Hendrix PG, Briers
TW, De Broe ME: Immunohistochemical localization of
placental alkaline phosphatase, CEA and CA125 in normal
and neoplastic human lung. Cancer Res. 46, 866-876 (1986).
27. Maslow WC, Muensch HA, Azama F, Schneider AS: Sensitive
fluorimetry of heat-stable alkaline phosphatase (Regan
enzyme) activity in serum of smokers and non-smokers.
Clin.Chem. 29, 201-215 (1978).
28. McLaughlin PJ, Twist AM, Evans EC, Johnson PM: Serum
placental type alkaline phosphatase in cigarette smokers.
J.Clin.Pathol. 37, 826-828 (1984).
29. Fishman L, Inglis MR, Fishman WH: Preparation and
characterization of human intestinal alkaline phosphatase
antigens. Clin Chim.Acta 38, 75-83 (1972).
30. Van Belle H, De Broe ME, Wieme RJ: L-p-bromotetramizole, a
new reagent for use in measuring placental or intestinal
isoenzymes of alkaline phosphatase in human serum.
Clin Chem. 23, 454-459 (1977).
31. Sussman HH, Small PA Jr., Cotlove E: Human alkaline
phosphatase. Immunochemical identification of
organ-specific isoenzymes. J.Biol.Chem. 243, 160-166
(1968).
32. Millan JL, Stigbrand T,: Sandwich enzyme immunoassay for
placental alkaline phosphatase. Clin Chem. 27, 2014-2018
(1981).
33. Seargent LA, Stinson RA: Evidence that three structural
genes code for human alkaline phosphatases. Nature 281,
152-154 (1979).
34. Berger J, Garattini E, Hua J-C, Udenfriend S: Cloning and
sequencing of human intestinal alkaline phosphatase.
Proc.Natl.Acad.Sci.USA 84, 695-698 (1987).
35. Lehman F-G: Preparation of monospecific antisera for
immunoassay of human placental (Regan) and intestinal
alkaline phosphatase. J.Immunol.Methods 36, 137-148
(1981).
36. Slaughter GA, Cosco M, Cangro M, Harris H: Detection of
enzyme polymorphism by using monoclonal antibodies .
Proc.Natl.Acad.Sci.USA 78, 1124-1128 (1981).
37. Loose JH, Damjanov I, Harris H: Identity of the
neoplastic alkaline phosphatase as revealed with
monoclonal antibodies to the placental form of the enzyme.
Am.J.Clin.Pathol. 82, 173-177 (1984).
38. Pollet DE, Nouwen EJ, Schelstraete JB, Renard J, Van de
Voorde A, De Broe ME: Enzyme -antigen immunoassay for
human placental alkaline phosphatase in serum and tissue
extracts, and its application as a tumor marker.
Clin.Chem. 31, 41-45 (1985).
39. Pollet D: PhD. thesis, UIA, Antwerp, Belgium (1986).

40. Neville MA: Tumor markers and their clinical value.
    Tumour Biology 7, 83-90 (1986).
41. Tonik SE, Ortmeyer AE, Shindelman SE, Sussman HH:
    Elevation of serum placental alkaline phosphatase levels
    in cigarette smokers. Int.J.Cancer 31, 51-53 (1983).
42. McLaughlin PJ, Travers PJ, McDicken IW, Johnson PM:
    Demonstration of placental and placental-like AP in
    non-malignant human tissue extracts, using monoclonal
    antibodies in an enzyme immunoassay. Clin.Chim.Acta 137,
    341-348 (1984).
43. Mach JP, Carrel S, Forni M, Ritschard J, Donath A,
    Alberto P: Tumor localization of radiolabeled antibodies
    against CEA in patients with carcinoma. A critical
    evalution. The New England J.Med. 303, 5-10 (1980).
44. Pateisky M, Philipp K, Shodler WD, Czernenka K, Hamilton
    G, Burchell J: Radioimmunodetection in patients with
    suspected ovarian cancer. J.Nucl.Medecine 26, 1369-1376
    (1985).
45. Jackson PC, Pitcher EM, Davies JO, Rhys-Davies E,
    Sadowski CS, Staddon GE, Stirrat GM, Sunderland CA:
    Radionuclide imaging of ovarian tumours with a
    radiolabeled ^{125}I-monoclonal antibody (NDOG2).
    Eur.J.Nuclear Medecine 11, 22-28 (1985).
46. Schroff RW, Woodhouse CS, Foon KA, Oldham RK, Farrell MM,
    Klein RA, Morgan AC Jr.: Intratumor localization of
    monoclonal antibodies in patients with melanoma treated
    with an antibody to a 250 kDa melanoma-associated antigen.
    J.Natl.Cancer Inst. 74, 299-306 (1985).
47. Goldenberg DM, Kim E, Deland F, Spremulli E, Owens M,
    Elson N, Glokerman JP, Primus FJ, Corgan RL, Alpert E
    Clinical studies on the radioimmunodetection of tumors
    containing alfa-fetoprotein. Cancer 45, 2500-2505 (1980).
48. Goldenberg DM, Kim EE, Deland FH, Van Nagell JR Jr.,
    Javadpour M: Clinical radioimmunodetection of cancer with
    radioactive antibodies to human chorionic gonadotropin.
    Science 208, 12484484-1286 (1980).
49. Jeppson A, Wahren B, Millan JL, Stigbrand T: Tumor
    localization by use of monoclonal and polyclonal
    antibodies to placental alkaline phosphatase.
    Brit.J.Cancer 49, 123-128 (1984).
50. Epenetos AA, Hooker G, Durbin H, Bodmer WF, Snook D,
    Begent R, Oliver RTD, Lavender JP: Indium-111 labeled
    monoclonal antibody to placental alkaline phosphatase in
    the detection of neoplasms of testis, ovary and cervix.
    Lancet II, 350-353 (1985).
51. Jemmerson R, Millan JL, Klier G, Fishman WH: Monoclonal
    antibodies block the bromelain-mediated release of human
    placental alkaline phosphatase from cultured cells. FEBS
    Lett. 179, 316-320 (1985).
52. Garnett MC, Embleton MJ, Baldwin RW: Studies on the
    mechanism of action of an antibody-targeted drug carrier
    conjugate. Anti-cancer Drug Design 1, 3-12 (1985).
53. Segal DM, Hurwitz E: Binding of affinity cross-linked
    oligomers of IgG to cells bearing Fc-receptors. J.Immunol.
    118, 1338-1347 (1977).

54. Levy R, Miller RA: Tumor therapy with monoclonal antibodies. Fed.Proc. 42, 2650-2654 (1983).
55. Bradwell AR, Fairweather DS, Dykes PW, Keiling A, Vaughan A, Taylor J: Limiting factors in the localization of tumours with radiolabeled antibodies. Immunol.Today 6, 163-170 (1985).
56. Gordon J, Stevenson GT: Antigenic modulation of lymphocytic surface immunoglobulin yielding resistance to complement-mediated lysis. Immunol. 42, 13-17 (1981).
57 Cabilly S, Riggs AD, Pande H, Shively JE, Holmes WE, Rey N, Perry LJ, Wetzel R, Heyneker HL: Generation of antibody activity from immunoglobulin polypeptide chains produced in E.coli. Proc.Natl.Acad.Sci.USA 81, 3273-3277 (1984).
58. Wood CR, Boss MA, Kenten JH, Calvert JE, Roberts NA, Emtage JS: The synthesis and in vivo assembly of functional antibodies in yeast. Nature 314, 446-449 (1985).
59. Weidle UH, Borgya A, Mattes R, Lenz H, Buckel P: Reconstitution of functionally active antibody directed against creatine kinase from separately expressed H and L chains in non-lymphoid cells. Gene 51, 21-29 (1987).
60. Neuberger MS: Expression and replication of immunoglobulin heavy chain gene transfected into lymphoid cells. EMBO J. 2, 1373-1378 (1984).
61. Land H, Grez M, Hansen H, Lindenmaier W, Schutz G: 5'-Terminal sequences of eukaryotic mRNA can be cloned with high efficiency. Nucleic Acids Res. 9, 2251-2266 (1981).
62. Maxam AM, Gilbert W: A new method for sequencing DNA. Proc.Natl Acad.Sci.USA 74, 560-564 (1977).
63. Ollo R,Rougeon F: Mouse immunoglobulin allotypes: post-duplication divergence of gamma2a and gamma2b chain genes. Nature 296, 761-763 (1982).
64. Ohno S, Mori N, Matsunaga T: Antigen-binding specificities of antibodies are primarily determined by seven residues of V_H. Proc.Natl.Acad.Sci.USA 82, 2945-2949 (1985).
65. Max EE, Seidman JG, Leder P: Sequence of five potential recombination sites encoded close to an immunoglobulin kappa constant region gene. Proc.Natl.Acad.Sci.USA 76, 3450-3454 (1979).
66. Von Heije G: Signal sequences. Limits of variation. J.Mol.Biol. 184, 99-105 (1985).
67. Pepe VH, Sonenshein GE, Yoshimura MI, Shulman MJ: Gene transfer of immunoglobulin light chain restores heavy chain secretion. J.Immunol. 137, 2367-2372 (1986).
68. Huylebroeck D, Maertens G, Verhoeyen M, Lopez C, Raeymakers A, Min Jou W, Fiers W: High level transient expression of Influenza virus proteins from a series of SV40 late and early replacement vectors. Gene, submitted.
69. Maniatis T, Hardison RC, Lacy E, Lauer J, O'Connell C, Quon D: The isolation of structural genes from libraries of eukaryotic DNA Cell 15, 687-701 (1978).
70. Gluzman Y: SV40-transformed Simian Cells support the replication of early SV40 mutants. Cell 23, 175-182 (1981).
71. McCutchan JH, Pagano JS: Enhancemant of the infectivity

of SV40 DNA with DEAE-dextran. J.Natl.cancer Inst. 41, 351-357 (1968).

72. Kaufman RJ, Sharp PA: Construction of a modular DHFR cDNA gene: analysis of signals utilised for efficient expression. Mol.Cell.Biol. 2, 1304-1319 (1982).
73. Scahill SJ, Devos R, Van der Heyden J, Fiers W: Expression and characterization of the product of a human immune interferon cDNA gene in CHO cells. Proc.Natl.Acad.Sci.USA 80, 4654-4658 (1983).
74. Graham FL, Van der Eb AJ: Transformation of rat cells by DNA from Adenovirus 5. Virol. 54, 536-539 (1973).
75. Marquart M, Deisenhofer J, Huber R: Crystallographic refinement and atomic models of the intact immunoglobulin molecule Kol and its antigen-binding fragment at 3A and 1,9A resolution. J.Mol.Biol. 141, 369-391 (1980).
76. Michaelsen TE, Frangione B, Franklin EC: Primary structure of the hinge region of human IgG3. J.Biol.Chem. 252, 883-889 (1977).

Acknowledgements: We thank B. Van Oosterhout and W. Drijvers for editorial assistance and D. Ginneberghe for assistance with the transfections. This research was supported by grants from the Belgian Algemene Spaar- en Lijfrentekas and from the "Geconcerteerde Onderzoekacties".

VACCINIA VIRUS AS AN EXPRESSION VECTOR

Robert Drillien
Transgène 11 rue de Molsheim, 67000 Strasbourg, France
Laboratoire de Virologie 3 rue Koeberlé 67000 Strasbourg

Introduction

In recent years vaccinia virus (VV) has become a popular vector for the synthesis of various proteins in mammalian cells. Up to now, the main emphasis has been put on using VV to obtain protein antigens capable of inducing a protective immune response against an infectious disease. Less well known is its value as a vector for the expression of other potentially useful proteins such as enzymes, hormones, immunomodulators. Our intent here is not to give an exhaustive review of the many applications being explored particularly in the area of vaccines but rather to introduce the field to the outsider, to underline some of the more recent refinements and to illustrate the vector system with some examples developed in our own laboratory (recent reviews on the topic can be found in ref. 1 and 2)

General features of the VV vector

VV, the prototype orthopoxvirus was the first poxvirus to be used as a vector. The possibility of similarly engineering other poxviruses such as Fowlpox or Capripox is also being studied. Several distinctive features of the poxviruses are important to consider for vector design. The viruses, large brick shaped particles in the case of VV contain a linear double stranded DNA genome (about 190 kbp) that is non infectious in contrast to the DNA from many other animal viruses. DNA replication and transcription take place in the cell cytoplasm where they depend largely on virus encoded enzymes. Assembly of viral particles also occurs in the cell cytoplasm and infection spreads either through virus budding from the cell surface or cell lysis which is the ultimate result of infection. The large size of the poxvirus genome and its lack of infectivity preclude in vitro construction of recombinant viral DNA. Instead, viral DNA fragments cloned on bacterial vectors are engineered so that a foreign DNA insert is flanked on both sides by a normally contiguous viral segment. These recombinant molecules are then introduced by transfection into cells previously infected with VV where they can undergo homologous double recombination events with an intact VV genome. Foreign genes recombined into the viral genome must be integrated into a region non essential for viral replication. Several regions of the genome, particularly the genes encoding the viral thymidine kinase (3), the hemagglutinin gene (4), the inclusion body gene (5) and a gene cluster located at the so called left end of the genome (6) have been characterized or are expected to be useful sites for insertion. In order to be expressed, foreign genes must also be under the control of promoters specifically recognized by the virus encoded RNA polymerase. A large amount of information that will not be dealt with here has accumulated on VV transcriptional promoters as to the time they are

251

A. O. A. Miller (ed.), Advanced Research on Animal Cell Technology, 251–260.
© 1989 by Kluwer Academic Publishers.

turned on (early or late), the efficiency with which they are transcribed and the mimimum length needed for their activity. Suffice it to say that poxvirus transcription still holds in store some surprises as illustrated by the recent finding of discontinuous late mRNAs containing poly A 5' ends (7, 8) despite the fact that no conventional splicing has been recognized for viral RNA. Thus foreign genes are inserted into the VV genome without introns. Exogenous genetic information up to 25 kbp has been introduced into VV recombinants (9). Theoretically it should be possible to insert much more since viable deletion mutants lacking up to 20 kbp have been isolated (10, 11, 12). So far recombinants expressing simultaneously three foreign genes have been constructed (13) and clearly many more genes could be expressed from the same recombinant viral genome. A variety of techniques have been established to select and or identify VV recombinants. Among the most straightforward are nucleic acid hybridization of the appropriate probe to viral plaques (14) or in situ immunoassay for the detection of antigens produced by recombinant plaques (15). Another widely used method relies on the ability to interrupt the VV thymidine kinase gene (tk) with foreign DNA (3) thereby giving rise to virus that is capable of multiplying in the presence of the mutagenic analogue of thymidine (5' bromodeoxyuridine). The assay is attractive since it allows one to select against the multiplication of non recombinant virus although spontaneous tk negative mutants may appear at a low frequency. Selection of tk negative recombinants requires that the host cells lack tk activity. Other techniques avoiding the use of a mutagen and the need for particular cell lines have been set up. One uses the β galactosidase gene of E. coli and the detection of the corresponding enzymatic activity with chromogenic substrates (16, 17). The foreign gene of choice can be physically linked to the β galactosidase gene in which case blue plaques will be picked or the foreign gene can be designed to replace the β galactosidase gene in a previously constructed recombinant virus and then white plaques will be looked for. Another method employs the insertion of the neomycin resistance gene to select virus capable of multiplying in the presence of the antibiotic G418 (18, 19). Procedures that combine several of the preceeding techniques or improve the frequency of appearance of recombinants by adding additional selective pressures such as selection against a virus with a temperature sensitive or drug dependent phenotype have also been successfully employed (20, 21). Recently, recombinant VV molecules have been constructed using single stranded phage vectors (22). The method enables gene transfer to the virus to be carried out following site directed mutagenesis of the vector without any additional steps. It is expected that further improvements of the VV vector will evolve as more and more information is gained from fundamental studies of the virus. The characterization of drug resistance genes in the VV genome and the ability to select for amplification of those genes (23) are just one of the many means that can be suggested for increasing expression. Already, coinfection with one recombinant virus containing a bacteriophage T7 promoter and another the T7 RNA polymerase has proved an efficient way to substantially increase the level of synthesis of a foreign gene (24). Likewise a late promoter for inclusion body formation in cowpox virus appears to be a promising tool to increase the transcription of a foreign gene (25).

Genetically engineered vaccines

The first application suggested by the two groups that pioneered the use of VV vectors was vaccination against infectious diseases. This was inspired by the success of the smallpox eradication campaign in which VV, the non pathogenic close relative of smallpox was employed. Candidate recombinant vaccines based on VV have been constructed for use against a number of infectious diseases caused particularly by viruses but also to a limited extent bacteria and parasites. In cases where these candidates are intended to replace an already existing vaccine they must still prove that they are equally or more efficient, devoid of undesirable effects and less costly. Where the recombinant vaccines are intended to fill a gap, their acceptance should be easier particularly if the disease being targeted is life threatening or economically important. Whatever pratical applications occur much is being learned in the process using VV recombinants as tools to discover which antigens provide protective immunity, what is the best way to display the antigen to the immune system (secreted, inserted in the cell membrane, etc.) and what type of immunological response should one look for against a particular disease (ref 2). Furthermore the large capacity of the VV vector enables one to insert several genes encoding protective antigens to look for possible synergy.

Work in our laboratory has focused on several viral diseases particularly rabies (for a review see ref.26), measles and AIDS. The cDNA encoding the surface antigen of the rabies virus, a 67kd glycoprotein, was integrated into the VV genome. The rabies glycoprotein was shown to be synthesized, processed and transported to the infected cell surface just as in a natural rabies infected cell. Inoculation of live VV recombinants encoding the rabies protein into a number of mammalian species induced the synthesis of neutralizing antibodies and protected them from subsequent rabies virus challenge. Moreover, the recombinants are still effective when taken orally which suggests the possibility of including the vaccine in baits (27). If successful this would provide a cheap and pratical means of wildlife vaccination.

Measles virus encodes two surface proteins (the hemagglutinin and the fusion protein) that are thought to be required for efficient vaccination. Present day measles vaccines are attenuated strains of measles virus that consequently include all measles proteins. A VV based recombinant measles vaccine could be cheaper and more effective in the warm climates of many developing countries where measles is a serious cause of childhood death. We have constructed two VV recombinants encoding either the hemagglutin protein or the fusion protein and shown that just as in the case of the rabies glycoprotein they are synthesized, processed and transported to the infected cell surface. Mice vaccinated with the recombinants develop measles neutralizing antibodies and resist a challenge infection with a lethal cell associated measles strain (28). Further work is being conducted to evaluate the efficiency of protection with recombinants encoding both glycoproteins or alternatively internal proteins of measles virus.

Antigens of the human immunodeficiency viruses that could provide protection against AIDS have still to be identified. VV vectors are particularly useful tools in this search. Many of the structural proteins in particular the envelope protein (29, 30, 31) as well as non structural proteins of HIV 1 have been expressed using VV in our laboratory and others. Studies so far indicate that the viral envelope protein is a

relatively poor antigen as compared to envelope proteins from other viruses. Modifications of the coding sequence have been carried out in order to improve antigenicity and some encouraging results have been obtained although clearly further work is required.

Vaccinia as a mammalian cell expression system

When considering VV as a mammalian cell expression vector for producing proteins under in vitro conditions one has to keep in mind several of the distinctive properties of the virus. First of all, synthesis of the desired product will be maintained for at most several days since VV is a cytopathic virus and will eventualy kill the host cell. Chronic infection with VV has been achieved (32, 33) however the usefulness of such systems for expression vectors is unknown. Secondly, cell cultures containing the VV encoded foreign product will also contain considerable quantities of live virus that one may wish to remove. Suitable techniques are available and are constantly being improved. Among the main advantages of using VV for mammalian cell expression are the following: the ease with which recombinants can be constructed and quickly assayed, the wide host range of VV enabling convenient screening of many cell types for the ability to synthesize an active product, the level of synthesis of proteins comparable or higher than with most other mammalian cell vectors, the correct post translational processing of foreign proteins (in our experience we have documented the proper glycosylation, phosphorylation, acylation, carboxylation and cleavage of various VV expressed proteins).

Work in our laboratory has been concerned with the in vitro synthesis of biologically active coagulation factors VIII and IX. These proteins are required for patients suffering from haemophilia and are as yet only available from human blood. We demonstrated a few years ago that active coagulation factor IX can be synthesized in mammalian cells under the control of a VV vector. The protein is correctly carboxylated and secreted into the cell medium where it can be recovered. The VV recombinant expressing factor IX was used to infect a variety of different cell types to determine which could make the largest amount of product with the greatest activity (34). The cDNA for factor VIII is an extremely long molecule (about 9 kb) and the primary translation product undergoes a series of post translational modifications necessary for its full activity namely glycosylation and specific proteolytic cleavages. VV recombinants having integrated factor VIII cDNA have also been constructed and shown to induce the synthesis of active coagulation factor with an efficiency that depends somewhat on the cells infected (35). Modifications of the original cDNA which delete coding sequences that are subsequently removed by proteolytic enzymes are now being attempted to further understand factor VIII maturation and possibly improve the levels of expression (36).

Conclusions

VV vectors have already proved their potential as tools for gene expression in mammalian cells. As might be expected with any system several drawbacks are apparent. For instance when considering its use as a live recombinant vaccine much concern has been expressed about the uncontrolled dissemination of VV in nature or the rare accidents known to occur after smallpox vaccination. It is therefore important to study the possibility of further attenuation of the vector or targeting the virus to particular

species or cell types. Current research carried out on some of the other animal poxviruses could provide an ideal solution for restricting the host range. A perfectly attenuated live VV vector for human use may no longer be an effective carrier of foreign antigens thus some kind of comprimise may have to be reached. At the present rate of progress one may soon expect effective and safe VV recombinant vaccines to be available against a number of infectious diseases. Exploiting VV vectors for the production of proteins in mammalian cells is still in an early stage. Naively one would wish to retain all the properties of the virus particularly its replication and transcription machinery that make it such an efficient expression system and discard the features that lead to the production of large quantities of virus and eventually cell lysis. Such an attempt would probably be a failure however a number of ways of improving expression are already being experimented and much scope remains.

References

1. Mackett, M. and Smith, G. L., J. Gen. Virol., 67, 2067-2082, 1986
2. Moss, B. and Flexner, C., Ann. Rev. Immunol., 5, 305-324, 1987
3. Mackett, M., Smith, G. L. and Moss, B., Proc. Natl. Acad. Sci. USA 79, 7415-7419, 1982
4. Shida, H., Virology, 150, 451-462, 1986
5. Patel, D. D., Pickup, D. J. and Joklik, W. K., Virology, 149, 174-189, 1986
6. Perkus, M. E., Panicali, D., Mercer, S. and Paoletti, E., Virology, 152, 285-297, 1986
7. Bertholet, C., Van Meir, E., ten Heggeler-Bordier, B. and Wittek, R. Cell,50, 153-162 1987
8. Schwer, B., Visca, P.,Vos, J. C. and Stunnenberg, H. G., Cell, 50, 163 -169, 1987
9. Smith, G. L. and Moss, B., Gene, 25, 21-28, 1983
10. Moyer, R. W. and Rothe, C. T., Virology, 102, 119-132, 1980
11. Lake, J. R. and Cooper, P. D., J. Gen. Virol., 48, 135-147, 1980
12. Drillien, R., Koehren, F. and Kirn, A., Virology, 111, 488-499, 1981
13. Perkus, M. E., Piccini, A., Lipinskas, B. R. and Paoletti, E., Science 229,981-984, 1985
14. Panicali, D. and Paoletti, E., Proc. Natl. Acad. Sci. USA, 79, 4927-4931, 1982
15. Mackett,M. and Arrand, J. R., EMBO J., 4, 3229-3234, 1985
16. Chakrabarti,S., Brechling, K. and Moss, B., Mol. Cell. Biol., 5, 3403-3409, 1985
17. Panicali, D., Grzelecki, A. and Huang, C., Gene, 47, 193-199, 1986
18. Franke, C. A., Rice, C. M., Strauss, J. H. and Hruby, D. E., Mol. Cell. Biol., 5, 1918-1924, 1985
19. Mars, M., Vassef, A. and Beaud, G., Ann. Inst. Pasteur./Virol. 137E, 273-290, 1986
20. Fathi, Z., Sridar, P., Pacha, R. F. and Condit, R. C., Virology, 155 97-105, 1986
21. Kieny, M. P., Lathe, R., Drillien, R., Spehner, D., Skory, S., Schmitt, D., Wiktor, T., Koprowski, H. and Lecocq, J. P. Nature, 312, 163-166, 1984
22. Wilson, E. M., Hodges, W. M. and Hruby, D. E., Gene, 49, 207-213, 1986
23. Slabaugh, M. B. and Mathews, C. K., J. Virol., 60, 506-514, 1986
24. Fuerst, T. R., Earl, P. L. and Moss, B., Mol. Cell. Biol., 7, 2538-

    2544, 1987
25. Patel, D., Ray, C. A. and Pickup, D. J., Abstract VII Int. Cong. Virol.
    Edmonton 1987
26. Kieny, M. P., Desmettre, P., Soulebot, J. P. and Lathe, R., Prog. Vet.
    Microb. Immunol. 3, 1986
27. Blancou, J., Kieny, M. P., Lathe, R., Lecocq, J. P., Pastoret, P. P.,
    Soulebot, J. P. and Desmettre, P. Nature, 322, 373-375, 1986
28. Drillien, R., Spehner, D., Kirn, A., Giraudon, P., Buckland, R., Wild,
    F. and Lecocq, J. P. submitted
29. Kieny, M. P., Rautmann, G., Schmitt, D., Dott, K., Wain-Hobson, S.,
    Alizon, M., Girard, M., Chamaret, S., Laurent, A., Montagnier, L. and
    Lecocq, J. P., Biotechnology, 4, 790-795, 1986
30. Hu, S. L., Kosowski, S. G. and Dalrymple, J. M., Nature, 320, 537, 1986
31. Chakrabarti, S. Robert-Guroff,R., Wong-Staal, F., Gallo, R. C. and
    Moss, B., Nature, 320, 535, 1986
32. Pogo, B. G. T. and Friend, C. Proc. Natl. Acad. Sci. USA, 79, 4805-
    4809, 1982
33. Paez, E., Dallo, S.and Esteban, M. Proc. Natl. Acad. Sci. USA, 82, 3365
    -3369, 1985
34. De La Salle, H., Altenburger, W., Elkaim, R., Dott, K., Dieterlé, A.,
    Drillien, R., Cazenave, J. P., Tolstoshev, P. and Lecocq, J. P., Nature
    316, 267-270, 1985
35 Pavirani, A., Meulien, P., Harrer, H., Schamber, F., Dott, K.,
    Villeval, D.,Cordier, Y., Wiesel, M. L., Mazurier, C., Van de Pol, H.,
    Piquet, Y.,Cazenave, J. P. and Lecocq, J. P., Biotechnology, 5, 389-
    392, 1987
36. Pavirani, A., Meulien, P., Harrer, H., Dott, K., Mischler, F., Wiesel,
    M. L., Mazurier, C., Cazenave, J. P. and Lecocq, J. P. Biochem.
    Biophys. Res. Comm., 145, 234-240, 1987

Drillien, R. : <u>Vaccinia virus expression vectors</u>

Miller :   Aren't there poxviruses which multiply intranuclearly and might therefore transform cells ?

Drillien : No, there are no poxviruses that multiply in the nucleus. There are poxvirus events say RNA synthesis that have been described within the nucleus but careful autoradiographic studies seem to indicate there is no intranuclear repli- cation. There might be a dependence on nuclear functions for poxvirus replication, that's another story. As regards to transformation by poxviruses there are certainly poxviruses that are not transforming viruses in the true sense. You have for instance shope fibroma. This virus will induce non malignant tumors which will regress spontaneously. So, these are not true transforming viru- ses. Poxviruses indeed and even vaccinia virus, encode an EGF-like protein. So poxviruses do encode a growth factor. No, after millions of vaccinations, no cancers have been associated with pox vaccination.

Spier :   There is a perennial sort of verbal battle going on in terms of being able to use vaccinia as a human vaccine or modified vaccine. The measles situation that you have highlighted seems to indicate that there is a terrific potential for measles vaccine based upon vaccinia in the Third World and I am sure there are other viruses that would be equally usefully carried into Third World popu- lations. Presumably you have been talking to various people and you have a view as to how you are going to get this particular kind of vaccine accepted by the various authorities.

There is what is the present state of the play ? I don't suppose you have got to the bottom line but could you tell us what

the present state is and what you think the prospects are ?

Drillien : That is like in any idea that is a little bit new. When trains came along, people were totally frightened and nobody wanted to get into these apparatuses and then little by little, trains got accepted and everybody is using trains. So, I think it is the same old story. It takes a long time to be accepted and sometimes, once you know the trains got finally accepted, there were airplanes that came along. My feeling is that pox recombinant viruses can certainly contribute enormously.

There are many political problems as you know, many regulatory problems and it might just be slowed down by one of these problems. Yes. I think certainly WHO at this time has a very open mind to these studies and is actualy encouraging work in this area. So, I think it is satisfying.

Chowdhury : In mammalian cells how long does this expression go on ?

Drillien : That is an important point. Poxvirus infection eventually leads to lysis of the cell. So, after twenty four hours, at most 48 hours, the cell is going to lyse and expression will end at a reasonable multiplicity. So, that is a very important point, yes.

Horaud : Is it possible to have a persistent infection with poxviruses ?

Drillien : Yes, people have persistently infected three leukaemia cells. As you might expect, the virus becomes dormant in these cells and there is very little viral expression. So, although you have a system where the poxvirus machinery is present, it is really shut down and you cannot take advantage of the amplification you get in the viral

infection for expression. At this stage, this is really not an interesting system for recombinant work.

Horaud :   Is something known about the efficiency of the replication of vaccinia virus ? Does the virus replicate locally ?

Drillien :   Yes, certainly. I think the best demonstration is Enzo PAOLETTI's work where he immunized with a recombinant expressing influenza hemagglutinin or something like that and then he came back with the herpes virus glycoprotein in a vaccinia recombinant (rabbit in this case) and he got antibodies against herpes virus glycoprotein. So, definitely you can go trough several cycles.

Anonymous :   What is the FAO position concerning the use of rabies vaccine recombinants in the wild ? I've heard something is happening in Argentina.

Drillien :   Yes, I am aware. That is really bad publicity. There were trials that were approved by the Pan American Health authorities but were not approved by the Argentinian Health authorities.

Horaud :   The people in charge of the project did not ask permission to the local regulatory agency in Argentina, to perform the experiment. So, they performed their experiments in the Buenos Ayres area. When these were under way, the Argentinian authorities heard about them. It was also a sort of media operation in which ecologists and other people started to say that this constituted an enormous danger. This operation was clearly a faux pas.

Van Meel :   Is there a way to manipulate the expression of antigens in order to make them secreted by the infected cells or remaining seated on the membrane ?

260

Drillien : Yes, you can use the chemical approach. This is a problem
that I have suffered from a little bit personally. Supposing you have your antigen which normally is secreted with
all its signal sequence and all that sort of things, there
is no problem : that antigen is going to be secreted. If
you have an antigen that is not normally secreted and you
put a signal on it, telling that it ought to be secreted
the chances are it will be going to the endoplasmic
recticulum and might even go to the Golgi. Sometimes it
even gets out of the cell but as you know probably, there
are more signals to tell a protein to stay inside, that
will trap it within one of these compartments I mentioned.
It is really a question of trial and error and it also
depends on the sequence you are considering. This is not
peculiar to vaccinia virus but also occurs with other
animal viruses : secretion also depends on the sequence
you consider.

EXPRESSION AND INTEGRATION OF EXOGENOUS DNA SEQUENCES TRANSFECTED INTO MAMMALIAN CELLS

F. COLBERE-GARAPIN[1], M.L. RYHINER[1] & A.C. GARAPIN[2]

1: Unité de Virologie Médicale ; 2: Unité de Biologie Moléculaire du Gène, INSTITUT PASTEUR, 75724 PARIS CEDEX 15, France

1. INTRODUCTION

It has been previously shown that the frequency of cointegration into cellular DNA of two genes borne on the same plasmid molecule varies considerably from cell line to cell line (Colbère-Garapin et al. 1985 & 1986). In monkey kidney Vero cells, cointegration of the entire human hepatitis B surface (HBs) antigen (Ag) gene and the aminoglycoside 3'-phosphotransferase (APH3') gene occurs only in 15 % of the transformed clones whereas in murine LM cells under the same conditions, both genes are cointegrated into 100 % of the clones (Colbère-Garapin et al. 1985 & 1986). In Vero cells, DNA rearrangements and deletions in the exogenous DNA could occur at any one of the 3 stages of the transformation process 1 - during the first days after transfection by degradation of free plasmid DNA molecules, 2 - during recombination with cell DNA and/or the early replication cycles following integration or 3 - later after integration if the transgenome is unstable. We therefore investigated the penetration, persistence and integration of cotransferred genes carried by double-stranded (ds) DNA molecules in monkey kidney Vero cells and 3 cell lines of human origin, HeLa, GM4312A and HepG2, in comparison with similarly transfected murine LM cells. Furthermore, we have compared the expression of transfected genes borne on single-stranded (ss) and ds DNA molecules in order to find out whether transient gene expression could be obtained from ss DNA, and whether stable coexpression of two genes could occur if distinct ss DNA molecules carrying each of the genes were co-transferred.

2. MATERIALS AND METHODS

2.1. Enzymes, bacterial strains, plasmids and phage. Bacteriophage T4 DNA ligase and restriction endonucleases were purchased from New England Biolabs, Appligene and Boehringer. Escherichia coli strain 803 su III r_k^- m_k^- used for transformation was kindly provided by K. Murray. Construction of plasmids pAG60 (Colbère-Garapin et al. 1981), pAG400-2 (Colbère-Garapin et al. 1985), pAG452-2 and pAG480-15 (Colbère-Garapin et al. 1986) has been described. BMK 71-18 [Δ (lac-pro AB), thi, sup E ; F'lac i 9, ZΔMI5, pro A$^+$B$^+$] (source : B. Müller-Hill) was kindly supplied by Dr. H.J. Fritz. Phages pAG511-2 and pAG511-6 are derivatives of M13 mp19 (Messing et al. 1977) into which the HBs gene was inserted. The replicative form of M13 mp19 (Kramer et al. 1984) was linearized with Bam H1 and ligated to a 2400 bp Bgl II DNA fragment carrying the HBs gene with two HBs promoters in tandem. The recombinant phage pAG511-2 carries the HBs gene such that it is transcribed into functional HBs mRNA, while pAG511-6 only encodes if transcribed an anti-sense HBs RNA. The hygromy-

A. O. A. Miller (ed.), Advanced Research on Animal Cell Technology, 261–275.
© 1989 by Kluwer Academic Publishers.

cin B (Hy) phosphotransferase gene (HPH) under the control of the herpesvirus hominis 1 thymidine kinase promoter confers resistance to Hy to cells. This hybrid gene, carried by a 1940 bp Pvu II fragment, was inserted into M13 mp19 at the Sma I site of the polylinker. Two recombinant phages were obtained : pAG510-17 and pAG510-14. The former encodes HPH mRNA and the latter, if transcribed, encodes an anti-sense HPH RNA. Single-stranded DNA was purified on benzoylated naphthoylated DEAE-cellulose (Gamper et al. 1985).

2.2. Eukaryotic cells and cell transformation. HeLa, TK⁻ mouse LM cells clone ID and monkey kidney Vero cells clone VC10 were cultivated in DMEM supplemented with 10 % calf serum. Human hepatoma Hep G2 cells (Aden et al. 1979) were cultivated in MEM containing 10 % fetal calf serum (FCS). GM4312 A cells were obtained from the NIGMS Human Genetic Mutant Cell Repository and are derived from human Xeroderma pigmentosum cells transformed by simian virus 40 (SV40). They were cultivated in DMEM supplemented with 10 % FCS. Cell transformation with plasmid DNA was carried out as described (Colbère-Garapin et al. 1985) using the calcium precipitation technique. Twenty-four hours after transfection, cell culture medium containing G418 (Gibco) or Hy (Eli Lilly) was added to the cells. The final concentration used was 150 µg G418/ml for Vero, HeLa and LM cells, 100 µg G418/ml for GM4312 A cells, 300 µg G418/ml for Hep G2 cells, and the hygromycin B concentration used was 150 µg Hy/ml.

2.3. Isolation of plasmid DNA after transfection. Between 24 hr and 3 weeks after transfection, plasmid DNA was isolated from the cells by the method of Hirt (1967). High molecular weight DNA was precipitated in 1 M NaCl and pelleted. Plasmid DNA which remained in solution was treated with ribonuclease from bovine pancreas (Calbiochem) at a concentration of 10 µg/ml for 30 min at 37°C. Proteinase K (Merck) was then added at a concentration of 200 µg/ml and the extract was further incubated at 50°C for 30 min. After a phenol-chloroform isoamyl alcohol (24-24-1) extraction, plasmid DNA was precipitated with ethanol.

2.4. Analysis of DNA from transformed cells. Cellular DNA was extracted as previously described (Colbère-Garapin et al. 1981). After cleavage with restriction enzymes, DNA fragments were analyzed according to Southern (1975). DNA fragments used as probes specific for the Ap, Tc and HBs genes were isolated as described (Colbère-Garapin et al. 1986) and ^{32}P-labeled by nick-translation (specific ^{32}P activity 3 x 10^8 cpm/µg). A Bam HI-Bam HI DNA fragment isolated from plasmid pLG89 (Gritz and Davies, 1983) was used as a probe specific for the HPH gene. Filters were hybridized at 68°C and autoradiographed. Some autoradiographs were scanned with a Gelman gel scanner.

2.5. Detection of the HBs antigen. HBsAg detection in cell supernatants was performed by solid phase radioimmunoassay using an Ausria II kit (Abbott Laboratories).

3. RESULTS

3.1. Penetration and persistence of free double-stranded plasmid molecules in monkey kidney Vero and mouse LM cells. We compared the efficiency of plasmid penetration and persistence in monkey Vero and mouse LM cells. Twenty-four hours after transfection with pAG60 carrying the APH3' selective marker or pAG452-2 carrying the hygromycin B resistance gene, low molecular weight plasmid DNA was extracted from the transfected cells by the method of Hirt (1967) and analyzed by Southern blotting using a

^{32}P-labelled probe including the hygromycin phosphotransferase (HPH) gene and some plasmid sequences. The results shown in figure 1 indicate that plasmid penetration and persistence during the first 24 hr after transfection is comparable for both cell types. About 25 % of the transfected unintegrated molecules are found to be linearized. One week after transfection, 500 plasmid molecules are found per cell in both Vero and LM cells (Skala et al. 1986).

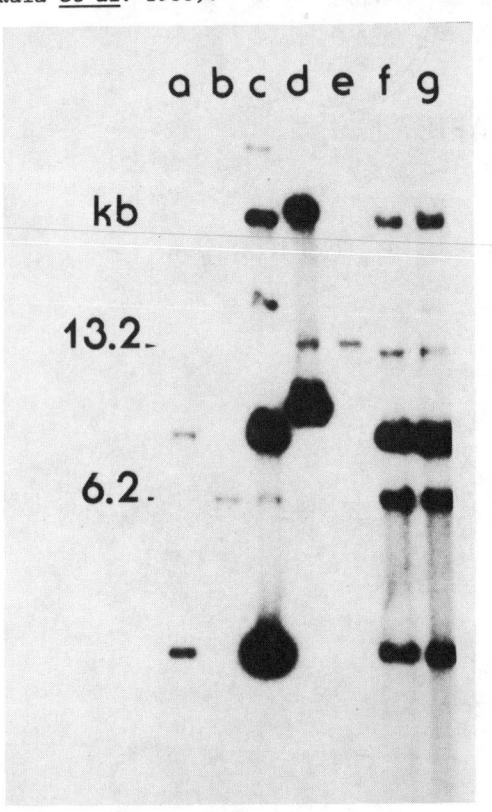

FIGURE 1. Plasmid molecules recovered from Vero and LM cells 24 hr after transfection. Uncleaved pAG60 extracted from E. coli (lane a), pAG60 linearized with Sma I (lane b), pAG452-2 (lane c) , a 13.2 kb plasmid (lane d), and the same plasmid linearized with Hind III (lane e) were used as markers. Plasmid pAG452-2 was extracted from Vero (lane f) and LM cells (lane g) 24 hr after transfection. After transfer according to Southern (1975) filters were hybridized with a probe specific for plasmid sequences.

3.2. Integration of exogenous (double-stranded) DNA sequences into cellular DNA of four primate cell lines. The plasmid pAG400-2 was used to study integration of exogenous DNA into cellular DNA. A single integration event of both the APH3' and the HBs genes of this 9 kb plasmid would require integration of 5.2 kb of continuous plasmid sequences. After selection of transformants in G418, we found that the HBs gene was deleted or rearranged in the majority of Vero cell clones (figure 2). The

264

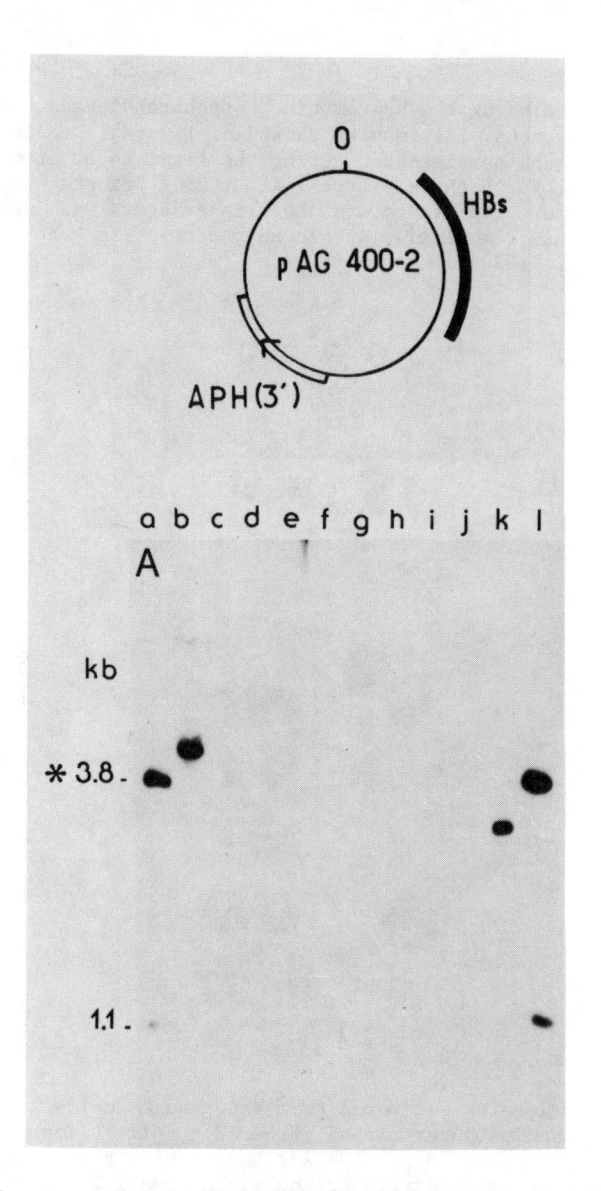

FIGURE 2. Detection by Southern blot hybridization of HBs gene sequences in clones of Vero cells transformed by pAG400-2. After cleavage with Eco RI, DNA fragments were analyzed in 0.7 % agarose gel electrophoresis and transferred onto filters. The HBs specific probe in pAG400-2 is shown above the filter by a heavy line next to the plasmid. The 3.8 kb Eco RI fragment which carries the coding and 3' non-coding regions of the HBs gene is marked by an asterisk. DNA fragments of plasmid pAG400-2 cleaved with Eco RI were used as Mr markers (lanes a and l). Cellular DNAs were cleaved with Eco RI (lanes b to k).

average length of exogenous DNA was 4.7 kb for Vero cell clones, and 4.5
kb for human HeLa and GM4312 A cell transformants (Colbère-Garapin et al.
1986). Human hepatoma Hep G2 cells integrated on the average 6.4 kb of
plasmid DNA, i.e. about 2 kb more DNA than the three other primate cell
lines used (figure 3). This resulted in a 4-5 times higher gene cointegration frequency in Hep G2 than in Vero cells. Mouse LM cell clones,
studied comparatively, integrated on the average 57 kb of exogenous DNA.
Similar results were obtained with different transfection techniques and
pairs of genes (Colbère-Garapin et al. 1986).

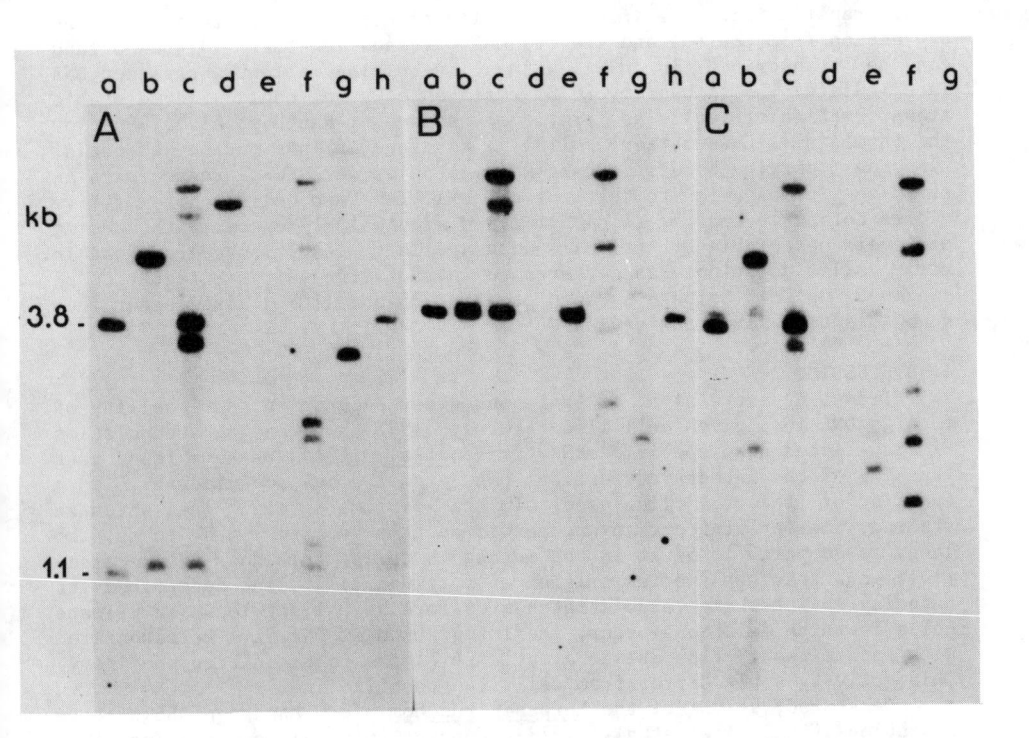

FIGURE 3. Detection by Southern blot hybridization of HBs, ampicillin,
and tetracyline resistance gene sequences in clones of Hep G2 cells
transformed by pAG400-2. Specific ^{32}P-labelled probes for the HBs, the
Ap^R and the Tc^R genes were hybridized to the DNA of filters A, B and C,
respectively. DNA fragments of plasmid pAG400-2 cleaved with Eco RI were
used as Mr markers (lanes a and h). Cellular DNAs were cleaved with Eco
RI (lanes b to g).

3.3. <u>Transfer and expression of single-stranded DNA into animal cells</u>. We studied the expression of transfected genes carried by single-stranded (ss) compared to double-stranded (ds) DNA molecules during the first days after transfection in cell lines of rodent and primate origin. The HBs gene was inserted into the M13 phage genome (see Materials and Methods). Two recombinant phages pAG511-2 and pAG511-6 were selected which carried the (-) and (+) strands of the gene, respectively. The hygromycin B resistance gene was also cloned into the M13 genome and two recombinant phages were isolated : pAG510-7 and pAG510-14 carrying the (-) and (+) stran ds of the gene, respectively. After transfer into monkey kidney Vero cells, the HBs gene carried by the (-) ss DNA was transiently expressed, although at a rate lower than that obtained with the same number of ds molecules (35.2 %). The homologous (+) DNA strand was also transiently expressed (11.8 % of the rate obtained with the ds DNA). If the (+) DNA strands were transcribed in transfected cells, it would give rise to anti-sense RNA which could not be translated into HBs antigen. Our results indicated then that about 12 % of the ss DNA molecules are duplicated - before or after integration - during the first few days after transfection. The simplest explanation for the difference of HBs expression from the (-) and the (+) DNA strands is that (-) DNA strand can be transcribed in mRNA without requiring a double-stranded DNA intermediate. The ss hygromycin B resistance gene could induce stable transformation of cells as previously shown by Rauth <u>et al</u>. (1986) for the thymidine kinase marker. Following co-transfer into murine LM cells, two genes carried by distinct ss DNA molecules were co-expressed with an efficiency comparable to that of ds DNA. In Vero cells, none of five clones cotransfected by ss pAG511-2 and ss pAG510-17 excreted the HBs Ag at levels detectable by radioimmunoassay. This result suggests that in Vero cells deletions and rearrangements observed after integration of exogenous ds DNA (Colbère-Garapin <u>et al</u>. 1985 and 1986) also occur when ss DNA is used for transfection.

4. DISCUSSION

Transient expression of exogenous genes depends on the stability of foreign DNA in a given cell line (Alwine, 1985) and on the accumulation of transfected gene specific mRNA (Skala <u>et al</u>. 1986). We have found that the size of the integrated foreign DNA also varies considerably as a function of the cell line used. In the 4 primate cell lines which we studied, the integrated transgenome was on the average 4.5 to 6.4 kb long, as compared to 57 kb in the murine LM cells. This does not necessarily means that the species origin of a given cell line determines its capacity to integrate large stretches of foreign DNA. At least in primate cells in which deletions occur, the integration of DNA may be associated with random restriction due to a nuclease of the recombination machinery. This activity would differ from cell line to cell line.

Denatured, ss DNA of the <u>herpes virus hominis</u> 1 has been shown to be infectious (Sheldrick <u>et al</u>. 1973). However, partial renaturation of ss molecules may have occurred during transfection or once inside the cell. The same notion may be evoked concerning the infectivity of the (+) and (-) strands reported for some parvoviruses. In experiments we performed with ss DNA, only one of the two DNA strands was used. The duplication of a unique ss DNA molecule may, however, occur in the cell, before or after integration. If the transient gene expression observed following transfection were the result of the phage DNA duplication, one would expect the (+) and (-) strands to induce the same level of transient expression.

The fact that in transient expression the (-) strand induced 3 times more HBs Ag than the (+) strand strongly suggests that the (-) strand is able to be transcribed directly into mRNA without going through a ds DNA intermediate.

ACKNOWLEDGEMENTS

We wish to thank Dr. S. Michelson for critical reading of the manuscript. We are grateful to S. Perrot for her excellent technical help, to P. Picouet for photography and N. Perrin for typing the manuscript. This work was supported by grants from CNRS (ATP N° 955.339), M.I.R. (grant N° 84V816) and CCAR (N°84 1938/RD.G.G.E.C).

REFERENCES

Aden D.P., Fogel A., Plotkin S., Damjanov I. & Knowles B.: Controlled synthesis of HBs antigen in a differentiated human liver-carcinoma-derived cell line. Nature, 1979 **282**, 615-616.

Colbère-Garapin F., Horodniceanu F., Kourilsky P. & Garapin A.C.: A new dominant hybrid selective marker for higher eukaryotic cells. J. Mol. Biol. 1981, **150**, 1-14.

Colbère-Garapin F., Horaud F., Kourilsky P. & Garapin A.C.: Comparative expression of the hepatitis B surface antigen gene in biochemically transformed human, simian and murine cells. J. Gen. Virol. 1985, **66**, 1741-1752.

Colbère-Garapin F., Ryhiner M.L., Stephany I., Kourilsky P. & Garapin A.C.: Patterns of integration of exogenous DNA sequences transfected into mammalian cells of primate and rodent origin. Gene, 1986, **50**, 279-288.

Gamper H., Lehman N., Piette J. & Hearst J.E.: Purification of circular DNA using benzoylated naphthoylated DEAE-cellulose. DNA, 1985, **4**, 157-160.

Gritz L. & Davies J.: Plasmid encoded hygromycin B resistance : the sequence of hygromycin B phosphotransferase gene and its expression in E. coli and Saccharomyces cerevisiae. Gene, 1983, **25**, 179-188.

Hirt B.: Selective extraction of polyoma DNA from infected mouse cell cultures. J. Mol. Biol. 1967, **26** 365-369.

Kramer W., Valerij Drut S.A., Jansen H.W., Kramer B., Pflugfeloder M. & Fritz H.J.: The gapped duplex DNA approach to oligonucleotide-directed mutation construction. Nuc. Acid. Res. 1984, **12**, 9441-9456.

Messing J., Gronenborn B., Muller-Hill B. & Hofschneider P.H.: Filamentous coliphage M13 as a cloning vehicle : insertion of a Hind III fragment of the lac regulatory region in M13 replicative form in vitro. Proc. Natl. Acad. Sci. USA, 1977, **74**, 3642-3646.

Rauth S., Song K.Y., Ayares D., Wallace L., Moore P.D. & Kucherlapati R.: Transfection and homologous recombination involving single stranded DNA substrates in mammalian cells and nuclear extracts. Proc. Natl Acad. Sci. USA, 1986 **83**, 5587-5591.

Sheldrick P., Laithier M., Lando D. & Ryhiner M.L.: Infectious DNA from herpes simplex virus : infectivity of double-stranded and single-stranded molecules. Proc. Natl Acad. Sci. USA, 1973, **70**, 3621-3625.

Skala H., Garapin A.C. & Colbère-Garapin F.: The transient expression of genes following transfection : dependence on the cell line. Ann. Virol. Inst. Pasteur, 1986, **137E**, 13-26.

Southern E.M.: Detection of specific sequences among DNA fragments separated by gel electrophoresis. J. Mol. Biol. 1975, **98**, 503-517.

Colbère-Garapin, F. : <u>Integration of exogenous DNA sequences trans-</u>
<u>fected into mammalian cells</u>

Spier :   Florence, could you tell us the levels of expression of
           hepatitis surface antigen in micrograms per 10^6 cells per
           day.

Colbère-Garapin : It is about one microgram under the best conditions
           but I think it is about twenty times less than what is
           obtained in CHO cells because in CHO, the gene is ampli-
           fied.

Spier :   And that is a consistent expression on a daily basis ?

Colbère-Garapin : Yes. What you can do with these cells, which is in
           my opinion quite interesting, is that you can feed the
           cells without trypsinizing them and I showed you the graph
           of HBs antigen expression after trypsinization. If you
           trypsinize the cells, the level of expression is very low.
           You have to feed the cells without trypsinizing them and
           under these conditions I have been able to keep some cells
           for about one year. Under these circumstances you can keep
           a very high level of antigen expression.

Horaud :   This happens in VERO cells ?

Colbère-Garapin : Yes.

Sherman : May I ask you about that last experiment, when you claim
           that your experiments demonstrated that you could get
           expression of single stranded DNA. I am not quite sure how
           you did that experiment.

Colbère-Garapin : We transfected with the single stranded DNA. Some
cells were transfected with the plus DNA of the gene, some
cells were transfected with the minus form of the gene and
some cells were transfected with double stranded DNA.

Sherman :   Right, and you were just measuring the amounts of expression ? Under these conditions I can't really see how you
can conclude, based on that, that you are getting expression of single stranded DNA.

Colbère-Garapin : If it was duplex DNA, you would get the same level
with the plus and the minus strand. It would be expressed
equally.

Sherman :   But those are in different test tubes and different
cells...

Colbère-Garapin : This is an average from several experiments...

Sherman :   For instance, when you did the experiment, let's say with
the plus strand construct, when you are going to measure
the complementary messenger RNA, did you look at those
same cells and measure for the same strand ? In other
words, if you are getting transcription of one strand
only, of the single stranded DNA, then you will only have
the one strand transcribed, you would not see the complementary one transcribed.
If you made double stranded DNA, might you not have, seen
complementary, both strands or do you have promoters on
both sides ?

Colbère-Garapin : I don't know if I have promoters on both sides.

Sherman :   If you don't have promoters on both sides, I don't see how
you can draw the conclusion you draw.

Colbère-Garapin : This means that about 12 % of the molecules are
duplicated because theoretically, the plus strand couldn't
be transcribed into functional messenger RNA. This is
impossible anyway. It will give antisense RNA. Because
there is an expression, part of the molecules are duplicated and than transcribed. It will be the same thing for
the plus strand. So you have to make the difference.

Kedinger : It does not mean, I am sure that promoters could be single
stranded ?... I hope.

Colbère-Garapin : I did not say from where the transcription started.
I just said that there was expression.

Kedinger : In one of your slides, you showed that there could be
inversion of one gene compared to the other one. How
frequent is that event ?

Colbère-Garapin : You mean when we transfect the plasmid with the two
genes in tandem ?

Kedinger : Right, and in one of the cases you showed they were head-
to-tail instead of being head-to-head.

Colbère-Garapin : For HepG2 cells we used two plasmids with different
orientations of the HBs gene.

Baserga : Did you measure CAT or did you just put the sample in the
liquid scintillation counter ?

Colbère-Garapin : I didn't perform this part of the experiment myself.
I said that. That is somebody else's work but a lumino-
meter was used.

Baserga : Because the last time I tried to measure ATP with the
luciferase system, I had to buy a new scintillation

counter : the amounts of light produced, litterally stripped the electrode. Your system I guess is a different one.

Colbère-Garapin : You can use an X-Ray film.

Drillien : To come back to that single stranded DNA experiment, if you interpret the experiment in another way, supposing the freshly replicated DNA strand has a lower probability of being transcribed, could this possibly explain your results ? It is biased towards one of the strands that has been replicated.

Colbère-Garapin : The replication would be preferential for one strand rather...

Drillien : Suppose the transcription is occuring better on the strand that was not freshly replicated, the parental strand.

Colbère-Garapin : If you make the hypothesis that the strand is replicated, the situation should be the same for both strands anyway.

Drillien : Are you sure ?

Colbère-Garapin : Why ? It shouldn't.

Drillien : It might be something else besides transcription of single stranded DNA...
As far as the L cells are concerned, what is the stability of plasmids in L cells ? Is that peculiar to mouse cells or to L cells only ?

Colbère-Garapin : In my hands, the clones were quite stable. I know that PELLICER has said that about 50 % of his clones were unstable but in my hands 80 to 90 % of the clones were stable.

Baserga : If you keep the selection pressure...

Colbère-Garapin : Even if you don't keep it.

Drillien : Have you tried other mouse cell lines besides the L cells ?

Colbère-Garapin : I did not test all of these other cells. It may depend on the transfection technique whether you use carrier or not, how much DNA is used to transfect and so on...

Anonymous : We use carrier all the time.

Baserga : We use carrier all the time too, but if you put a gene like NEO or GPT and you then relieve the pressure, it disappears after the first passage.

Colbère-Garapin : I haven't done that experiment. Some have said that the clones are more stable when you don't use carrier. I don't know whether this is true or not.

Petricciani : You suggested in your data and other existing data show that a number of variables related to the integrated exogenous DNA, transfect among these species and cell type.
I wonder if you could speculate about the relative importance of species and make a comment - if you would - about the possible difference or the extent to which differences are demonstrated between aneuploid cell lines such as the ones you used and normal human diploid cells.

Colbère-Garapin : I have no experience with the human diploid cells because they have a short life, so you cannot isolate clones with these cells. I would be very careful concerning the influence of the species specificity on the

phenomenon that we observe. I don't say that this is primate-specific. But it is true that for the four out of five cell lines that we have studied, all of them integrated less foreign DNA than the mouse cells themselves and probably on the average, primate cells are less permissive to stable transformation than rodent cells. Certainly it would be necessary to study many more cell lines to know whether this is true or not.

Chowdhury : This is not a question but I want to bring your attention on a new method of transfection which can be directed at specific tissues or cells. In this particular case, the DNA is linked to asialo orosomucoid and it is directed to the liver which is the only cell in the mammal having the capacity to take up the galactose-terminated glycoproteins. This method has not only been used in vitro for Hep G2 cells but also used in vivo to transfect living liver and have CAT activity expressed, albeit transiently.

Colbère-Garapin : There is the prospect of obtaining a stable expression in Hep G2 cells or only a transient expression ?

Chowdhury : It is transient in both at this time.

Hope : One question actually directed at to Baserga. Did I hear him rightly that you said that if you removed the gene G418 selective presure, you would rapidly loose expression in your experience ?

Baserga : These are Balb/C 3T3 cells.

Hope : All right, because I was going to say from the user of these cells on a large scale, G418 is very expensive and if one is to make 20.000 L of medium with G418 in it, it doesn't work. I have personally dealt with G418 selective cells which were grown for periods considerably in excess

of ten generations with no loss of expression in the absence of G418.

Baserga :   Which cells ?

Hope :   They were L cells.

COOPERATION OF ONCOGENES IN MALIGNANT TRANSFORMATION

D. STEHELIN

Institut Pasteur, Unité d'Oncologie Moléculaire, F - 59019 LILLE Cedex

SUMMARY

A molecular approach to Cancer was hampered for a long time by the lack of identification of genetic targets involved in the multistep processes leading to the outgrowth of tumours. A dramatic new field of research was begun about twelve years ago with the discovery of (cellular) proto-oncogenes. Such genes appear constitutively present in our chromosomes, they have been well conserved during the evolution of Metazoans and serve important functions in cell growth, regulation, development and/or differentiation. They can undergo spontaneous or induced alterations leading to active "oncogenes" able to sustain abnormal cell growth and possibly induction of tumours. More than 30 proto-oncogenes have now been isolated and characterized. Several abnormalities have been deciphered, leading to activated oncogenes expressing mutated products and/or abnormally elevated levels of gene products. Combinations of such oncogenes have been shown to exert cooperation in the transformation process of certain tissues. Despite the complex patterns of alterations involved at the molecular levels, an impressing amount of knowledge has been accumulated, some of which may already be useful to design tools suitable for clinical use.

I will briefly review some of the landmark experiments in oncogene research as well as some prospects for possible medical applications.

A. O. A. Miller (ed.), Advanced Research on Animal Cell Technology, 277–299.
© 1989 by Kluwer Academic Publishers.

Burnet said a long time ago that the more he knew about the fine structure of the mammalian cell and the complexity of intracellular and intercellular interactions, the more he was convinced that detailed understanding of the malignant process and practical control of cancer based on understanding would never be possible.

This is not encouraging for a young scientist starting cancer research and I wonder if we should ever say "never" !

Burnet was however certainly right in the fact that man represents a rather complicated machine. It must be recalled that our organism is made out of some 60 thousands billions cells, 200 different tissues, billions of new cells are made every hour, about half a billion km of new DNA is made per day by each of us (this represents about the distance from earth to sun).

So, if there are potential abnormalities, they will be dugged in our organism. Each haploid cell in our body contains some $3x10^9$ nucleotide pairs one fraction only of this number being phenotypically expressed. In these cells there are some 50.000 structural genes and of course each type of cell expresses its specific programme.

We now know that there are potential cancer genes (protooncogenes) in our cells, a fact which was not evident twelve years ago. I recall that the first way cellular oncogenes were deciphered was by the use of retroviruses and I was fortunate enough to discover the first of these cellular oncogenes while working in Bishop's laboratory in California.

We were able to show that Rous sarcoma virus was probably initially not a highly transforming virus but became so by stealing a tiny little fragment of cellular chromosome, putting then this sequence under its own viral control. Consequently when such a virus infected cells, it was then able to put their growth under control of its own promoter. This resulted in the appearance of the transformed state of the cells.

Many different oncogenes have been isolated from retroviruses.

The other means by which cellular oncogenes have been found is mainly by the use of initiated cells like in the 3T3 test (these are mouse cells which are no longer "normal" : they are somehow initiated. If one puts tumor DNA in them, it is possible in some instances after transfection to obtain foci on the plates and one can pick such foci and try to fish for the oncogene which appeared in these cells.

A third way which is now largely being used is based on the chromosomal translocation characteristic of many cancers. The chances are - and we shall see two examples - that at the rupture point there may be a protooncogene that becomes activated following the translocation. So, by cloning these translocation rupture points, several people have been able to fish out genes that can now be called protooncogenes.

Over 30 such oncogenes have been found. They share the following common properties : these protooncogenes are normal genes. They exist usually as 1 copy per haploid genome. Using one such oncogene as a probe, it is often possible to fish out other family members. So, for many of these oncogens, we now know that there are families of related genes in our chromosomes. These genes are expressed in certain normal tissues. They have been conserved throughout the evolution of vertebrates (in all metazoans yet tested) for over 500 millions years.
From this last observation, three important consequences derive.

First, these genes have been kept because they most probably fulfilled important functions in cell growth and/or differentiation.

Second if a protooncogene is identified in an animal species, it is relatively easy to clone within a few months the homolog of this gene present in the human genome.

Third, if such genes are implicated in natural cancers, such cancers may proceed by a similar mechanism both in man and in the animal.

One can wonder whether all the oncogenes products represent the same kind of molecules (big family of kinases for instance). This is not so. It is possible to classify the different oncogenes in several classes (Fig. 1). An oncogene like sis which is PDGF's β chain codes for a growth factor or has an homology with growth factor. Some oncognes specify growth factor receptors like erbB1 which codes for the EGF receptor, CSF-1 and other ones, some gene products are cytoplasmic, at the inner face of the membrane, tyrosine kinases and this is the src family that has now but ten members. Some like the ras family and related genes have GTPAse activity, some are cytoplasmic like the serine-threonine kinases mil, raf or mos. Finally, gene products have a nuclear localization and may be gene activators. That is the case for myb, myc, fos, erb-A. Erb-A was recently shown by Vennström's group in Heidelberg, (in collaboration with Jacques Ghysdael from our laboratory) to be one of the receptors for the thyroid hormones.

If you have at hand oncogenes or oncogene probes, it is important to try and localize them on human chromosomes for the reason I have just told : many cancers have translocations or chromosomal abnormalities and it is interesting to know where these oncogenes are located to study if one could make the association between the presence of a protooncogene at a given place in a chromosome and a specific break of this chromosome in a given transformed or cancer cell. On the whole, oncogenes are scattered throughout the genome (Fig. 2). None have been found yet on chromosome Y but all the other chromosomes have oncogenes. Oncogenes located on a

Protooncogene localization within the cell	Type of protooncogene	Protooncogene product
External	sis	PDGF βchain (homology with a growth factor)
Transmembrane	erb B1 fms ros	EGF receptor CSF-1 receptor Insulin receptor (homology with a growth factor receptor)
Intracellular : innerside of the cytoplasmic membrane	src yes fes/fps ras	Tyrosine kinase activity GTPAse activity (homology with G proteins)
Cytoplasm	mos mil/raf	Serine/threonine kinase activity
nucleus	myc myb fos erb-A jun	Gene activators ? Receptor for thyroid hormones T_3, T_4 Transcription regulator AP1
nucleus	ets rel	Not yet classified

FIGURE 1. Classification of protoncogenes according to their cellular localization within the cell.

given chromosome do not belong to the same class. Expressed in other terms, oncogene products such as those displaying a nuclear localization have their genes scattered throughout the genome.

What are the mechanisms or factors acting on protooncogenes which activate them into cancer genes ? Fig. 3 gives an idea of their diversity.

Chromosome	Band	Protooncogene
1	p 11 - p 13	N-ras*
	p 36.1	fgr
	q 21 - q 23	ski
2	p 23 - p 24	N-myc*
3	p 25	raf (mil)-1**
5	q 33	fms (Rec. CSF-1)
6	q 22 - q 23	myb**
7	p 12 - p 14	erB-1 (Rec. EGF)**
	q 21 - q 31	met*
8	q 11.2	mos
	q 24.1	myc**
9	q 34.1	abl
11	p 15	Ha-ras-1
	q 23.3	ets-1**
12	p 11 - p 12	Ki-ras-2
14	q 21 - q 31	fos
15	q 26.1	fes
17	p 13	p 53*
	q 11 - q 12	erbA-1**
	q 11 - q 22	erbB-2 (neu)*
20	q 13.3	src
21	q 22.3	ets-2
22	q 11	bcr*
	q 13.1	sis (PDGFβ)

FIGURE 2.   The human protooncogene map.
** Protooncogenes discovered or codiscovered in our laboratory in Lille.
   mil : Jean Coll ; myb, ets-1 : Dominique Leprince and Anne Gégonne ;
   myc, erb-A and erb-B : Martine Roussel and Simon Saule.
* No known viral equivalent.

Activation results from
- modifications of the <u>NATURE</u> of the gene product
   <u>Mutations</u> : ras, erbB.2, HER, neu.

- modifications of the <u>AMOUNTS</u> of the gene product
   <u>Translocations</u> : c-myc, abl
   <u>Amplifications</u> : c-myc, N-myc, erb.B2, HER, neu
                   erb.B.1, myb, ras, abl.
   <u>RNA accumulation</u> : c-myc

FIGURE 3. Modes of protooncogene activation.

Activation of the protooncogene may result from <u>modification</u> of the very nature of the gene product namely the protein. This can result from point mutations, deletions, substitutions, additions or whatever nucleotide modifications we know of. Alternatively, <u>increased dose</u> of the product can also lead to activation. This may result from translocations, amplification or abnormal accumulation of RNA. This means that ten times more activator can be enough to induce a cell to grow anarchically. When I say increase, we now know that in some cases it can be a decrease. In retinoblastoma for example, - what have been called with misuse I believe anti-oncogenes-, there may be negative regulators of growth functions. This may be an example of special cases where environmental factors regulate the expression of oncogene products.

Whether changes occur in the <u>nature</u> or in the <u>dose</u> of the oncogene product, this can result in the stimulation of abnormal cell growth. Of course these two processes can coexist and exert cumulative effects.

I'd like to give a few examples of some of the best known activation mechanisms.

The first and also the best studied case was probably the activation of the ras gene, specially the Harvey ras gene for which has clearly been shown the existence of two critical regions out of the 189 amino acids (Fig. 4). These are positions 12 and 61. Glycine in position 12 or glutamine in position 61 can be changed by almost any other amino acid except I believe proline. This unvariably will result in activation which can be scored by the famous NIH 3T3 test. Clearly, DNA mutated at position 12 or 61 or at both positions together, is able to induce the appearance of foci in these cells. This is a very convenient way to analyze the influence a given type of mutation has on activation.

Position 13 has been seen <u>in vivo</u> in some leukaemias to be potentially involved in the leukaemogenesis process but this has not yet been firmly established.

This means that in the Harvey ras gene product, there are two critical positions. Having established this, it was interesting to see if chemical carcinogenesis or radiations or other means known to promote transformation could be used to alter positions 12 or 61.

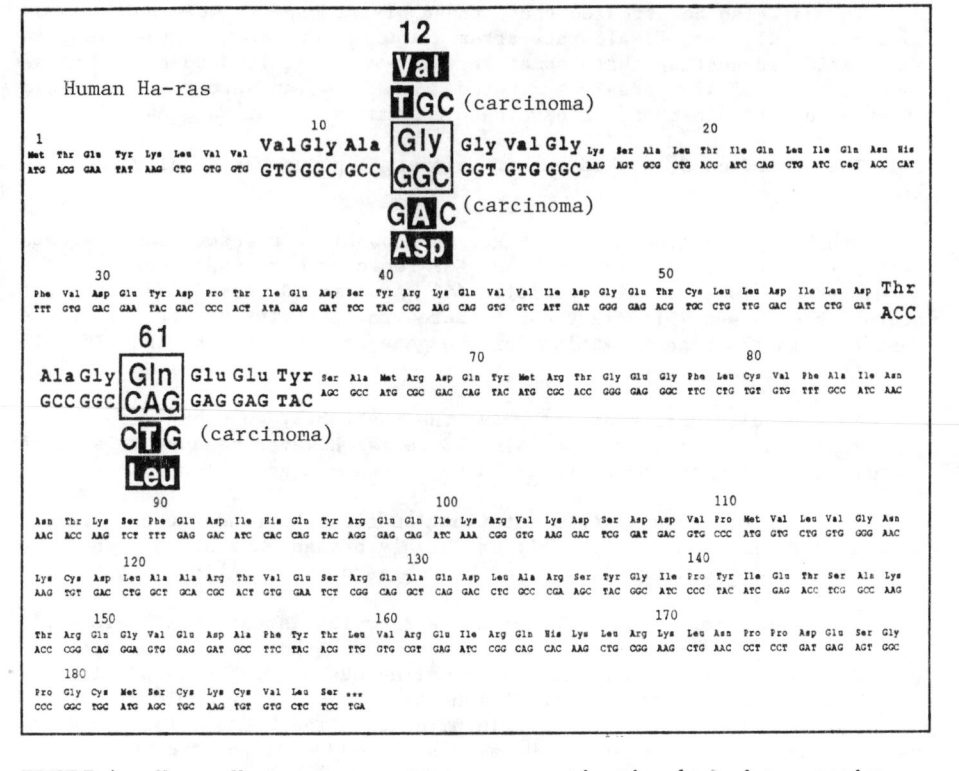

FIGURE 4. Human Harvey ras protooncogene activation by point mutations.

That this was indeed possible was shown very nicely by Barbacid's group. Using Sprague Dawley rats, he was able to show that injection of nitrosomethyl urea into newly born female animals invariably resulted in mammary carcinomas. In 60 cases out the 60 experiments performed, modification of the ras gene occured at position 12.

In these experiments, activation due to modification at position 61 was never observed. This is easy to understand by people familiar with chemical carcinogenesis and has to do with the type of nucleotides involved and the mechanism of modification.

If another carcinogen is taken such as dimethylbenzanthracene, again breast carcinoma are obtained. This time, it is the position 61 which is systematically modified but never position 12.

This was the first unambiguous demonstration that in specialized cells, oncogenes were the target for chemical carcinogens and second, that different modifications could be obtained within the same gene depending upon the nature of the carcinogen being used.

Interestingly enough, both modifications led to the same type of tumour which by the way may not always be the case.

It must also be stressed that these breast cancers were only developing in newly born female rats after allowing for sexual maturation, an observation suggesting that something else was required to allow these tumours to manifest themselves. Indeed, animals which were ovariectomized after such a treatment never developed a mammary breast carcinoma.

The second case, worked out in Weinberg's group was equally compelling.

ErbB2, also called neu or HER.2 is probably a transmembrane growth factor receptor. It looks like the EGF receptor but isn't one. In the normal gene exists a T in position 664. Associated specially with rat neuroblastomas and glioblastomas, there are transitions from T to A resulting in the transformation of the gene as detected by the NIH 3T3 transfection assay test.

Such results show that not only the ras family gene but also other gene families can be scored on NIH 3T3 cells. However because this test is negative for many oncogenes it is by no means a general one.

In addition, because this mutation suffices to render the molecule transforming in these cells, it is highly probable that the oncogene product is also responsible for rat neuroblastomas and glioblastomas.

The reason for me introducing these examples is that what Weinberg's group studied was less likely to happen at the homologous position in man. At this position there is also a valine but the codon is GTC instead of GTG which means that the transition from T to A here would introduce an aspartic acid residue and not glutamic acid. The NIH 3T3 test remains negative with such a molecule. However a second mutation from C to G in the same codon makes now the test positive.

This observation is potentially interesting because it suggests that men having managed to survive for many decades could have succeeded in doing so by protecting themselves at critical positions by requiring two mutations against one for the rat.

It would be interesting to see whether other long-lived animals also show this kind of "protection" mechanism.

This is all I want to tell about activation by mutations and I won't talk about activation resulting from overtranscription because the mechanisms are so far still very poorly understood.

Translocations : there are now several well documented cases of activation of oncogenes at the breakpoint of chromosomal translocations the best example being that of the Burkitt lymphoma, a special cancer of the jaws. It is characterized in all cases hitherto studied by translocation of a fragment of chromosome 8 (q24) which contains the protooncogene c-myc to either chromosome 14 or chromosome 2 (sometimes also chromosome 22).

Another well documented case is that of chronic myelogenous leukaemia. It involves a reciprocal translocation between chromosome 9 (which carries protooncogene abl normally poorly expressed in myeloid cells) and chromosome 22 (which carries the gene bcr, active in normal cells of the myelord lineage).
As a result of the chromosomal rearrangement, the Abelson (abl) gene in the transformed cellis now found next to the ber gene on chromosome 22, under the latter constitutive control. So, once bcr-Abelson chimaeric RNA molecules are made, giving rise to the corresponding chimaeric protein, there is activation of the Abelson protein by the bcr first two exons. And we know that there are at least two things which result from the translocation. The first is the fact that the strong bcr promoter which is expressed constitutively in normal conditions in such myeloid cells becomes positioned upstream of the Abelson gene which normally is poorly expressed. Second, in order to function, Abelson's gene needs to have something changed at its 5'- end and that is exactly what this chimaeric molecule is doing. There results a very high level of p 210 tyrosine kinase activity whereas the original product of the Abelson gene has a very low tyrosine kinase activity.

Here we have a very good example which tells us about at least one step of modification responsible of chronic myelogenous leukaemia. This is probably not the only step. Indeed, in some acute leukemia happens another phenomenon : the probable activation of the N-ras gene.

Cancer arising from amplification of certain oncogenes has been shown in several instances and especially during the late stages of carcinogenesis (amplification does not seem to happen often in early tumours).

Many advanced tumours show deregulation of the DNA replication mechanism. As a result, some sequences in the chromosome, instead of being present as one copy per genome, become amplified several hundred times, giving rise to the so-called Homogeneously Stained Regions (HSR. Fig. 5).

If the amplified sequence contains an oncogene, its transcription will clearly be more abundant.
Alternatively, if there are mutations leading to oncogene activation, the chances of a successful hit leading to malignancy, increase accordingly.

These are a few simplistic considerations of how given oncogenes can become activated. Retroviruses of course are potent oncogenic means. They are however exceedingly abnormal because they make 500 times more RNA than the normal gene and they have had the time to accumulate mutations such as deletions, substitutions etc within the viral oncogene. As a consequence, their potent oncogenic properties which result from multiple changes, bear little resemblance with the original cellular gene that went into the virus.

Each modification which takes place in a given cell is stably inherited by all daughter cells. Some of them in turn undergo additional modifications which again are passed on to cells of the next generations.

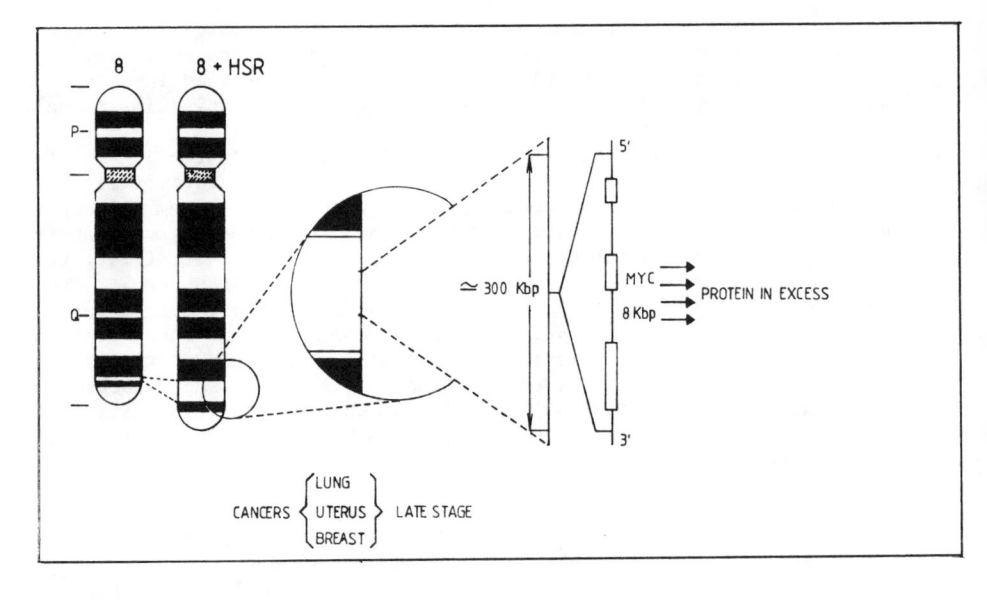

FIGURE 5. Amplification of c-myc giving rise to HSR.

An accumulation of metabolic abnormalities evolves from this sequence of events. Together they ultimately can cooperate so as to destabilize a normal cell and make it become malignant.

These multistep processes can happen during a long period of time extending from several months or years to several decades, leading to the increase of tumours observed after the fourties. In the case of papillomas or of cancers of the cervix, this may take as long as 30 years for the multiple modifications to happen before one actually gets the cancer.

As shown in Fig. 6, these different phases can be delimeated, called initiation, promotion and progression.

In molecular terms, it is however extremely difficult to know what is really an initiator, a promoter and a progressor.

Like in many other laboratories, ours was interested in these steps. Questions such as - If there were oncogenes involved, would it be possible to show the same oncogenes playing always the role of initiator, others being always progressors or would there be more than one oncogene involved ? - are clearly of prime importance.

I shall try to answer these questions.

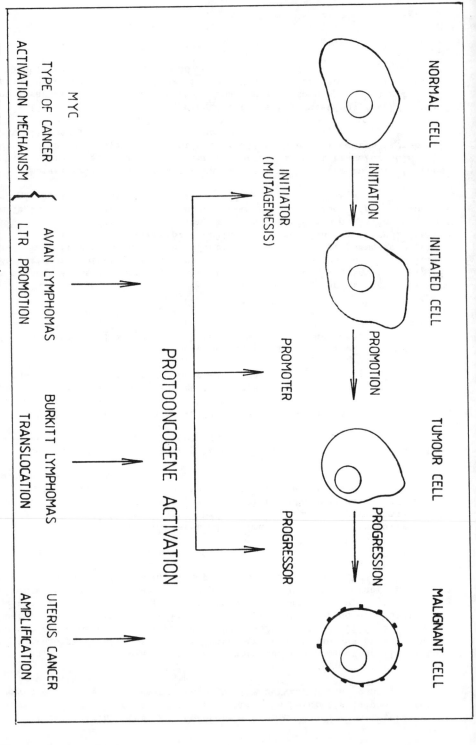

Fig. 6. Multi step tumour progression.

It is striking that an oncogene like myc can be an initiator at least in avian lymphomas but is a promoter in Burkitt's lymphoma and in many other cancers including cancer of the cervix, breast cancer etc..., it is a progressor.

So, the same oncogene can play sometimes the role of an initiator, of a promoter or of a progressor. Moreover, the mechanisms by which activation proceeds do not seem to be the same in those different cases. Whether this is true in the whole or not is yet under debate but it may be that if a given oncogene acts as an initiator, it will do so by a given mechanism, if it acts as a promoter it could be another mechanism and so also if it is acting as a progressor.

This is the case for myc.

When it acts as an initiator in avian lymphomas, it is governed by LTR promotion. If it works as a promoter as in Burkitt lymphomas, this results from a specific translocation yielding RNA over accumulation. As a progressor like in cancer of the cervix, it does so as a result of DNA amplification (Fig. 6).

In conclusion, the same gene can be involved in different steps using for that purpose different molecular mechanisms.

At this point, one must realize that if somebody comes with a given tumour, it is impossible for the molecular biologist to determine the initial mechanisms having led to the manifestations of malignancy, these mechanisms having happened a long time ago.

What is nice today is the possibility to dispose of reconstituted in vitro and in vivo systems which allow you to characterize the first steps delineating which type of activation is actually taking place at what moment and in which type of cell.

That is actually what many groups are doing and I shall try to analyze the best reconstitution systems worked out by Weinberg, Wigler, Barbacid and others, using one or the other of the three following systems : transfection of NIH 3T3 cells with the ras protooncogene, which is an easy way to transform cells, transfection of normal embryo fibroblasts and finally injection of such transfected cells in the nude mice in order to see if one can get tumours or not. These involve increasingly efficient ways to transform cells.

Normal cellular ras protooncogene does not perform in any of these three systems. Mutated in position 12 or 61, it will transform NIH 3T3, will do the same very poorly or not at all with normal embryo fibroblasts but will not give any tumour in the nude mice.

If instead of one or the other of these two mutations, an LTR viral promoter is put in front of the normal ras protooncogene. Then foci of transformed cells will appear in the NIH 3T3 test and some rat embryo fibroblasts will be transformed.

In conclusion, mutation or LTR promotion allows activation of the cells.

What happens if the abnormal ras cellular protooncogene is simultaneously mutated either at position 12 or 61 and promoted by an LTR ? The answer is that there is more effect than by using either one or the other activation alone.

These results suggest that within a single oncogene, there can be several ways to activate that gene - here at least two - and that the effects produced by both activations together will be more important than those produced by each activation alone.

This situation is reminiscent of multistep carcinogenesis acting on the same gene.

Another way to evaluate cooperation is to use two distinct oncogenes which are both sligthly modified. Such is the case with ras activated at position 12 and a sligthly activated myc gene. In those tests, the myc gene alone performs negatively. It does not give rise nor to foci nor to tumour. However cotransfection of normal rat embryo fibroblasts with mutated ras and activated myc results in the appearance of transformed cells which upon injection into the nude mice induce the formation of tumours (Fig. 7).

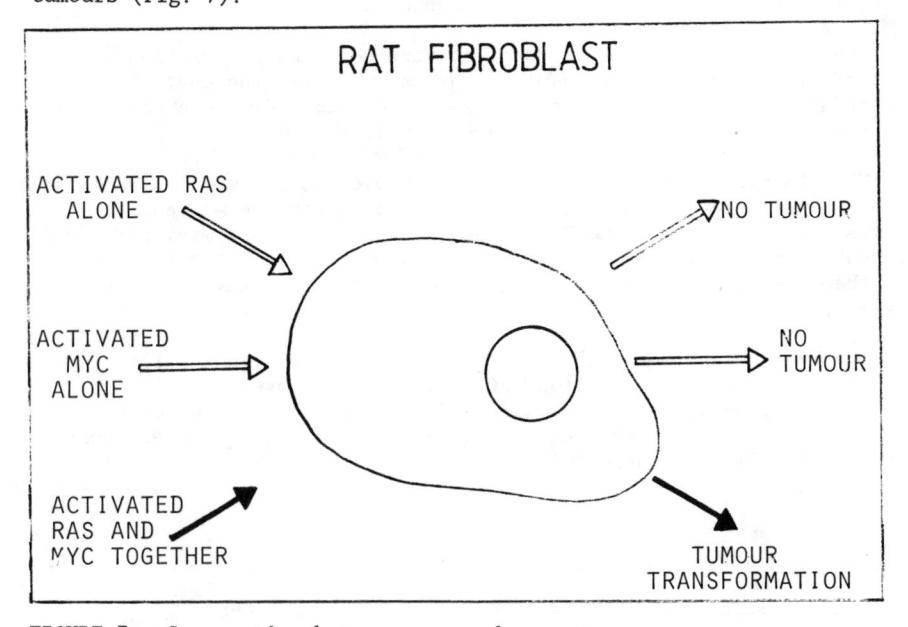

FIGURE 7. Cooperation between ras and myc oncogenes.

In other words, it is possible to obtain cooperation of the effects of these oncogenes which now transform cells in a way that none of them alone could do. The technique used in these experiments clearly allows one to monitor these effects in vitro.

Whether these results obtained by in vitro experiments are meaningful, immediately raised the necessity to develop similar in vivo tests.
This was very elegantly done by Leder's group using transgenic mice and
the MMTV promoter - enhancer which can be more or less specifically
activated by glucocorticoids[1].

Having put myc or ras under the MMTV - glucocorticoid activator
promoter, mice were then made transgenic with that construction.

In the case of mice having constitutively only the MMTV - myc construction, the percentage of tumour - free females versus the number of
days the animals have been living is high during a considerable period
(Fig. 8).

This means that even though all the cells of these transgenic
animals contain an MMTV promoter activated gene, they nevertheless can
live happily for months. However, when they become older, some of these
animals come down with tumours. Control, untreated animals stay free of
tumors at that age (not shown). These experiments clearly show that the
predisposition of mice to tumours has been enhanced.

The very type of the tumour which appear - mammary gland
adenocarcinoma - is probably related to the glucocorticoid-specific MMTV
enhancer promoter used in these experiments.

The same overall time-dependent pattern of tumour formation can be
observed with mice made transgenic with ras put unter the control of the
MMTV enhancer-promoter. At an older age, such animal die more frequently
from tumours than did MMTV-myc-transgenic carriers.

What happens following crosses between myc and ras mice ? When this
is done transgenic mice are obtained which have both the myc and the ras
genes activated by the same MMTV promoter. In this case many more mice
come down now with tumours and this happens much earlier in the animal's
life then when the animals had either one of the two activated protooncogene alone.

The results tell us that in vivo, by using transgenic mice, it is
possible to obtain cooperation of these two oncogenes. Mice become
increasingly predispoed to mammary adenocarcinomas. These experiments
also show that to some extent the ras and myc in vitro transformation
system of NIH 3T3 is meaningful.

---

(1) The specificity is not absolute. These sequences are activated best in breast cells for these
contain many glucocorticoid receptors.

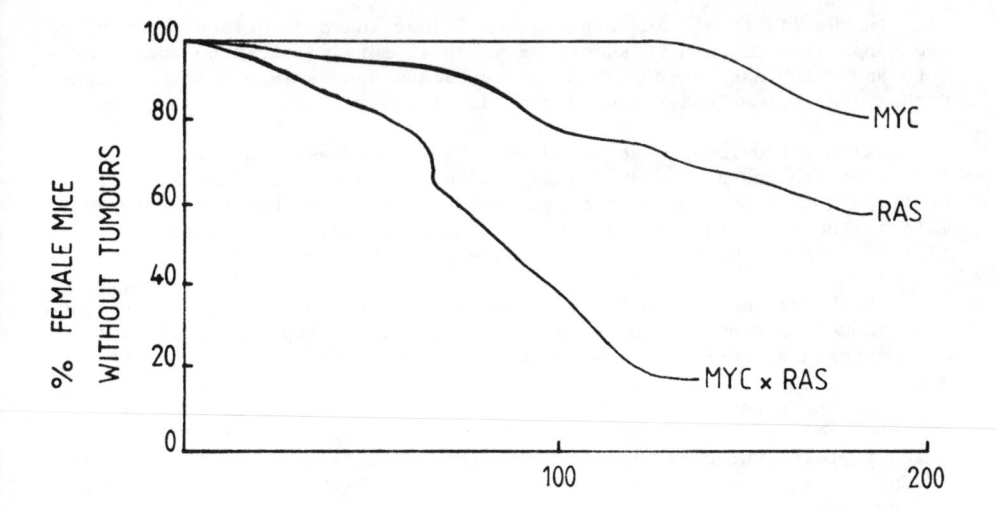

FIGURE 8. Frequency of tumours in female mice made transgenic with MMTV-myc(-ras).

It was interesting at that point to know what would happen if one of the two oncogenes ras or myc were put under control of another promoter. This is what did Adams from Australia who replaced the MMTV - glucocorticoïd promoter by the immunoglobulin - enhancer promoter. Again the mice were pretty happy for about 80 days and then began to come down with tumours. However in this case, tumours were not of the mammary carcinoma type but of the B - lymphocyte lineage.

This clearly showed that for a gene like myc, a change in the tissue specificity of the enhancer - promoter, changes the type of tumour accordingly.

It is by no means certain that these conclusions can be generalized namely that the same results will be obtained using other known onco- genes. Indeed, in some tissue, many other oncogenes just do not exhibit anything detectable even when they are placed under the control of very strong promoters.

In the light of the experiments I just have described to you, I would now like to turn to something which is more specific to our laboratory and which has been sorted out by Simon Saule, Jean Cole, Fabienne Denhez, Christine Dosier and Patrick Martin.

Using retroviruses, we studied other putatively cooperative oncogenes. We showed a few years ago with the group of Thomas Graf in Heidelberg that some retroviruses captured not one but two oncogenes. This is the case for acutely transforming leukaemia viruses. The Mill Hill n° 2 virus (MH2) is a good example of that class of viruses.

This virus makes multiple polyclonal tumors of macrophagetype and also induces sarcomas. In vitro it transforms macrophages, fibroblasts and neuroretina cells. It has in its genome the mil-raf oncogene together with myc.
The oncogene products are made from two distinct mRNAs in two distinct proteins. Mil yields a fusion protein with a viral determinant, a gag-mil p 100 protein which is localized in the cytoplasm while the myc gene product is made from the subgenomic RNA as a p 62-63 protein localized in the nucleus.

Understanding the contribution made by each of these two oncogene products in transformation relies on mutants expressing in the correct way only one or the other gene product.

We isolated mutants which expressed only mil or only myc gene products and then tested them on different cell types in order to find an in vitro biological test for each of these two oncogenes.

Using chicken neuroretinal cells and plating them in dishes, it is possible to monitor their life time (duration of the culture).

MH2 (wild type) - infected neuroretina cells are quickly induced - within a few days - to resume cell growth. Uninfected control cells on the other hand, stick to the plate and ultimately die.

What is (are) the gene(s) responsible for induction of cell growth ? Is it myc or mil or both ? With our mutants this was an easy study to perform.

All the mil expressing vectors were able to perform almost as well as the wild type virus whereas the myc mutants like clone 16 or clone 25 were not doing much more than uninfected cells. This means that stimulation of the growth of neuroretinal cells is obtained through expression of the mil gene product.

We then tried to develop another test which would score myc mutants and possibly not mil mutants. This could be done using chick embryo fibroblasts since we knew these cells can be transformed by the wild type virus that has both oncogenes, also by myc expressors but not by mil expressors. The mil gene indeed does not transform fibroblasts.

Disposing of biological tests to detect the effects of each oncogene product, we thus started the experiments by trying to demonstrate cooperation between these two genes.

If on the one hand neuroretinal cells are tested for proliferation and on the second hand for transformation using the soft agar cloning technique, it is possible to see that the wild type virus is able to induce proliferation and simultaneously transform such cells.

Mil expressors induce proliferation but do not induce foci in agar. Myc alone doesn't seem to do much under the same circumstances. However careful analysis of the data seems to indicate that the wild type virus is able to do something that the mil expressor is unable to do, which tells you that maybe cells induced to proliferate by the mil gene product may become now sensitive to further transformation using superinfection with myc expressors.

In conclusion, seven day-old chicken embryo neuroretina cells are quiescent. Induced with v-mil, they will start to grow. At that point they are not yet transformed under these conditions. If they are now superinfected with myc expressor, transformation occurs. This happens very quickly.

These in vitro tests are very nice indeed because they allow one to rapidly evaluate within the myc or the mil gene the part responsible for cancer.

Does cooperation work in several cell types and if yes, is the order of oncogene product action important ? The answer to these two questions is yes : the two same genes are able to cooperate for the transformation of macrophages.

These experiments were made with Thomas Graf in Heidelberg. Taking normal chicken macrophages, they can become transformed by v-myc expressors. At that point they are mildly tumorigenic, they are not metastatic and they depend for their growth in culture upon the presence of the CMG factor. In absence of the latter, cells simply won't grow. Now, if these cells are superinfected with the mil expressor mutant, not only do all the cells grow very well in culture, but they are now tumorigenic, highly metastatic and have become independent of CMGF : they have become autocrine for this factor.

We first thought that maybe the mil gene product is CMGF is not true. Mil does something in the cell that leads to growth factor independence. We still can't explain in molecular terms this action.

What these experiments clearly show is that in culture the two same oncogenes can cooperatively transform two completely distinct cell types.

We first started this talk by saying that the myc oncogene could cooperate with the ras gene. Now we clearly showed that myc is able to cooperate with mil, which tells you that a given oncogene can choose different partners to transform different cell types.

Fibroblasts are the target for myc and ras while myc and mil choose neuroretinal cells and chicken macrophages.

It is important indeed to try and establish the cell specificity of each of the 30 different oncogenes or so known today as well as the molecular mechanisms underlying activation and the type of combinations leading to cooperation.

What we do not know yet is if we can find tumors that exhibit abnormal forms of the two protooncogenes myc and mil in spontaneous tumors in man. We can expect this to happen but have'nt seen that so far. This has already been shown for myc and ras.

There is another virus, the avian acute leukemia virus also called avian erythroblastosis virus which induces very strong in vivo erythro-blastosis. Infection by this virus kills 100 % of the birds within three to four weeks. In vitro, it transforms erythroblasts and fibroblasts and again that virus has two oncogenes : erb-A and erb-B.

Here again there are two independent strategies : it is possible to obtain the erb-A product as a gag-erb-A fusion product gp75 which is of nuclear origin and the erb-B gene product which arises from subgenomic RNA as a transmembrane glycoprotein. It is necessary to remember here that the erb-A gene product is a truncated from of a thyroid hormone receptor.

Again several people in different laboratories made mutants which expressed either erb-A or erb-B gene products.

The results can be summarized as follows : erb-B gene product is a transmembrane glycoprotein and has kinase activity. In the virus, the N terminus which specifies the binding for the growth factor has been lost. It is supposed that removal of the N terminus from the cellular protooncogene makes the gene product constitutively active. The receptor for the growth factor is no longer present and the conformation of that transmembrane glycoprotein is such that its protein kinase activity is permanently switched on. At the C-terminus of the viral gene as compared to the cellular protooncogene another piece is also lacking. We know that this segment in the cellular protooncogene is a negative regulator of protein kinase activity.

So, with erb-B oncogene the virus has accumulated two distinct modifications each of which is activating the protein kinase constitutively.

The erb-A gene product also found in a truncated form in the virus.

The hormone binding domain is situated at the C-terminus. It is this C-terminus which is truncated in the virus. Again it is supposed that erb-A gene product molecule becomes constitutively active. As the other end the absence of another fragment leads to the disappearance of anothis negative regulator in the virus. In addition to this, then are two more places in the virus which have also been mutated and whose mutation lead to loss of negative control. In the erb-A viral product we have at least

four modifications all of which are known to be activators of the hormone receptor. So, the avian erythroblastosis virus has managed to put together two oncogenes and to make altogether six distinct modifications inducing such a dramatic situation as erythroblastosis.

It must be remembered that virus mutants which only express erb-B make a kind of weak erythroblastosis : erythroid cells have a hard time to mature but they ultimately do so. erb-A alone will make erythroblasts grow nicely a little bit more than they should but does not prevent differentiation. Both genes together block differentiation at a very early stage when the immature cells have reached the CFU stage. At that stage they will grow very actively and that probably results from these two genes being present together, their togetherness promoting erythroleukemia much more readily than when each oncogene is present alone.

The above examples show us which genes can cooperate. They also show by a comparison of the viral gene products with their normal counterpart the structure modifications which are putative positive regulators or inducers.

I shall speak now of some potential applications . Modest applications of oncogenes technology can be found in the field of diagnosis more precisely in the diagnosis of activated protooncogenes.

If one knows what kind of position is dangerous for oncogenes like position 12 in Harvey ras, it is possible to devise a specific test. it is known that ras genes in the whole are involved in more than 20 % of all human tumours. Ras mutation is not specific of one tumour type but has been detected in many different tumours.

Some ras-related genes may be more specific of a given disease and may serve to orient therapy. In the case for instance for polyps of the colon the following is being tested : in these cells if ras is induced at position 12, maybe immediate resection should be made if the prognosis is shown to be very bad. If these polyps do not have the modifed ras gene, prognosis could be shown to be better.

Another possible application is in the case of the family - inherited cancer called retinoblastoma (there are very few such cancers). Retinoblastoma in its family form is probably regulated by the absence of an inhibitor coded by a gene on chromosome 13 at position q14. The absence of one allele in the embryo causes a potential risk for retinoblastoma. Cancer is induced by the absence of the two alleles which occurs sporadically in predisposed children. There are now probes to test this absent allele. Such probes could possibly serve in the future for early genetic counselling and prevent the birth of children affected by this chromosome abnormality.

Having probes to detect certain oncogenes such as activated neu in some breast cancers could allow therapy modulation, that is the anticipated chemotherapeutic treatment of these cancers which, if neu is activated, may develop unusually fast.

These few examples show that there are many potential applications where the knowledge gained on oncogenes can advantageously be used.

In conclusion, most researchers involved in molecular oncology hope that the future opened by the discovery of oncogenes may be more promising that Burnet anticipated.

Stehelin, D. : Cooperation of oncogenes in malignant transformation

Sherman :   I wonder whether you would comment on whether you need to
             have constant expression of these various oncogenes or
             whether these oncogenes are acting as trigger and even if
             you were able to design a therapy that would suppress the
             action of these oncogenes that wouldn't result in a cure.

Stehelin :   This is a very good question. It is very hard to answer
             that question in vivo because you just don't know what
             happens. But there are many experiments that have been
             done in vitro in cell cultures where you can take normal
             fibroblasts per example and you can induce these fibro-
             blasts with the Ts viral oncogene. You can switch the Ts
             to a non permissive temperature. After a few days, a few
             weeks, a few months, and this is a way - sort of - to shut
             off the gene at least if you designed the experiment
             correctly and I think that experiments performed along
             these lines - of course in many cases where you shut off
             the expression of the gene, the cells become normal again
             - but if you wait long enough and you put these cells in
             culture, by having allowed them to grow for many weeks,
             you will see that if you shut down the expression of that
             oncogene, some cells are now modified well enough to
             become anarchic and induce tumorigenesis. I don't know if
             I answered your question but I think there is a message
             behind that which is : if you allow a cell to undergo many
             more passages than that cell should, it is dangerous.

Sato :   It seems to me that the use of oncogenes is a very inte-
             resting way of getting at the physiology of the tissues. I
             am not sure that anyone has known that EGF and thyroid
             hormone are involved in erythropoeisis. If one could
             accumulate more information like this, one might be able
             to use these markers to pull out the pluripotent stem
             cells that everybody has been looking for.

Van Hamme : First, is it possible to obtain differentiation of neuroretina cells and second, if yes, do you observe different
effects of mil alone and myc alone or myc and mil on the
differentiation of these cells ?

Stehelin : This is a very good question indeed. Actually, there are
several cell types in the neuroretina. There are glial
cells, Muller cells, pigmented cells and neuronal cells.
We are looking for the effects of mil and myc on these
types of cells. It is very difficult to make a general
statement because this has been shown in vitro and we have
no clonal populations to make certain that individual
clones would not have benefited from other factors released by other cells around these ones.
Roughly, mil and myc both seem to amplify the compartment
of undifferentiated cells. It does not seem to prevent
cell maturation and myc has been known in many examples
not to prevent maturation. In certain cases such as
pigmented cells, myc is probably a maturation factor : myc
put on these neuroretina cells, the cells being left
afterwards during one month in culture, you will see the
appearance of pigmented cells that just begin to pop out
although you never get this with mil.
Mil prevents that, myc allows that. It is hard to make
generalizations.

Cassiman : When you do these transfections, these double transfections, do you have any idea what percentage of the total
population will respond to the double transfection ?

Stehelin : Yes, I think we are working under very special conditions.
They are optimal which is that we use viral vectors and
viral vectors are very efficient to infect I would say
close to 100 % of the cells.

Cassiman :   I don't mean the number of cells transfected but the cells responding to the transfection, transforming or forming tumors... Is that 80 % of the population ?

Stehelin :   It is hard to have a good number for that but it is probably most of these. It is very hard because it is difficult to evaluate cell death in the very early steps. But I would say that under these conditions many of these cells respond and this is judged by immunofluorescence using antibodies directed against the viral products. So, you can see even the next day already how many cells are positive although you get cell death but I would say certainly more than 50 %.

Chowdhury : You told us about many of the potential applications of the oncogenes. But I hope you didn't really mean that you want to leave behind a colonic polyp that does not have an oncogene detectable in it... Because oncogenes are not found in all potential malignant say - villous adenomas - and one polyp can have it and maybe another does not and sometimes the oncogenes are detectable in advanced cancer which you don't want to have.

Stehelin :   You are absolutely right. It is really hard to give many experiments in three quarters of an hour. I made probably too many understatements or made things a little bit too simple. But it is interesting to consider for example that in familial polyps, ras seems to be an event before cancer really pops out. And with skin papillomas, according to some, ras seems to be present once it is a cancer but not in the precancerous form. So that tells you that the same oncogene may be again a promoter in one case and a progressor in another. I am not trying to make the story simple. It is very complicated.

Part VI.

Applications of Animal Cell Cultures

MOLECULAR DISSECTION OF THE CELL CYCLE

R. BASERGA

Department of Pathology and Fels Research Institute, Temple University School of Medicine, Philadelphia, PA 19140 USA

ABSTRACT.
There are essentially three approaches to the identification of genes that play a major regulatory role in the control of proliferation in animal cells: 1) the identification of proto-oncogenes, i.e. the cellular equivalents of retroviral transforming genes; 2) the molecular cloning of the wild-type alleles complementing temperature-sensitive (ts) defects in cell cycle specific ts mutants, and 3) the identification by differential screening of cDNA libraries of growth regulated genes, i.e. genes that are inducible by growth factors or are cell cycle regulated. Introduction of these genes under the control of appropriate promoters can influence the growth characteristics of the cells so that they can be grown in the absence of defined growth factors.

INTRODUCTION.
1. As discussed in this research workshop the growth of cells in culture can be improved by manipulating the substrate, or the environment (stimulatory and inhibitory growth factors) that control the extent of proliferation of animal cells. Another way to approach the problem of increasing the growth of cells in culture is, instead of manipulating the environment, to manipulate the genetics of the cells, for instance, by making cells independent of growth factors which are often expensive and tedious to purify. Cell lines have occasionally arisen spontaneously that grow in the complete absence of growth factors in the medium, usually because they produce their own growth factors. However, one cannot depend on spontaneous mutations occurring in cells in cultures and it is desirable to know how to manipulate the cells so that they can grow efficiently in the absence of the known growth factors. In theory this is not too difficult to accomplish. For instance, it is well known that c-myc can replace platelet derived growth factor (PDGF) in stimulating cellular DNA synthesis in quiescent Balb/c3T3 cells in culture (Kaczmarek et al. 1985). Microinjection of the c-myc protein into quiescent Balb/c3T3 cells stimulated cellular DNA synthesis provided that platelet poor plasma was added to the cells after microinjection. Platelet poor plasma (PPP) by itself had no stimulatory effect on cellular DNA synthesis in quiescnet Balb/c3T3 cells. Kaczmarek et al. (1986) also reported similar results by microinjecting, instead of c-myc, a gene containing the coding sequence of the p53 protein under the control of a strong viral promoter. Here again the microinjection of the p53 gene, presumably overexpressed by the controlling viral promoter, resulted in stimulation of cellular DNA synthesis but only when PPP was added to the cultures after microinjection of the chimeric p53 gene. Overexpression was also the end result of the microinjection of the c-myc protein since it can be calculated that in those circumstances about 1 million molecules of c-myc proteins were microinjected into the nuclei of the quiescent cells.

303

A. O. A. Miller (ed.), Advanced Research on Animal Cell Technology, 303–314.
© 1989 by Kluwer Academic Publishers.

These experiments suggest that either the p53 gene, or the c-myc gene, when properly integrated into the cellular genome, could make cells grow in the absence of at least one growth factor (PDGF). Indeed, in my laboratory Xiao Xia Gai has developed a cell line in which the p53 gene, under the control of the polyoma promoter, is overexpressed and this cell line can grow efficiently in 5% horse serum, which does not support the growth of the parent cell line, Balb/c3T3 (unpublished data). Interestingly, in the experiments described above with the microinjection of the c-myc protein, or the p53 gene, PPP could be replaced by purified insulin-like growth factor 1 (IGF-1). It should be possible, therefore, to grow appropriate cell lines in serum-free media completely devoid of growth factors, provided that the cell line will express either c-myc or p53 and IGF-1.

However, if we really wish to develop growth-factor independent cell lines that have the characteristics we desire we may have to use different genes for different cells. I am basing this statement on the facts that there is increasing evidence of genomic redundancy in animal cells, and that certain genes are expressed in a cell type-specific manner. At any rate one can simply compare fibroblasts (usually requiring PDGF) and lymphocytes (in which Interleukin-2 replaces PDGF) to realize that different cells may have totally different requirements for growth. To identify the genes that properly overexpressed would allow cells to grow in the absence of growth factors, we therefore have to understand what are the genes that control the proliferation of animal cells.

2. Identification of growth regulatory genes: There are three major approaches to the identification of growth regulatory genes, i.e. of genes that play a major role in the proliferation of animal cells. These approaches are: 1) the identification of proto-oncogenes, i.e. the cellular equivalents of retroviral genes.
2) the molecular cloning of the wild-type alleles that complement ts defects in cell cycle specific ts mutants, and
3) the identification by differential screening of cDNA libraries of growth regulated genes, i.e. genes that are inducible by growth factors or are cell cycle regulated.
Though all three approaches are important and necessary, none of them by itself is sufficient.

a. Cellular Oncogenes: It seems reasonable to assume that cellular oncogenes must have something to do with the regulation of animal cell proliferation since when they are altered, either by point mutations, deletions, or insertions into a retroviral vector, or when they are overexpressed, they can transform cells in culture (for reviews see Bishop, 1983 and 1987, and Land et al. 1983). We can say, therefore, that an alteration in the structure or function of a cellular oncogene can result in an alteration in the regulation of cell growth.

b. Temperature-sensitive Mutants of the Cell Cycle: It is obvious that the genes of cell cycle-specific ts mutants like cellular oncogenes, also have something to do with cell proliferation since when their products are defective (at the non-permissive temperature) the cell stops at a precise point in the cell cycle. Molecular cloning of the wild-type alleles of the genes responsible for these ts mutants is therefore of extreme importance since it will identify a class of genes that are essential for cell cycle progression.

c. Growth Regulated Genes: By growth regulated genes we mean genes whose expression is induced by growth factors, or mitogens, or varies during different phase of the cell cycle. A number of genes have been shown to be cell cycle dependent, i.e. to be expressed in a cell cycle dependent manner (for a review see Kaczmarek 1986). A good illustration are the core histones (Plumb et al. 1983). Other growth regulated genes have been identified as cDNA clones by differential screening of cDNA libraries (Linzer and Nathans, 1983; Cochran et al. 1983; Hirschhorn et al. 1984; Edwards et al. 1985; Lau and Nathans, 1985). This approach identifies genes that are growth regulated, not that are growth regulatory. However, the importance of growth regulated genes in cell proliferation is strengthened by the finding that several bona fide oncogenes are expressed in a cell cycle dependent manner (Kelly et al. 1983; Greenberg and Ziff, 1984; Reich and Levine, 1984; Campisi et al. 1984). The fact that these proto-oncogenes are inducible by growth factors does not necessarily mean that they regulate the cell cycle. However, these proto-oncogenes when properly activated, as mentioned above, are known to transform cells and therefore they must have some influence on growth regulation. One can turn this statement around and say that if oncogenes can be induced by growth factors then one can look at other growth factor inducible cellular genes and perhaps identify possible candidates for potential new oncogenes or, at any rate, genes that play a major role in the control of cell proliferation. For instance, the p53 gene, whicn is now generally accepted as a bona fide oncogene (Eliyahu et al. 1984), was originally demonstrated to be growth regulated. Only subsequently it was shown to be capable of substituting c-myc in transformation experiments. Interestingly enough no p53 sequences, or sequences even remotely related to the p53 sequence, have ever been detected in a retroviral transforming gene. It seems, therefore, that some potentially very important genes have never been picked up by retroviruses and, therefore, that our search for growth regulatory genes has to extend beyond the genes that are found in retroviruses.

3. Growth Regulated Genes: I will discuss here three growth-regulated genes that have been identified and molecularly cloned in my laboratory.
a. Calcyclin. Calcyclin was originally isolated as a growth regulated cDNA clone called 2A9 (Hirschhorn et al. 1984). The human cDNA clone of 2A9 was isolated from an Okayama-Berg library and was fully sequenced. It consists of 434 bp with an open reading frame of 270 nucleotides coding for a putative polypeptide of 90 amino acids (Calabretta et al. 1986). The open reading frame has 55% homology with the coding sequence of both α and β subunits of the rat S-100 protein (Kuwano et al. 1984). The homology between 2A9 and the subunits of S-100 is particularly striking in the regions which, in the S-100 sequence, code for the calcium binding sites (Van Eldik et al. 1982). Here the homology reaches 91%. There are also homologies between the deduced amino acid sequence of 2A9 and that of the p11 subunit of a protein complex that serves as the major cellular substrate for tyrosine kinases (Gerke and Weber, 1985; Glenney and Tack, 1985). Calcyclin is inducible by serum, PDGF and EGF, but not by insulin and platelet poor plasma (Calabretta et al.1986). The 2A9 gene has been isolated and completely sequenced (Ferrari et al.1987). The promoter of the calcyclin gene has a dual regulation. It is inducible by PDGF but it also contains negative regulatory elements that are down regulated by epidermal growth factor (Ghezzo et al. in preparation).

b. Vimentin: The human cDNA clone isolated from the Okayama-Berg library that was called 4F1 turned out to be, after sequencing, the vimentin gene (Ferrari et al. 1986). Vimentin is one of the important components of the

intermediate filaments family which also include desmin and keratins with which, indeed, vimentin has substantial homologies.

c. 2F1. The human cDNA clone isolated as a 1.5 kb clone and called 2F1, was fully sequenced and turned out to be a human ATP/ADP translocase (Battini et al. 1986). The ATP/ADP translocator is one of the most important proteins in the energy metabolism of the cell. It sits at the mitochondrial membrane where it transfers ATP and ADP from the mitochondrial to the cytosol and vice versa.

4. Importance of Late G_1 Genes: In a recent review (Baserga and Ferrari, 1987) I have pointed out that it may be possible that the most important and specific genes controlling the proliferation of cells may be those that are expressed in late G_1 or at the G_1-S boundary. The evidence has been summarized in that review. It includes the lack of specificity of early events, and the ability of certain genes that are supposedly growth regulated, to be induced also by differentiating agents which actually turn off cell proliferation. In addition one platelet derived growth factor can be replaced quite easily by very aspecific stimuli like, for instance, a pulse of cycloheximide. Based on the evidence presented in the review by Baserga and Ferrari (1987) we formulated the following hypothesis: All events, after stimulation of G_0 cells by growth factors, are part of a common response of cells to a variety of environmental signals including proliferative stimuli, while events occurring in late G_1 are specific for cell proliferation. It is, though, an important part of our hypothesis that in ordinary circumstances the proliferation specific events cannot take place unless the cell has been previously primed by its initial response.

This hypothesis is strongly supported by the fact that at least two DNA viruses, SV40 and adenovirus 2, can bypass the temperature sensitive blocks in G_1 specific ts mutants of the cell cycle (for review, see Baserga 1985). These experiments indicated that DNA oncogenic viruses could bypass the G_1-ts block and induce DNA synthesis in the absence of events that are required by serum stimulated cells. It suggested the hypothesis that these DNA viruses activate late G_1 genes or genes that are at the G_1-S boundary and that control the duplication of the genetic material that characterizes S phase. This hypothesis turns out to be substantially correct. Liu et al. (1985) looked at the RNA levels of a panel of cell cycle genes, including three cell cycle dependent oncogenes in tsAF8 cells, and in 3T3 cells infected with adenovirus. Liu et al. (1985) concluded that the cell cycle genes activated by adenovirus are a subset of the cell cycle genes activated by serum and that, with one exception, the cell cycle genes activated by adenovirus were late G_1 or S-phase genes. These results suggested that activation of late G_1 genes may be sufficient for the starting of S-phase.

5.   CONCLUSIONS

By manipulating the genetic make-up of the cells it should be possible to produce cell lines with the desired characteristics, but which are capable of growing in media completely devoid of growth factors. The choice of the genes necessary for the growth of cells in media completely devoid of growth factors, must be based on a rational selection that will allow the growth of cells without interfering with the characteristics that make them important to the cell biologist. The recent developments in the area of cell cycle studies seem to indicate that late G1, or G1-S phase boundary genes may play

very important and specific roles in the control of animal cell proliferation. One could hypothesize that using these genes it would be possible, so to speak, to short-circuit the cell cycle and create cell lines that eliminate most of the G1 period and can grow in the absence of growth factors.

REFERENCES

1. Baserga, R. The Biology of Cell Reproduction. Harvard Univ. Press. Cambridge, MA 1985
2. Baserga, R, Ferrari S: Haematologica 72:1-4, 1987
3. Battini E, Ferrari S, Kaczmarek L, Calabretta B, Chen S-T and Baserga R: J. Biol. Chem. 262:4355-4359, 1987
4. Bishop JM: Ann. Review of Biochemistry 52:301-354, 1983
5. Bishop JM: Science 235:305-311, 1987
6. Bravo R, and Celis JE: J. Cell. Biol. 84:795-802, 1980
7. Calabretta B, Battini R, Kaczmarek L, deRiel JK, and Baserga R: J. Biol. Chem. 261:12628-12632, 1986.
8. Campisi J, Gray HE, Pardee AB, Dean M and Sonenshein JE: Cell 361:241-247, 1984
9. Cochran BH, Reffel AC and Stiles CD: Cell 33:939-947, 1983
10. Edwards DK, Parfett CLJ, and Denhardt DT: Mol. & Cell Biol. 5:3280-3288, 1985
11. Eliyahu DA, Raz A, Gruss P, Givol D, and Oren M: Nature 312:646-649, 1984
12. Ferrari S, Battini R, Kaczmarek L, Rittling S, Calabretta B, de Riel JK, Philiponis V, Wei J-F, and Baserga R: Mol. Cell. Biol. 6:3614-3620, 1986
13. Ferrari S, Calabretta B, deRiel JK, Battini R, Ghezzo F, Lauret E, Griffin C, Emanuel BS, Gurrieri F, and Baserga R: J. Biol. Chem. 262: 8325-8332, 1987
14. Gerke V, and Weber K: EMBO J. 4:2917-2920, 1985
15. Glenney JR Jr, and Tack BF: Proc. Natl. Acad. Sci USA 82:7884-7888, 1985
16. Greenberg MD, and Ziff EB: Nature 311:433-438, 1984
17. Hirschhorn RR, Aller P, Yuan Z-A, Gibson CW and Baserga R: Proc. Natl. Acad. Sci 81:6004-6008, 1984
18. Kaczmarek L: Lab. Investigation 54:365-376, 1986
19. Kaczmarek L, Hyland JK, Watt R, Rosenberg M and Baserga R: Science 1313-1315, 1985
20. Kaczmarek L, Oren M and Baserga R. Exp. Cell Res. 162:268-272, 1986
21. Kelly K, Cochran BH, Stiles CD, and Leder P. Cell 35:603-610, 1983
22. Kuwano R, Usui H, Maeda T, Fukui T, Yamanari N, Ohtsuke E, Ikehara M, and Takahashi Y: Nucleic Acids Res. 12:7455-7465, 1984
23. Land H, Parada LF, and Weinberg RA: Science 222:771-778, 1983
24. Lau LF, and Nathans D: EMBO J 4:3145-3151, 1985
25. Linzer DIH and Nathans D: Proc. Natl. Acad. Sci. 80:4271-4275, 1983
26. Liu H-T, Baserga R and Mercer WE: Mol. & Cell. Biol. 5:2936-2942, 1985
27. Maniatis T, Fritch EF and Sambrook J: Molecular Cloning, a Laboratory Manual, Cold Spring Harbor Lab. Cold Spring Harbor, NY 1982
28. Plumb M, Stein J and Stein J: Nucleic Acids Res. 11:2391-2410, 1983
29. Reich NC and Levine AJ: Nature 308:199-201, 1984
30. Van Eldik LJ, Zendegui JG, Marshak DR and Watterson DM: Int. Rev. Cytol. 77:1-61, 1982

This work was supported by grants GM33694 from the National Institutes of Health and CD-214 from the American Cancer Society.

Baserga, R. : <u>Molecular dissection of the cell cycle</u>

Sherman : Renato, this is maybe completely wrong, do you see any
relationship between these two critical - what you foresee
as critical phases - the early G0-G1 transition and the
late G1 and the ideas of components in progression. Is it
possible that the G0 genes are the competent genes and
that G1 genes are the progression...

Baserga : It depends of how you made your model. If you make it
very tight it usually does not last very long. It dies in
a very few hours. If you make it very loose then it can
continue. By competence it is understood that you have to
add one growth factor and then the other otherwise it
doesn't work. As such it is not competent any longer.
Competence prevention theory seems to indicate you need
one growth factor, no question about that. The
explanation as far as I am concerned is that these early
events simply prepare the cell for a second signal. But
it is not something like one, two, three, four, it is more
like... if you have a walk in the bank to mortgage your
home, you go to the bank, you know that you need two keys
to open the door one does not do it by itself, the key of
the bank doesn't do it by himself. And I think this is
what happens. You need two keys to open the bolts of DNA
synthesis.

Sherman : But are the two sets of genes that you are looking at,
which might change differentially, be the consequence of
the two growth factors ?

Baserga : We have been looking at the regulation of the thymidine
kinase promoter and of the cyclin promoter which are part
of the DNA synthesizing machinery. We were hoping that

they would be about the same.   That would have made
everything easy because then the only thing to do is to
isolate the protein that binds to these two promoters and
you have the two keys.

Well, they behave differently.   But the thymidine kinase
promoter does need the activation of the early genes
c-myc, c-fos, p53. Now whenever we will be able to find
the products of c-myc and p53 on the thymidine kinase
promoter gene, I do not know. But it does need activation. Cyclin does not.   The cyclin promoter is merely
activated by EGF only. Thymidine kinase needs PDGF

Spier :   I have been fascinated by these protein phosphorylations
            which seem to take place and yet throughout your talk, you
            didn't mention them at all. Do you think they are comple-
            tely irrelevant ?

Baserga :   No, of course not. Even in a plenary lecture of about one
            hour you cannot talk of everything. But if you want to
            hear me think out loudly...

            I think that phosphorylation has more to do with cell
            growth than with cell proliferation. Cell growth being
            growth in size. Now, a cell to divide has to grow in size
            for if a cell does not double you obtain two smaller cells
            and if you do that very frequently you are not going to
            have many cells left. The cell has to double, to divide
            and keep the same size. When you come into cellular DNA
            synthesis in total absence of cell growth, adenovirus
            infection for instance, I think the control of DNA syn-
            thesis is independant of cell growth. Any reasonable cell
            will not accept the signal for cellular DNA synthesis
            unless it has first accepted the signal for cell growth or
            it is a kind of suicide so to say.   I think that under
            physiological conditions, the two are coordinated but I
            think that phosphorylation has more to do with cell
            growth. But growth in size rather than proliferation.   I
            have no evidence for it all right.

Spier :   Can we just come back and think about what is happening
             with these protein phosphorylations. They are obviously
             very quick reactions and obviously cascade throughout the
             cell because you get the tyrosine kinases phosphorylating
             proteins which phosphorylate proteins. Why should the
             phosphorylation of proteins be - as you are referring -
             specifically or even generally connected with increase in
             cell mass or increase in cell size ?

Baserga :   Suppose you are making two copies of everything instead of
             one copy. Double in size. A good method is for instance
             to phosphorylate certain enzymes and make them a little
             bit more active. There has been persistent reports for
             instance that when you stimulate cells to proliferate, RNA
             PolI and RNA PolII are phosphorylated. This has been
             going back to years ago. So, it is possible that these
             phosphorylations seem to have the effect of doubling the
             number of molecules in a cell. I am far from ready to
             defend this theory...

Kedinger :   Have people looked for viral mutants which do not stimu-
             late cell growth or DNA synthesis ? In other words, what
             are the viral factors which are involved in cell growth ?

Baserga :   You mean, if there are genes that are induced by viruses ?
             Yes, this has been done actively but is not yet published.
             It was one of these thesis that the postdoc stretched the
             truth a little bit and so the chef de laboratoire has
             retracted it.

Miller :   During mitosis there is no protein synthesis.

Baserga :   Not quite - nuclear proteins are synthesized during
             mitosis.

Miller :   It is said that there are stored forms of messenger RNA
             accumulated in the cytoplasm. Wouldn't these be the early
             events that might make the cell ready to receive these
             signals needed for DNA replication ?

Baserga :   This is in exponentially growing cells ?

Miller :   Yes.

Baserga :   In fact you can do without G1 as everybody knows, you can
             eliminate G1 if the cell has everything ready, it comes
             out of mitosis and goes straight into S phase. And that
             is probably because everything that is needed is there.
             For one thing that I always try to make it clear, G0, G1
             and G2 do not exist, they are simplified notations. K_M
             and V_{MAX} do not exist, K_M and V_{MAX} exist but if you want
             to describe an enzyme, K_M and V_{MAX} are useful notations.
             What exists are mitosis and DNA synthesis, not S phase.
             So we use these notations such as G1 simply to say that
             the cells come out of mitosis and now take eight hours to
             enter S phase. But that is simply langage. The truth of
             the matter is that everything that counts is cellular DNA
             synthesis, doubling in size and mitosis. The rest is
             conjecture.

Remacle :   In the same line and for kinase, could you comment about
             the role of calcium. It is difficult in such short a time
             of course...

Baserga :   I was very skeptical about the role of calcium until
             growth regulated genes turned out to be a calcium binding
             protein. So calcium suddenly acquired a lot of importance
             in my eyes. However, it seems to me so stupid that a cell
             should confine what is its most important activity which
             is reproduction to such - shall we say - an aspecific
             thing as an ion. You need it and there is no question in

my mind, but I still hope that a cell is intelligent enough to rely on specific bits of information.

Everybody makes a great confusion between things that are necessary for cell proliferation and things that control cell proliferation. Everybody, when they see one gene going up, they say : "Ah ! c-myc controls cell prolife-ration. The c-fos goes up ten minutes before c-myc and now it is c-fos which controls cell proliferation. I have been around for too many years and I have made even a list of things that control cell proliferation : cyclic AMP, cyclic GMP, histone deacetylation, histone phosphoryla-tion, histone dephosphorylation, ribosomal RNA synthesis, etc... I have a list of sixty four molecules by now and they are all necessary because without them you don't have cell proliferation but I want to say that up to now no one has found the real thing that controls cell proliferation. There are only molecules which are necessary which is a different thing.

Spier : Just one comment. You are thinking that calcium per se must contain information, the concentration of calcium and the spatial localization of calcium contain other bits of information as well.

Baserga : True, but for instance, in keratinocytes, actually calcium instead of stimulating, inhibits. So, in WI-38 cells, you can replace EGF with calcium. I can grow WI-38 cells for seven days in culture, to confluence with calcium, dexame-thasone and IGF1 (insulin - like growth factor 1). The only one factor that I can't get rid is IGF1 but every other growth factor can be eliminated.

Taylor : A trivial comment on that. The specificity of the calcium signal of course is very often conveyed by the specific protein that binds to, rather then calcium as an ion per se and so, acquires a considerable degree of specificity through its binding protein and of course you have a calcium - binding protein that you are interested in.

Miller :   I remember hybridization studies whereby mouse cells were fused I think with guinea pig cells or some other sort of cells bringing therefore additional ribosomal genes. These experiments were made in Henry HARRIS' laboratory. Ribosomal RNA genes of one species were down regulated as if the other parent brought some negative element. If some chromosome were lost, again both sorts of ribosomal RNA were again being expressed.

Baserga :   The experiments of Henry HARRIS if I remember, were published many years ago. What he did was to fuse HeLa cells with chicken erythrocytes. Chicken erythrocytes have a nucleus but don't make any RNA whatsoever. Zero. And he noticed that as soon as he fused the HeLa with the chick, the chick nucleus incorporated uridine. Right away. However he could not show the presence of chick specific proteins. After 24 hours, the chick nucleus swells, it makes nucleoli and then, new chick specific antigens appear. By knocking out the nucleoli with an UV microbeam, the appearance of chick-specific antigens could be inhibited. Although some of his hypothesis concerning messenger RNA were not quite right, the experiments do remain. If you realize - and again I am thinking out loudly now, sometimes you have to make provocative statements - if you now realize that Tom CECH has shown that RNA is self splicing and further than that, that ribosomal RNA can furnish elements for the splicing of messenger RNA - all these events occur in Tetrahymena - one is actually tempted to imagine that maybe you have transcription of chromosomal genes but you don't have the messenger RNA appear in the cytoplasm unless the ribosomal RNA comes in and splices it. There are some who claim that ribosomal RNA made after stimulation of cells, is not quite the same as the ribosomal RNA that is made by quiescent cells. You know that there are about 200 copies of ribosomal RNA genes but he says that these copies are not exactly alike,

they are different. They are very similar but different. You can say that some specific ribosomal RNA must be necessary for the processing of some of these genes. Again, I think it is interesting to throw out some ideas.

USE OF MICROBEADS FOR CELL TRANSPLANTATION

J. ROY CHOWDHURY*, N. ROY CHOWDHURY*, A.A. DEMETRIOU§ AND J.M. WILSON¶

*Department of Medicine, Albert Einstein College of Medicine, Bronx, NY;
§Department of Surgery, Vanderbilt University Medical Center, Nashville, TN, and
¶Department of Biology, Massachusetts Institute of Technology, Cambridge, MA.

INTRODUCTION

Transplantation of isolated hepatocytes and other animal cells have many potential therapeutic and investigative applications. Liver cell transplantation, if successful in man, could be used in managing acute liver failure, which at this time carries dismal prognosis. It could also possibly be used for replacement of specific deficient liver functions in inherited liver diseases. Finally, hepatocytes or other transplanted cells may have potential application as vehicles for somatic gene therapy.

During the past few years various methods of liver cell transplantation have been described (1-3). However, until recently, long-term survival and function of the transplanted hepatocytes were not unequivocally demonstrated in vivo (4,5). Mutant rats with inherited bilirubin-UDP-glucuronosyltransferase deficiency (Gunn strain), exhibit life-long jaundice due to failure to glucuronidate bilirubin (6). Gunn rats have been used in several laboratories for testing the function of transplanted hepatocytes (4,5,7). Intrasplenic injection of isolated normal liver cells into Gunn rats result in only transient biliary excretion of relatively small amounts of conjugated bilirubin (7). Thus, the results of liver cell transplantation have been largely disappointing.

For several years, dextran microbeads (Cytodex, Pharmacia) have been used as substrata for culture of mammalian cells. We hypothesized that attachment of isolated hepatocytes to microbeads prior to transplantation by injection into body cavities might support the transplanted cells long enough for develpment of vascular supply. Isolated liver cells attached to collagen-coated dextran microcarriers (MC; Cytodex 3, Pharmacia) were intraperitoneally injected in rats. Effectiveness of this method in promoting the survival and function of the transplanted liver cells was demonstrated in mutant rats with deficit of specific liver functions, such as conjugation of bilirubin or albumin synthesis (8-10), and in rats with acute liver failure induced by 90% hepatectomy. Here we also report the application of this technique in the transplantation of small intestinal epithelial cells (enterocytes) into Gunn rats, and retrovirally transduced fibroblasts into Wistar-RHA rats. Methods for storing microcarrier-attached liver cells from human and other species have been also described.

PROCEDURE

Materials

Animals: Outbred Wistar rats (200-250 g) were purchased from Charles River Breeding Laboratories (Cambridge, MA). Syngeneic Wistar (RHA) rats and congeneic Gunn rats, which have identical genetic make-up with Wistar (RHA) rats except for the bilirubin glucuronidation locus, and genetically athymic rats were developed by C. Hansen of the National Institutes of Health, Bethesda, MD and maintained as inbred strains at the Albert Einstein College of Medicine. Genetically analbuminemic rats (NAR) were

A. O. A. Miller (ed.), Advanced Research on Animal Cell Technology, 315–327.
© 1989 by Kluwer Academic Publishers.

provided by S. Nagase of Sasaki Institute, Tokyo, Japan, and maintained at the Albert Einstein College of Medicine. The male Gunn rats were bred with female athymic rats, and the offsprings (F_1) were inter-bred. Approximately one out of sixteen of the second generation offsprings (F_2) were both hyperbilirubinemic and athymic (as recognized by sparse hair and failure to respond to tuberculin test after BCG vaccination); these were maintained in a germ-free barrier facility at the Albert Einstein College of Medicine. All rats were maintained on a standard chow (Rodent Chow 5001, Purina) in a 12-hr light/dark cycle.

Chemicals and equipment: Cytodex 3 (dextran microcarrier beads, coated with collagen, type I) was purchased from Pharmacia (Piscataway, NJ). Collagenase, type I, betaglucuronidase and all chemicals for perfusion were obtained from Sigma (St. Louise, MO). Tissue culture media were purchased from Grand Island Biological Company (Grand Island, N/). Millipore filters were obtained from Nalgene (Rochester, NY). For fast protein liquid chromatography (FPLC) of serum, an FPLC apparatus (Pharmacia) equipped with a Mono Q (I.D. 5 mm, height 50 mm) anion exchange column was used.

Methods

Isolation of rat liver cells: Asceptic methods were used throughout. Liver cells were isolated by collagenase perfusion through the portal vein in situ by the method of Berry and Friend (11), modified by Seglin (12), as described elsewhere (10).

Transplantation of isolated rat liver cells: Details of this procedure have been published (10). In brief, Isolated liver cells were attached to MC by incubation for 90 min in Dulbecco's modified Eagle's Medium (DEM) containing 10% fetal calf serum. After this the cells were washed twice with DEM without fetal calf serum and 1-2 x 10^7 MC-attached cells were injected into the peritoneum of Gunn or NAR rats. Bile from transplanted Gunn rats was collected by bile duct cannulation at various intervals. Bilirubin and its conjugates in Gunn rat serum and bile were quantitated by h.p.l.c. or by a diazo method as described (8,10). Serum albumin in NAR rats was quantitated by serum electrophoresis, by immune electrophoresis (8-10) or by FPLC.

Subtotal (90%) hepatectomy (13): Ninety percent of the liver mass (left, median, right lower and right upper lobes) was removed from Sprague-Dawley rats under ether anesthesia; leaving only the caudate lobe. MC-attached liver cells (1 x 10^7) were transplanted either 4 days prior to, or at the time of 90% hepatectomy. Control experiments included injection of microcarriers or liver cells alone. All rats were provided 5% dextrose ad lib after surgery. Blood glucose in tail blood samples was determined preoperatively and at various intervals post operatively. Survival of the animals was followed hourly for the first 12 post operative hours and every 4-6 hr thereafter.

Isolation and transplantation of human liver cells (14): Portions of normal human liver were obtained from fresh cadavers of organ donors at the Vanderbilt University Medical Center with appropriate permissions. The liver was perfused as above through the tributories of the portal vein visible at the cut surface. The cells were attached to MC and injected into athymic Gunn rats. Bilirubin and its conjugates in serum and bile were analyzed as above.

Cryopreservation of liver cells: Rat or human liver cells were attached to MC as described and frozen at -170°C in DEM containing 10% fetal calf serum and 10% dimethyl sulfoxide. The MC-attached cells were thawed at various intervals up to 6 months and transplanted into Gunn or

NAR rats. Function of the transplanted cells was determined as described above.

Isolation and transplantation of normal rat enterocytes into Gunn rats (15): Enterocytes were isolated as follows: normal Wistar (RHA) rat small intestine was resected under ether anesthesia, perfused with ice-cold normal saline, filled with 27 mM sodium citrate, pH 7.4, containing 2 mM dithiothreitol and clamped at both ends. After incubation of the segment in phosphate buffered saline at 37^0 for 15 min, the intestine was emptied and the procedure was repeated. Following gentle palpation on a chilled plate, the clamps were opened and the released cells were collected. An enterocyte enriched fraction was collected after centrifugation (500 x g x 10 min), suspended in DMEM, attached to MC and 1.5 x 10^8 cells were transplanted into congeneic Gunn rats as described for hepatocytes. Bile collected through choledochal fistula was analyzed for bililrubin and its conjugates on days 4,7,11 and 20 post-transplantation.

Transplantation of rat skin fibroblasts after retroviral transduction: A helper-free amphotropic producer was used for preparation of a retroviral vector containing human parathyroid hormone (PTH) cDNA. Fibroblasts were isolated from Wistar (RHA) rat skin and cultured in DEM containing 10% fetal calf serum. Cytodex 3 was sterilized in 70% ethanol as outlined by the manufacturer and thoroughly washed in DEM containing 10% fetal calf serum. The fibroblasts were attached as described above for liver cells and cultured for 2-5 days. The cultured cells were infected for 12 hours with viral stocks (1-4 x 10^5 cfu/ml) containing 8 ug/ml of polybrene. The MC attached cells (1 x 10^7) were then injected intraperitoneally into syngeneic Wistar RHA recipient rats.

Human PTH levels in sera of the recipient rats were determined by radioimmunoassay at various intervals after transplantation using a species-specific antiserum.

Infection of isolated hepatocytes with "BAG" virus: A helper-free amphotropic producer of the BAG virus was provided by Dr. C. Cepko (Harvard University, Boston, MA). The retroviral vector used to make this producer has been described (16); it coexpresses E. coli beta-gactosidase and the gene that confers resistance to neomycin in prokaryocytes and to G418 in eukaryocytes (Neo). Isolated liver cells were cultured 2-4 days on plastic tissue culture plates coated with collagen type I. Hepatocytes were infected for 12 hours with the viral stock as described for fibroblasts.

Southern analysis: Cellular DNA was isolated, aliquots were digested with the restriction endonuclease Kpn I, restriction fragments were resolved by electrophoresis in 1% agarose gels and blotted by the method of Southern (17). The blot was probed with the Bam/Hind III fragment of the neomycin gene that was labelled to high specific activity with ^{32}P-dCTP using the random primer method (18).

Cytochemical and immuocytochemical procedures: Activity of beta-galactosidase was detected in situ with the substrate 5-bromo-4-chloro-3-indolyl-beta-D-galactoside which forms a blue precipitate in infected cells (16). Immunocytochemical localization of UDP-glucuronosyltransferase and asialoglycoprotein receptor was performed using corresponding antibodies, horseradish peroxidase-conjugated staphylococcal protein A and diamino-benzidine cytochemistry as previously described (19).

RESULTS

Organization and morphology of the transplanted liver cells:
Seventytwo hours after the transplantation, the transplanted cells, MC and connective tissue formed a conglomerate on the anterior surface of the

pancreas (Fig 1). Light and electron microscopic examination revealed the presence of morphologically intact liver cells within the conglomerate (9). Capillaries containing blood cells were observed within the conglomerates as early as 72 hr after transplantation (Fig 2).

Transplantation of rat liver cells in Gunn rats

In allogeneic recipients, serum bilirubin showed a slight decrease up to 6 days; this was followed by a return to pre-transplantation levels. Bilirubin monoglucuronide and diglucuronide were detectable in bile upto 5 days (8). In congeneic recipients, bilirubin glucuronides were present in bile for 6 weeks (last day of the experiment and serum bilirubin progressively decreased to near normal levels in 3 weeks (Fig 3).

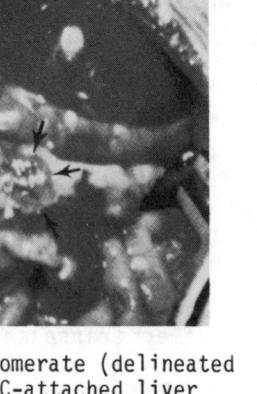

Figure 1. Conglomerate (delineated by arrows) of MC-attached liver cells on the anterior surface of the pancreas.

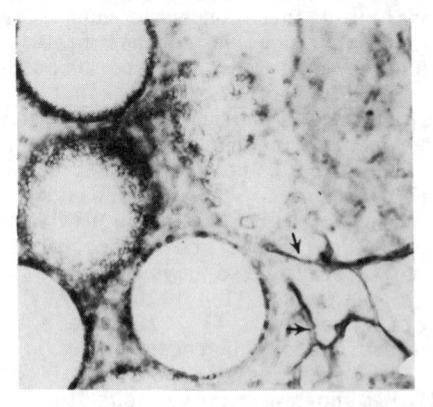

Figure 2. Light microscopic view of a section of conglomerates stained for nucleoside diphosphatase activity, showing MC with attached cells and capillary endothelium (arrow).

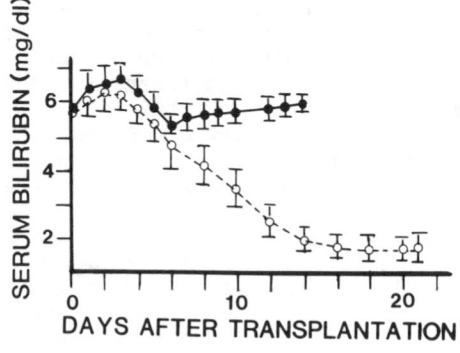

Figure 3. Serum bilirubin concentrations after transplantation of normal rat hepatocytes into allogeneic (solid line) or congeneic (interrupted line) Gunn rats. Each point represents mean of 6 experiments ± SEM.

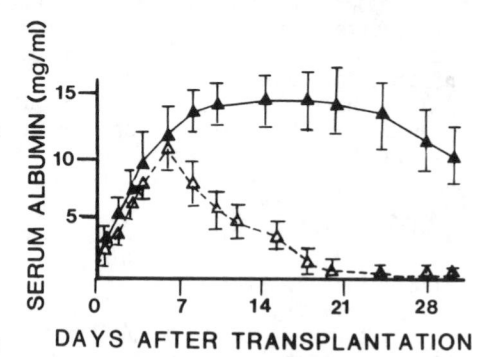

Figure 4. Serum albumin concentration in NAR recipients of transplanted allogeneic normal hepatocytes. Solid line: cyclosporin A-treated rats; interrupted line: untreated recipients. Data are mean ± SEM of 6 experiments.

Transplantation of normal rat liver cells in NAR rats
    After transplantation of allogeneic liver cells, plasma albumin
concentration progressively increased from 0.5 mg/ml to a peak of 10 mg/ml
in 6 days; this was followed by a decline (Fig 4). When immune systems of
the recipient NAR rats were suppressed with cyclosporin A (25 mg per kg of
body weight) given intragastrically daily for the first 5 days after
transplantation and was maintained close to that level until day 30, the
last day of the study (Fig 4). In normal Wistar rats, plasma albumin
concentrations ranged from 35 to 40 mg/ml.
    Ninety percent hepatectomy: All rats that received liver cell transplan-
tation at the time of 90% hepatectomy expired within 48 hr. The rats that
received transplantation of liver cells alone or MC alone 4 days prior to
transplantation, also all died within 48 hr (Fig 5). In contrast, the rats
that received MC-attached hepatocytes 4 days prior to 90% hepatectomy had
significantly improved survival (Fig 5); of this group 40% showed long term
survival. Examination of the peritoneal cavity by celiotomy or at autopsy
showed regeneration of the liver to near normal size in 7 days. However,
the livers consisted of single lobes.
    Blood glucose concentration in rats that had received MC-attached
liver cell transplantation 4 days prior to 90% hepatectomy, decreased only
slightly postoperatively (Fig 6). In contrast, the group that received
liver cells alone, exhibited pronouced hypoglycemia after 90% hepatectomy
(Fig 6).

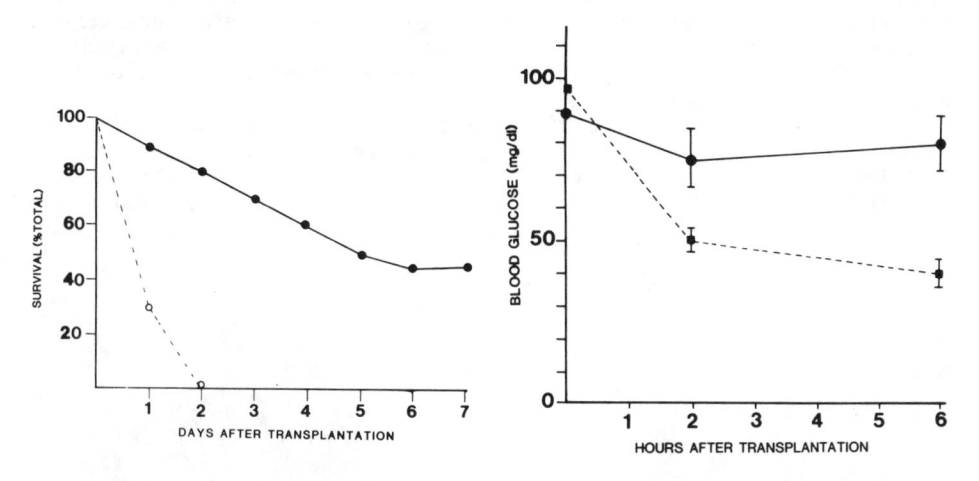

Figure 5. Survival of rats after
90% hepatectomy 4 days after
transplantation of MC-attached
liver cells (solid line) or liver
cells alone (interrupted line).

Figure 6. Blood glucose levels of
rats after 90% hepatectomy 4 days
following transplantation of MC-
attached liver cells (solid line) or
liver cells alone (interrupted line).

Transplantation of normal rat enterocytes into Gunn rats
    After transplantation of isolated small intestinal epithelial cells
into congeneic Gunn rats, bilirubin mono- and diglucuronide (total
conjugated bilirubin, 3% of normal) were detectable in the bile for 20 days
(last day of the experiment).
Transplantation of human liver cells into athymic Gunn rats

The conglomerate of MC and liver cells was similar to that formed after rat liver cell transplantation. Conjugated bilirubin was detectable in bile for 30 days (last day of the experiment). Serum bilirubin progressively decreased from 5 mg/dl to 1.5 mg/dl in 21 days, and remained at that level until the end of the experiment.

Cryopreservation of liver cells

Fifty to 60% of the frozen liver cells were viable upon thawing, as determined by trypan blue exclusion. When 1×10^7 viable cells were transplanted into NAR rats, serum albumin levels were comparable with those observed after transplantation of an equal number of freshly isolated liver cells.

Transplantation of fibroblasts transduced with retrovirus containing human PTH cDNA

After transplantation of fibroblasts with integrated retroviral genome containing human PTH cDNA, human PTH level in rat serum increased from undetectable levels to 175 pg/ml in 24 hr. This was followed by a decline to very low levels during the next 3 days (Fig 7). After this a second peak, smaller than the first was observed; this peak lasted for about 4 days (Fig 7).

Transduction of cultured hepatocytes with BAG virus

Maximal proviral integration occurred when the cells were infected on day 3 of culture on collagen-coated plates. Cultures of infected hepatocytes were analyzed in situ for retrovirus-mediated transduction and expression by cytochemical staining for beta-galactosidase. This procedure specifically labels cells that express virally directed beta-glactosidase; endogenous beta-galactosidase is not detected. Fig 8 shows an analysis of liver cells infected on day 3; approximately 25% of the cells express bacterial beta-galactosidase.

Immunocytochemical staining for UDP-glucuronosyltransferase and asialoglycoprotein receptor showed that these two hepatocyte-specific proteins (in the context of liver cells) were present in all viable cell in the culture, indicating that practically all cells surviving under these culture conditions were hepatocytes.

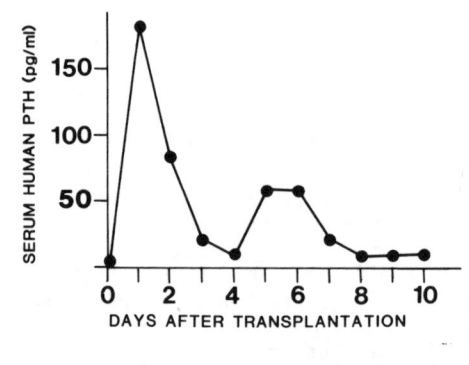

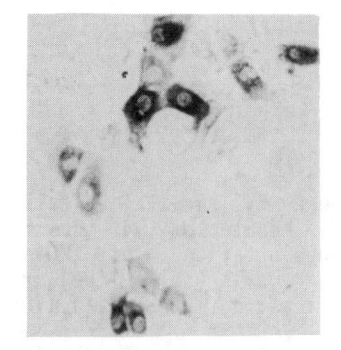

Figure 7. Human PTH levels in the serum of a rat transplanted with fibroblasts transduced with retrovirus containing human PTH cDNA.

Figure 8. Beta-galactosidase staining of cultured hepatocytes transduced with BAG virus. Darkly stained cells are positive for expression of bacterial beta-galactosidase.

DISCUSSION

Results of these studies clearly indicate that attachment of liver cells to collagen-coated dextran microcarriers prior to transplantation improves the survival and function of the transplanted cells.

It is interesting that transplantation of 1% to 2% of the total liver cell mass was adequate for reduction of serum bilirubin in Gunn rats. This suggests that there is an excess of bilirubin-UDP-glucuronosyltransferase activity in the normal liver. However, there may also be some proliferation of the transplanted hepatocytes (20).

Serum albumin concentrations observed after transplantation of normal liver cells into NAR rats, were more surprising. Serum albumin levels that are 20-30% of normal were reached after transplantation of only 1-2% of liver cell mass into NAR rats. The half-life of injected serum albumin has been reported to be longer than normal in NAR rats. In preliminary studies, we have observed that the rate of transcription of the albumin gene is 2 to 3-fold greater in the NAR rats than in normal rats, but the nuclear transcript is not efficiently processed to the cytoplasmic mRNA presumably due to a defect in an intron within the albumin gene. The mechanism of increased transcription rate for albumin in NAR rats is unknown; however, albumin synthesis by transplanted normal hepatocytes may be increased in NAR rats. Thus, increased synthesis and decreased degradation may both be responsible for the accumulation of serum albumin after transplantation of normal cells into NAR rats.

Liver cell transplantation has been used by many investigators to provide metabolic support and improve survival in animals with experimental acute liver injury (1,2,21). Many of these studies utilized D(+)galactosamine-induced liver injury as a model of acute liver failure (1,21). However, it is difficult to interpret the results obtained with this model, because the degree of liver damage varies according to the rat strain, age, route of administration and the capacity of the liver to regenerate in response to exogenous factors derived from transplanted cells or extracts. Similar problems are encountered when using ischemic liver damage as a model for acute liver insufficiency (22,23). The reported improvement of survival after liver cell transplantation was also obtained by administration of liver cell or splenic cell extracts, or tissue culture supernatants, and thus, may not represent function of viable transplanted cells. The 90% hepatectomy model used in our study has the advantage of being reproducible, uniformly lethal, and associated with a measurable biochemical abnormality, namely hypoglycemia. The improvement of survival was not obtained after injection of liver cells alone. Transplantation of MC-attached liver cells at the time of 90% hepatectomy also did not enhance survival, suggesting that organization and vascularization are important in the ability of the transplanted cells to provide metabolic support in acute liver failure. It is likely that after partial hepatectomy, the transplanted cells proliferate along with the residual liver (20). The ability of the transplanted cells to markedly improve survival in acute liver failure suggests that if this procedure is successful in the transplantation of human liver cells, it may have potential therapeutic benefit. The ability to preserve the MC-attached hepatocytes for future use should enhance this therapeutic potential.

We developed the athymic/Gunn rat hybrid to evaluate the effectiveness of this method in the transplantation of human liver cells. These rats, in addition to having a bilirubin conjugation defect, have inherited

deficiency of T-cells, and as a result, do not reject xenograft. Function of transplanted human cells in this animal, indicates that attachment to MC can also support the survival of human hepatocytes.

Bilirubin UDP-glucuronosyltransferase activity has been demonstrated in rat liver, renal cortex and small intestinal epithelium (19). However, in man, the kidney lacks bilirubin-UDP-glucuronosyltransferase activity, and only liver and small intestinal epithelial cells conjugate bilirubin. Therefore, we were interested in developing a method for transplantation of intestinal epithelial cells. Our studies indicate that attachment of enterocytes to MC prior to transplantation promotes their survival and bilirubin conjugating function. Thus the microcarriers may have a wide application in the transplantation of mammalian cells.

Production of human PTH by retrovirally transduced fibroblasts transplanted in rats indicates that it is possible to have expression of retrovirally introduced genes in vivo. We have been successful in maintaining primary hepatocyte cultures on MC for several days, and introducing new genetic material into them. Availability of this novel method for transplanting hepatocytes, together with the technique for efficient transduction of hepatocytes, opens the possibility of designing new approaches to liver-directed somatic gene therapy.

ACKNOWLEDGEMENT: The authors thank Dr. R.C. Mulligan of the Massachusetts Institute of Technology and Dr. P.M. Novikoff of Albert Einstein College of Medicine for help and guidance, and E. Thompson and Z. Khan for technical assistance. This work was supported by the following National Institutes of Health grants: DK34357 to JRC, DK39137 to NRC, DK38763 to AAD and DK17702. JMW was supported by the Howard Hughes Medical Institute.

REFERENCES
1. Sommer BG, Sutherland DER, Matas AJ, Simmons RL, Najarian JS: Hepatocellular transplantation for treatment of D-galactosamine-induced acute liver failure in rats. Transplant Proc 11:578-584, 1979.
2. Makowa L, Rotstein LE, Falk RE, Falk JA, Langer B, Nossal NA, Blendis LM, Phillips MJ: Reversal of toxic and anoxic induced hepatic failure by syngeneic, allogeneic and xenogeneic hepatocyte transplantation. Surgery 88:244-253, 1980.
3. Makowa L, Rotstein LE, Falk RE, et al.: Studies into the mechanism of experimental acute hepatic failure by hepatocyte transplantation. Can J Surg, 24:39-44, 1981.
4. Berg RGM, Ernst P, Van Maldegen-Drowkers C, et al.: Effect of viable isolated hepatocytes or hepatocyte fractions on survival following galactosamine-induced acute liver failure. Eur Surg Res, 17:109-118, 1985.
5. Woods RJ, Parbhoo SP: An explanation for the reduction of bilirubin levels in congenitally jaundiced Gunn rats after transplantation of isolated hepatocytes. Eur Surg Res, 13:278-284, 1981.
6. Gunn CH: Hereditary acholuric jaundice in new mutant strain of rats. J Hered, 29:137-141, 1938.
7. Vroemen JPAM, Blanckaert N, Buurman WA, et al.: Treatment of enzyme deficiency by hepatocyte transplantation in rats. J Surg Res, 39:267-275, 1985.
8. Demetriou AA, Whiting JF, Feldman D, Levenson SM, Roy Chowdhury N, Moscioni AD, Kram M, Roy Chowdhury J: Replacement of liver function in rats by transplantation of microcarrier-attached hepatocytes. Science 233:1190-1192, 1986.

9. Demetriou AA, Levenson SM, Novikoff PM, Novikoff AB, Roy Chowdhury N, Whiting JF, Reisner A, Roy Chowdhury J: Survival, organization and function of microcarrier-attached hepatocytes transplanted in rats. Proc Natl Acad Sci, USA 83:7475-7479, 1986.

10. Demetriou AA, Whiting JF, Levenson SM, Roy Chowdhury N, Schechner R, Michalski S, Feldman D, Roy Chowdhury J. New method of hepatocyte transplantation and extracorporeal liver support. Ann Surg 204:259-271, 1986.

11. Berry MN, Friend DS: High yield preparation of isolated rat liver parenchymal cells: a biochemical and fine structural study. J Cell Biol, 43:506-520, 1969.

12. Seglen PO: Preparation of isolated rat liver cells. Methods Cell Biol, 13:29-83, 1976.

13. Reisner A, Roy Chowdhury J, Capizzi A, Michalsky S, Levenson SM, Roy Chowdhury N, Demetriou AA: Improved survival in rats with acute liver insufficiency induced by 90% hepatectomy by intraperitoneal transplantation of microcarrier-attached hepatocytes. Hepatology (abstr.) 6:1760, 1986.

14. Roy Chowdhury J, Moscioni AD, Demetriou A, Lahiri P, Shouval D, Roy Chowdhury N: Comparison of bilirubin conjugates produced by dog, rat and human hepatocytes transplanted into athymic Gunn rats. Hepatology (abstr.), in press, 1987.

15. Arnaout W, Barbour R, Moscioni AD, Brown LL, Roy Chowdhury J, Demetriou AA: Intraperitoneal transplantation of microcarrier-attached enterocytes in Gunn rats. Hepatology (abstr.), in press, 1987.

16. Price J, Turner D, Cepko C: Retrovirus-mediated transduction of mammalian cells. Proc Natl Acad Sci, USA, 84:156-160, 1987.

17. Maniatis T, Fritsch EF, Sambrook T: Molecular Cloning: A Laboratory Manual. Cold Spring Harbor Laboratory, Cold Spring Horbor, New York, 1982.

18. Feinberg AP, Vogelstein B: Technique of labeling DNA at high specific radioactivity. Anal Biochem, 132:6-13, 1983.

19. Roy Chowdhury J, Novikoff PM, Roy Chowdhury N, Novikoff AB: Distribution of uridine diphosphoglucuronate glucuronosyltransferase in rat tissues. Proc Natl Acad Sci, USA, 82:2990-2994, 1985.

20. Demetriou AA, Felcher A, Schechter C, Capizzi A, Chang J, Davenport T, Roy Chowdhury J, Levenson SM: Proliferation of intraperitoneally transplanted microcarrier-attached rat liver cells. Gastroenterology (abstr.), 92:1720, 1987.

21. Baumgartner D, LaPlante-O'neill PM, Sutherland DER, Najarian JS: Effect of intrasplenic injection of hepatocytes, hepatocyte fragments and hepatocyte culture supernatants on D-galactosamine-induced liver failure. Eur Surg Res 15:129-135, 1983

22. Sommer BG, Sutherland DER, Simmons RL, Najarian JS: Hepatocellular transplantation for experimental acute liver failure in dogs. J Surg Res 29:319-325, 1980.

23. Sommer BG, Sutherland DER, Simmons RL, Najarian JS: Hepatocellular transplantation for experimental acute liver failure in dogs. J Surg Forum 30:279-281, 1979.

Chowdhury, J.A. : <u>Use of microbeads during liver cells</u>
<u>transplantation</u>

Szpirer :   The integration site of the virus is not controlled.   The
virus integrates everywhere.   You are not afraid of that ?

Chowdhury : Yes.   The virus does integrate   everywhere and there is   a
possibility that   it could   be oncogenic.   Now there   are
ways of handling that as I am sure you know, that is   make
a deletion in the LTR (Long Terminal Repeat) on one   site.
As the virus replicates, it copies the LTR from that   side
to the other site.   It will be inactivated at that   point.
This is called   a CIN   virus for   Cell Inactivated   Virus.
These constructions have been designed for this particular
purpose.   If these experiments succeed and if we ever   use
it in humans, we are going to use the CIN viruses.

Drillien :   If it integrates, it integrates anywhere.

Chowdhury : It integrates anywhere but it doesn't have the active   LTR
and it is not going to drive the gene which is in front of
it.

Drillien :   It can still interrupt an essential gene.

Chowdhury : It can, yes although that   has not been practically   shown
to actually happen and this   idea is always behind   everybody's mind.

Drillien :   It can be in an essential regulatory gene.

Chowdhury : If it kills, it looses   very little.   If it stimulates   to
divide and does lead to cancer then we have a chance.

Hulser :   I wonder about these multinucleated cells I saw on one of
            your last slides. Are these endomitotic cells or are they
            fused ?

Chowdhury : The multinucleated cells that you get from the liver, as a
            rule, do not divide because in liver as you know, there
            are many binucleated or tetranucleated cells. These are
            not dividing cells. They stay under these conditions for
            ever. Under our culture conditions, we know that some of
            the cells must be dividing because there is an uptake of
            tritiated thymidine. So, I cannot tell you for sure
            whether these cells are dividing or not.

Drillien : The cells that you are putting in culture before injecting
            them into the peritoneal cavity, are they purified hepato-
            cytes or a mixture of liver cells ?

Chowdhury : The cells that I showed ready for the injection in the
            analbuminemic rats and Gunn rats or the hepatectomized
            ones, are not really cultured, they are just maintained in
            the medium for attachment.

Drillien : They don't contain macrophages or other cells ?

Chowdhury : They contain the total cell complement obtained after
            collagenase perfusion which means 85 % hepatocytes, 15 %
            of other cells including macrophages, Kuppfer cells and
            epithelial cells. The cells that we showed later for the
            virus infection studies, those were cultured. They had to
            be cultured for two to three days in order to infect them
            and under these culture conditions, we investigated
            morphometrically the percentage of hepatocytes and it so
            happened that after three days of culture on that type I
            collagen-coated plate or beads, the cells that are still
            attached and alive are just about 100 % hepatocytes.   The
            other cells do not survive and as a result do not get

rejected because the antigen processing cells are lacking from this mixture.

Remacle : Is this the explanation that you propose for the better efficiency of hepatocytes bound to microcarriers versus hepatocytes alone ?

Chowdhury : That is difficult to say. The observation that the other cells failed to thrive was made just about one week before I came here and it is difficult for me to say more than this. No, I don't think that this is the reason because the antigen processing cells are obviously present in this mixture since their rejection can be prevented by prior UV irradiation.

Miller : All right, you eliminate the dendritic cells but you still have the HLA groups on your cells.

Chowdhury : For reasons that are not totally explained, liver is well known to be an immuno privileged organ and for whatever reasons ... I don't really understand. If you transplant skin, kidney, heart, whatever, the liver needs much less cyclosporin, much less immunosuppression and for some reasons it does not like to be rejected. So, although HLA groups are present, just after UV irradiation, we, for sure do not destroy them, we nevertheless see a persistance of the graft.

Miller : Maybe Dr Prunieras can comment about skin graft now or afterwards ?

Prunieras : You allude to the fact that keratinocytes in culture can be forced to reexpress HLA-DR antigens under the action of interferon gamma. This is right in culture but we don't have any information regarding the actual role of this expression in _in vivo_ rejection phenomena. It may or may not be involved. I mean to me the question is still unresolved.

Spier :   If you look at the Damon process for encapsulating cells, where you actually create a barrier between the cells that you put in and the surrounding cells, you shouldn't get the immunological problems. This has been tried for Langerhans cells for pancreas transplants. You may be interested in trying it for your liver transplantations.

Chowdhury : We tried this many years ago and didn't have significant results.

Merten :   I think your hepatocytes must stick to the microcarier. Perhaps therefore they did not work in the Damon process because there, normally, the cells don't stick to the surface.

Chowdhury : Yes, that is what I thought.

INFLUENCE OF EPIGENETIC FACTORS ON EPIDERMAL DIFFERENTIATION AND MORPHOGENESIS

Michel PRUNIERAS

Directeur de Recherche à l'INSERM
Laboratoires de Recherche Fondamentale de L'OREAL, 1 av. Eugène Schueller
93600 Aulnay sous bois, France

INTRODUCTION

The major and vital function of skin is protection : protection against environmental elements that can be hazardous if they penetrate into the blood stream ; protection against body fluid depletion. This latter point is examplified by the case of large burn patients who do not survive if their ion plasmatic concentrations are not adjusted and if their lost skin is not replaced.

The protection role of epidermis is achieved through the synthesis and organization of a thin layer of dead horny cells, the horny layer.

The horny layer results from two series of changes that take place in the epidermis during the progression of lower to upper epidermal cells.

The first series of changes represents differentiation of epidermal keratinocytes. It includes the accumulation inside the boundaries of the cell of a number of proteins among which keratin, envelope protein and keratokyaline have been the objects of recent interest.

In particular, the synthesis of the envelope protein which polymerizes as a thick band along the internal aspect of the cell membrane is an important step in the transformation of globular deformable basal keratinocytes into flat rigid corneocytes.

Thus, the synthesis of intracellular differentiation molecules correlates with the shape of the cells.

The second series of changes represents morphogenesis. Morphogenesis is equivalent to tissue organization. Obviously, the shape of the cells that compose a given tissue is of import to its organization. However, the way the cells are arranged inside the tissue depends also on amounts and qualities of intercellular substances. In the epidermis the composition of intercellular spaces varies. In the horny layer, the intercellular spaces are filled with lipids selectively enriched in ceramides. It may be hypothesized that the synthesis of ceramides is of import to the morphogenesis of the skin horny layer.

Differentiation and morphogenesis are both genetically programmed. To find out the epigenetic factors that can modulate this program it is necessary to create experimental situations in which alterations of differentiation and morphogenesis markers can be consistently induced.

A. O. A. Miller (ed.), Advanced Research on Animal Cell Technology, 329–343.
© *1989 by Kluwer Academic Publishers.*

Experimental approach

The influence of the dermis on epidermal differentiation and morphogenesis has been documented through recombination experiments (1). In such experiments, the skin is dissociated in epidermis and dermis. Epidermis from different region, polarity, age, genotype or species is reassociated with homo or heterotypic dermis. However, since the dermal tissue is composed of cellular and matrix elements, such experiments can tell only little about the respective role of each of these elements.

In addition, the dermis contains diffusible substances such as hormones, growth factors and vitamins the organizing influence of which can be assessed only through more refined experiments.

Cell culture systems, instead of tissue recombinations, allow dermal cells, matrix molecules and soluble factors to be studied independently. This is why we have concentrated on cell culture experiments to approach the problem of the influence of epigenetic factors upon epidermal differentiation and morphogenesis.

Efforts to develop keratinocyte culture systems can be directed along two mainlines. One is to favor growth, the other is to promote differentiation.

The most widely used method to favor growth is that of Rheinwald and Green (2, 3) which uses 3T3 feeder cells together with a growth factor enriched medium. We will first see the effect of this culture method on epidermal differentiation and morphogenesis.

In vivo markers of epidermal differentiation and / or morphogenesis

In human skin in vivo a number of markers have been found.

By photomicroscopy three can be recognized :
1. The epidermis is composed of 3 compartments including
   . tightly packed polarized basal cells
   . several layers of nucleated polygonal suprabasal cells and
   . orderly arranged horny layers of corneocytes
2. Upper nucleated suprabasal cells containing cytoplasmic keratohyaline granules compose a granular layer between lower nucleated keratinocytes and upper horny layer corneocytes
3. Horny layer corneocytes are devoid of nuclei.

By electronmicroscopy, 5 markers are of special interest
1. Hemidesmosomes at basal cell pole
2. Desmosomes between adjacent suprabasal nucleated keratinocytes
3. Keratohyaline granules are heterogeneous. They are composed of dense zone associated with less electrondense regions
4. There is thickening of the intracellular aspect of the plasma membrane inner leaflet in corneocytes situated above the granular layer
5. In corneocytes, keratin filaments organize according to the so-called keratin pattern.

By biochemical analysis ^{35}S-labelled proteins extracted by high salt buffers and separated by 2D-gel electrophoresis appear as two sets of three different spots. A first set is composed of three alkaline to neutral proteins of 67, 65 and 58 kd corresponding to keratins 1,2 and 5 of the Moll et al. (4) catalog. A second set comprises three acidic proteins of 56.5, 50 and 46 kd, corresponding to keratins 10, 14 and 17 of the above reference.

By immunochemical analysis using either polyclonal antibodies against purified proteins, or monospecific antisera against specific epitopic sequences or monospecific antisera obtained by elution or monoclonals, 4 protein constituents can be identified including keratin 67 kd, involucrin (precursor of the envelope protein) epidermal transglutaminase (acting in the synthesis of envelope protein) and filaggrine (a constituent of the histidine-rich protein of keratohyaline).

In vitro alterations of differentiation and morphogenesis markers in the growth promoting culture system of Rheinwald and Green

Epidermal keratinocytes cultured on plastic in the presence of 3T3 feeder cells exhibit progressive alterations in differentiation and morphogenesis markers. At 3rd subculture the most salient alterations are the following (5) :

At the photomicroscopic level, there are still three compartments but basal cells are loosely arranged and suprabasal layers are few ; horny layers are disorganized ; there are few (if any) keratohyaline granules; there is persistence of nuclei in corneocytes.

At the electronmicroscopic level, desmosomes and envelopes are present but few in number, there are no hemidesmosomes and when keratohyaline is visible it is not heterogeneous (dense zones are lacking).

At the biochemical level, 58, 50 and 46 kd keratins are expressed but 67, 65 and 56.5 kd keratins are not. In addition a 56 kd basic protein is expressed which corresponds to a keratin known to be associated with hyperproliferation.

Rationale for developping differentiation and morphogenesis promoting culture systems

In the in vivo situation, keratinocytes rest on a dermal bed, not on plastic.

The dermal bed is the basal lamina which contains laminin, fibronectin, collagen IV, heparan sulfate proteoglycans to cite only some of the constituents that have been so far identified in this structure.

The idea is therefore, to culture the cells on substrates containing at least one of these constituents or, even better, to use the genuine adult skin basal lamina itself as a substrate for culture.

It should be remarked, at this point, that the use of a pre-existing basal lamina as a substrate may not be needed provided i) the keratinocytes are the source of basal lamina constituents and ii) the culture substrate allows the keratinocytes to secrete basal lamina molecules. The culture substrate, be it plastic or else, would then become coated with a basal lamina equivalent, rendering unnecessary the use of natural basal lamina. The sole question would be to provide keratinocytes with a substrate that generates proper signals.

In following this general idea, we cultured human keratinocytes on various substrates including plastic, collagen films, collagen films coated with a basal lamina equivalent, dermal equivalents made of a mixture of collagen I and fibroblasts and dead de-epidermized dermis still covered with its genuine basement membrane.

However, in the in vivo situation, keratinocytes are not only attached to the basal lamina but they also are exposed to air. In ordinary cultures, like those made after the Rheinwald and Green method they remain covered with the tissue culture medium. Consequently, exposure to air was considered as a potentially significant epigenetic factor.

Brief description of the culture systems used

They have been reviewed recently (6). Keratinocyte suspensions were seeded at high density on :

1. Collagen films. Nitrocellulose filters were coated with either bovine collagen I or IV
2. Basement membrane equivalents. Collagen coated filters were used to culture bovine corneal endothelial cells. After stimulation with an eye-derived growth factor, these cells secrete basement membrane molecules that constitute a basement membrane equivalent (6) containing laminin and collagen IV, which overcoats the collagen. Keratinocytes were seeded after removal of bovine corneal cells.
3. Dermal equivalents. Fibroblasts were embedded in collagen gels. After retraction of the gels, keratinocytes were seeded on top (7).
4. Genuine basement membrane. Split thickness flaps of human skin were de-epidermized by soaking in PBS for five days at 37°C. After such treatment, the epidermis detaches above the basal lamina. The de-epidermized dermis (DED) was frozen and thawed 10 times to kill all cells that would have remained alive in the dermis . It was used as substrate, with the basement membrane on top.

It will be noted that in systems 2 and 4 the keratinocytes were cultured on basement membrane or equivalent whereas in systems 1 and 3 they were not.

Finally, to test exposure to air, coated filters, dermal equivalents and DED were placed on perforated stainless steel grids and lifted up to the interface between air and tissue culture medium (TCM)

Comparison of differentiation and morphogeneis in different culture systems

In a first series of experiments, it was shown that the cultures made on DED and exposed to air are dramatically close, morphologically, to normal skin. On vertical histological sections of cultures made on DED and air-exposed one can see that epidermal cells invaginate into the pits and holes of the dermal substrate ; intercellular bridges are seen, keratohyaline granules are conspicuous and the horny layer is coherent. In contrast, immersed cultures on plastic exhibit poor stratification, parakeratosis and no keratohyaline granules. In addition, cornified cells, when present do not arrange in a coherent horny layer (5).

At the ultrastructural level a number of markers were seen in DED cultures exposed to air and not in immersed cultures on plastic as summarized in Table I.

Table I

Ultrastructural markers of differentiation	In vivo	Plastic	DED
Tonofilaments	+	+	+
Desmosomes	+	+	+
Hemidesmosomes	+	−	+
Intracellular Membrane Coating Granules (MCG)	+	−	+
Heterogeneous keratohyaline	+	−	+
Cell membrane thickening	+	+	+
Gap junctions	+	−	+
Extracellular MCG	+	−	+
Keratin pattern	+	−	−
Interdesmosomal line	+	−	+

It can be seen that cultures on DED and exposed to air exhibit all but one of the markers seen in vivo. In contrast, cultures on plastic and immersed exhibit only 3 out of 10.
In conclusion, cultivation of keratinocytes on DED and exposure to air have a strong influence on the morphological differentiation and morphogenesis of the epidermis.

In a second series of experiments human keratinocytes were cultured according to Rheinwald and Green. At 3rd subculture the cells were either maintained on plastic or detached, suspended and seeded on the various substrates described above. After 1 to 3 weeks the expression of keratin polypeptides was studied by high salt buffer extraction and PAGE after labelling the cultures with ^{35}S-methionine.
The presence of keratohyaline granules was also taken as an indication of differentiation.

Results are summarized in Table II.

Table II

		keratohyaline	67 kd keratin
Plastic alone		−	−
Filter coated with film of coll. I or IV	I	−	−
	AE	−	−
Filter coated with base-ment membrane equivalent	I	−	−
	AE	−	−
Dermal equivalent	I	±	−
	AE	+	++
DED	I	±	−
	AE	++	++

I = immersed   AE = Air-exposed

It can be seen from this table that air-exposure + dermal substrate, either natural (DED) or artificial (dermal equivalent) induced keratohyaline synthesis and 67 kd keratin. On the contrary, neither collagens I nor IV nor basement membrane equivalent had any effect, be they in immersed or air-exposed situation.

Dermal substrates, air-exposure and vitamin A

Fuchs and Green have shown that at third subculture, conventional cultures made on plastic and immersed in TCM supplemented with fetal calf serum (FCS) do not express the 67 kd keratin (8). Keratohyaline granules are also lacking and corneocytes when present are not arranged in a horny layer. However, when FCS was delipidized, the 67 Kd keratin was re-expressed, keratohyline granules were synthetized and more cornified cells were produced.

Furthermore, if vitamin A was added to the delipidized serum, the synthesis of 67 kd keratin and keratohyaline granules was blocked and cornification went back.

These results are schematically represented in Table III.

Table III

	67 kd keratin	Keratohyaline	Cornification
Plastic + TCM + FCS	—	—	±
Plastic + TCM + delipidized FCS	+	+	+
Plastic + TCM + delipidized FCS + Vitamin A	—	—	±

Results indicate that the presence of vit. A in FCS blocks the synthesis of 67 kd keratin and keratohyaline.

If we now turn back to the results of our experiments it can be seen that keratohyaline and 67 kd keratin were synthesized in cultures made on dermal or dermal equivalent substrates and exposed to air even though the culture medium was supplemented with normal FCS.

The first interpretation is that exposure to air generates signals that override vitamin A control. But this cannot be the case since the same cultures made on basement membrane equivalent or collagen films and exposed to air did not respond to these hypothetical signals.

Another interpretation is that the thickness of the substrate regulates the amount of vitamin A which finally reaches epidermal keratinocytes when the cultures are raised to the air-liquid interface. Indeed, the raising of the culture results in feeding the cells through the substrate. Obviously, the thickness of it would be of great import if it acted as a filter, in retaining more or less vitamin A.

This points to the conclusion that the program of differentiation of skin keratinocytes normally contains all necessary information for the expression of keratins including the high molecular weight 67 kd. As a major epigenetic factor the dermis, by its filtering action, might regulate the amount of gene regulatory diffusible substances that reach the epidermal cells. Vitamin A may be the most important of these substances if not the only one possible.

Epidermal lipids and morphogenesis

As recalled in the first part of this paper the in vivo process of horny layer synthesis and organization comprises two major events :
1. synthesis of intracellular proteins that eventually enrich the keratinocyte and transform it into a rigid protein sac and
2. synthesis of intercellular cement rich in lipids. This intercellular cement represents 10 to 40 percent of the total epidermal volume.

The lipid composition of epidermis changes drastically from lower live to upper dead layers. This is due to first the degradation of nucleated cell membrane phospholipids and second the extrusion into the intercellular spaces of the inner content of lamellar bodies, a cytoplasmic organelle that merges with the plasma membrane of the keratinocyte at the time this latter cell becomes loaded with keratohyaline granules.

Changes in lipid composition according to epidermal cornification are schematically summarized in Table IV after P. Elias (9).

Table IV

| | changes in lipid composition during cornification (wt percent) | |
	lower layers	horny layer
Polar lipids	44.5 ± 3.4	4.9 ± 1.6
Neutral lipids	51.0 ± 4.5	77.7 ± 5.6
Ceramides	3.8 ± 0.2	18.1 ± 0.4

The role of intercellular lipids in the horny layer is thought to be double :
1. insure cohesion of corneocyte
2. function as barrier

Although there are still debates about the question of how much cohesion and how much protection the intercellullar lipids actually account for, it is widely accepted that these lipids at least significantly contribute to these tow points.

The following arguments have been put forward (9) : When rats are fed an essential fatty acid deficient diet the intercellular lipids are inadequate and the barrier function is altered ; regional variations in skin penetration relate to per cent lipid weight ; and, with regards to cohesion, the increase in horny layer thickness in ichtyosis relates to an abnormally high content in cholesteryl sulfate.

Therefore, it is most likely that the synthesis of those lipids which increase in the horny layers (neutral lipids and ceramides) has to do with the organization of epidermal outer layers, i.e. morphogenesis.

Epidermal lipids, differentiation and morphogenesis in culture

Epidermal human keratinocytes have been cultured as monolayers in the Rheinwald and Green system. At 3rd subculture, part of the cultures were trypsinized, suspended, seeded on DED and cultured at the air liquid interface.

After one to 3 weeks the lipids were extracted and analyzed by 2D-TLC on silicagel plates.

As noted above, the morphology of cultures made on DED and air-exposed are very similar to normal skin. In particular, a well organized thick horny layer is present. This is in contrast with immersed monolayer cultures on plastic in which the horny layer is either inexistant or loosely arranged. In addition, a prominent granular layer is visible.

After 2 weeks of culture lipid analysis revealed low amounts of phospholipids and high amounts of sterols and ceramides (10) a pattern of distribution which is close to the in vivo situation.

Results suggest that culture of keratinocytes on DED and exposure to air induce the synthesis of intercellular lipids that can play a significant role in epidermal morphogenesis.

SUMMARY AND CONCLUSION

In the present paper we discuss some experiments with epidermal keratinocyte cultures aiming at demonstrating the influence of epigenetic factors on epidermal differentiation and morphogenesis. Morphologically air exposed cultivation of keratinocytes on thick dermis or dermal equivalents induces histiotypic morphogenesis, with, in particular a conspicuous horny layer.

Biochemically exposure to air and thick collagenous substrates induce the synthesis of intracellular proteins such as the 67 kd keratin that is considered as a terminal differentiation marker as well as that of intercellular lipids associated with complete morphogenesis.

With regards to the 67 kd differentiation keratin, it is speculated that dermal thickness may be a significant epigenetic factor in controlling the amount of gene regulatory substances (such as vitamin A) that eventually reaches epidermal cells.

The reason why cultivation on dead de-epidermized dermis and exposure to air influence intercellular lipid synthesis remains unknown.

Acknowlegment
Experiments reported above have been performed in collaboration with Marcelle Régnier from the Centre International de Recherche Dermtologique, Sophia Antipolis, 06000 Valbonne, France.

REFERENCES

(1) WOODLEY, D.T., DEMARCHEZ, M., SENGEL, P., PRUNIERAS M. :
The control of cutaneous development and behavior. The influence of extracellular and soluble factors. In : "Dermatology in General Medicine", T.B. Fitzpatrick et al., eds., 3rd edition, Mc Graw-Hill, Inc, 1987, pp. 132-146

(2) RHEINWALD, J.G., GREEN, H. :
Serial cultivation of strains of human epidermal keratinocytes : the formation of keratinizing colonies from single cells. Cell 1975, 6, 331-335

(3) RHEINWALD, J.G., GREEN, H. :
Epidermal growth factor and the multiplication of cultured epidermal keratinocytes. Nature 1977, 265, 421-423

(4) MOLL, R., FRANKE, W.W., SCHILLER, D.L., GEIGER, B., KREPLER, R. :
The catalog of human cytokeratins : patterns of expression in normal epithelia, tumors and cultured cells. Cell 1982, 31, 11-23

(5) REGNIER, M., SCHWEIZER, J., MICHEL, S., BAILLY, C., PRUNIERAS, M. :
Expression of high molecular weight (67 K) keratin in human keratinocytes cultured on dead de-epidermized dermis. Exp. Cell. Res. 1986, 165, 63-72

(6) PRUNIERAS, M., REGNIER, M., WOODLEY, D.T. :
Methods for cultivation of keratinocytes with an air-liquid interface. J. Invest. Dermatol. 1983, 81, 28s-33s

(7) ASSELINEAU, D., BERNARD, B., BAILLY, C., DARMON, M. :
Epidermal morphogenesis and induction of the 67 kd keratin polypeptide by culture of human keratinocytes at the liquid-air interface. Exp. Cell. Res. 1985, 159, 536-538

(8) FUCHS, E., GREEN, H. :
Regulation of terminal differentiation of cultured human keratinocytes by vitamin A. Cell 1981, 25, 617-623

(9) ELIAS, P.E. :
The special role of the stratum corneum. In "Dermatology in General Medicine", T.B. Fitzpatrick et al., eds., 3rd edition, Mc Graw-Hill, Inc., 1987, pp.342-346

(10) PONEC, M., NUGTEREN, D. :
Lipid composition of cultured human keratinocytes. 17th annual meeting of the European Society for Dermatological Research, Abst.n° 28, 1987

Prunieras, M. : <u>Influence of epigenetic factors during morphogenesis
and differentiation of the epidermis</u>

Klepsch :   I am working for the European Commission. I am related to
an <u>in vitro</u> cell culture programme called BAP and as you
all probably know, we have other programs concerning the
reconstruction of skin.   One of them is headed by
Prof. DUBERTRET in collaboration with Prof. LAPIERE for
example and I think we have heard here some of his
coworkers, Dr Foidart. But I think your approach has
slight but important differences with theirs. They   start
with a matrix containing fibroblasts, the keratinocytes
being added later on.

Your approach looks more towards the protective functions
afforded by skin, analyzing for that purpose the differentiation process.   Such an approach is also considered in
the United States in order to develop means to assist
heavily burned people.

All these studies join together towards reaching a common
goal : the reconstruction of a particular organ : the
skin.

Prunieras : There are two ideas stemmed from the same route. But they
diverge at a certain point. They diverge for a very good
reason. When you have a culture of keratinocytes, even if
that culture does not express final differentiation and
morphogenesis, once it is transplanted, it will, because
then it finds we don't know what, proper messages, you
know. The cells are instructed and they reconstruct
something which is practically normal.

But in toxicopharmacology we do not transplant the cells,
we would have a lot of ethical problems.   So, the real
question to us is not to have a non-differentiated cell
culture to use because it would be totally meaningless.
So, what is the important factor ? The important factor
is the barrier and especially for the penetration of drugs
and this kind of things. How can you test anything which
would relate to the <u>in vivo</u> situation if you don't have

the same kind of a barrier ? So for us, this barrier is of fundamental importance. It is not fundamental for those who wish to transplant the cells because the job is made after transplantation.

Miller : I would like to point out in fact, human or chicken or mouse keratinocytes, since they lack HLA, can be allografted without rejection. Experiments have been done by Professor THIVOLET some time ago and none of the grafts he made has been rejected until now. Do you know Dr Punieras the time when the first graft was made ?

Prunieras : I don't like to discuss this too much. Simply, I could relate to the last congress in Berlin in which I was in charge of the meeting on this particular subject. The point was made last May of what we actually know. It seems that for a certain period of time, several weeks at least, the cells that are transplanted are not rejected. But the conditions that have been used for transplantation, I mean the depth of the bed on which the cells were transplanted, was not enough to make sure that they remained longer than that. Indeed there is an important phenomenon which takes place when you cut the skin around, to the surface but not down to the muscle fascia. If you remain a little higher, then you have a lot a remmants of hair bulbs, hair, shafts, you have remmants of sweat gland canals and from these remmants, repopulation of the surface of the host cell can take place. Eventually, after 2 or 3 months for example, the graft is progressively replaced by a tissue coming from the host. So, at the present time, there is no firm evidence that the take is permanent. That the rejection is delayed seems to be pretty well documented now. That the take is permanent, is still disputed.

Miller :   Dr Chowdhury do you want to comment because you spoke a
                little bit about the possibility of using cells as a
                vector to reconstitute organs. These skin cells might
                possibly be a good vector ?

Chowdhury :   We have not tried to reconstitute real organs by the
                microcarrier method since this would not have given real
                organs. For example, the liver that we are creating is an
                endocrine liver. It does not have a bile duct fortunately
                otherwise it should cause peritonitis but is is an endo-
                crine liver : it takes things from the blood and puts them
                back into the blood. No, we don't have means to recons-
                titute real whole organs like skin as Dr Prunieras men-
                tioned.
                But keratinocytes are being contemplated as vehicle for
                somatic gene therapy. People are already at work using
                the microcarrier in exactly the same technique as we used
                for the transplantation but that again is an endocrine
                skin, not a real skin.

Prunieras :   The idea has already been put forward two years ago, I
                think by Howard GREEN, to do the following : something
                which is very close to what you have described for the
                hepatocytes. You can obtain keratinocytes from an indi-
                vidual, culture them, try to reprogram them and graft them
                back to the donor, after having reprogrammed them for
                example to produce - I don't know - IL-1 for example, or a
                growth factor, or something. This is an idea which has
                been already proposed. I don't know whether anything has
                been done in laboratories actually but if something popped
                out in the litterature in the next future, I would not be
                surprised.

Chowdhury :   But for this why whould you need a real skin, why would
                you need the detailed organization that you describe ?

Prunieras : Oh, you don't, at this particular time you don't. No, No, you could use a regular sheet of epidermal cells. Absolutely.

Miller :   I think in the future we might be looking for other human cells lacking histocompatibility antigens in order to take them, reprogram them and reinsert them in the human body. I would like to stress the fact that such an approach has the advantage that the molecules which are produced have the correct three dimensional configuration so that you don't have to isolate these proteins by downstream processing, because proteins obtained in this way, depending upon their galenic form, may induce antibodies upon administration to the patient. I think that this reprogrammation of such HLA-minus cells will be a real breakthrough in the future.

Spier :   Are they really HLA-minus ?

Prunieras : Human skins in vivo seem to loose the HLA-DR expression.

Spier :   But this is after the process of differentiation and morphogenesis ? When you take an allograft and insert it into the peritoneum, what happens to it ?

Prunieras : It does not survive under these conditions but if you inject it into the muscle for instance, it forms a pearl, a keratinizing pearl which will eventually be rejected through a foreign body-type inclusion. It is not easy to distinguish you know the part which is due to immunology and that due to phagocytosis. So, this is not a good model. We would prefer to have a skin recombinant for example, applied to the skin and then study it according to the classic criteria.

Spier :   Yes, but what you were saying before about the way in which the keratinocyte eventually loosens as the skin underneath grows, that is the natural process of skin renewal which is going on all the time, what about your genetic engineered deviant, variant or whatever remains in place and doesn't get replaced is a good point.

Prunieras : The real point which is at stake with heavy burned patients is deep burn. The burn which eradicates the skin completely. So there is no way. You've got to have something here. This point has not yet been solved.

Spier :   I agree.

Miller :   When you consider the differentiation of these keratinocytes, these cells as they progressively move from the basal membrane upwards, the stickiness, the attachment of the several layers diminishing as you go towards the outside, is that constant ?
For instance if you put keratinocytes on microbeads, would it be absolutely unthinkable to imagine that by changing the speed of agitation, you preferentially detach those cells which have reached the most differentiated state ?

Prunieras : This is a very interesting question that you raised. The system of keratinocytes in culture is a very clear system in the sense that it constantly produces a number of cells that are floating in the medium. They are simply floating because they become detached. There are so many of them that we have used in the past a very simple way of evaluating the amounts of proteins produced in culture in function of the time simply by weighing the cells. We used to filter and weight the filter. It was very accurate. It is certainly the simplest and the cheapest way of evaluation. There is a lot of these cells which actually become detached. The process by which these cells detach is an extremely interesting subject which is

not yet well understood. We have just one experimental
model which is the exfoliation model in which staphylococcus aureus produces a protein of well defined
sequence. We have the cDNA of this protein and this
protein has the property of binding to a crucial something
- we still don't know what it is - which is situated at
the limit between the granular layer and the horny layer,
which means that when this protein arrives at this particular site, all the horny layer detaches. But this
happens exclusively in the newborn mouse or human implant,
during the first weeks of life. After that moment, you can
use as much as you want of this exfoliatin, nothing
happens. Recent studies of the receptor sites would point
to the keratohyalin layer and more precisely to fillagrin.
But this is still a matter of discussion. We have already
an antibody against this protein and active researches are
made in France in the laboratory of Professor MINE on this
subject. It is a very interesting question. There are
two important facts there. There is a protein which "does
it" but it cannot "do it" after a certain time, after
birth, which is a really exciting observation.

THE CHARACTERIZATION OF FIBROBLAST GROWTH FACTOR
AND ITS BIOLOGICAL EFFECT IN VITRO AND IN VIVO

D.J. GOSPODAROWICZ AND G. NEUFELD

CANCER RESEARCH INSTITUTE, UNIVERSITY OF CALIFORNIA MEDICAL CENTER
SAN FRANCISCO, CALIFORNIA

1. INTRODUCTION
        Over the last 4 years, both basic and acidic Fibroblast growth
factors (bFGF and aFGF) have been purified to homogeneity, their primary
structures determined, and their cDNA cloned and sequenced (reviewed in
ref.1-3). This information has had significant impact on our
understanding of a variety of mitogenic activities isolated from diverse
origin. It has become clear that growth factors isolated from ovary,
adrenal, kidney, eye, brain, placenta, macrophages, prostate, cartilage,
and various tumors, are structurally and biologically identical, or at
least, very similar to bFGF or aFGF (reviewed in ref.1). Availability
of the pure mitogens has led to the recognition of a wide spectrum of
activities for these two factors (4), most notably, their ability to
mimic the biological effect of the vegetalizing factor in early embryos
(5) and to act as angiogenic factors. Basic FGF or aFGF are
multifunctional, since they can both stimulate proliferation and induce
or delay differentiation. They stimulate other critical processes in
cell function as well. Thus far, so many new and varied functions have
now been described for FGF's that at present, one must consider these
peptides to be of special importance for the control of cell growth and
differentiation.

2. PRIMARY STRUCTURE, GENOMIC ORGANIZATION, AND mRNA EXPRESSION
    OF ACIDIC AND BASIC FGF
        Basic FGF has been purified from most mesoderm- or neuroectoderm-
derived tissues or cells which have in common a strong angiogenic
potential (1,2). Structural studies have shown that bFGF is a single
chain peptide composed of 146 amino acids (6) which can also exist in an
NH_2-terminally truncated form missing the first 15 amino acids (7). The
truncated form of bFGF is as potent as native bFGF, as demonstrated by
radioreceptor binding and biological assays, indicating that the NH_2-
terminal region of bFGF is neither involved in its binding to FGF cell
surface receptors nor in its biological activity (8). Related to bFGF
is aFGF, which shares a 55% total sequence homology with bFGF (9).
Acidic FGF is a 140-amino acid peptide which can also exist as an NH_2-
terminally truncated form missing the first 6 amino acids (10) The high
degree of homology between aFGF and bFGF suggests that they are derived
from a common ancestral gene.
        Evidence that a viral oncogene may code for a growth factor or
part of a growth factor receptor has recently emerged from studies on
the PDGF structure and that of the EGF receptor structure (reviewed in
ref.11). In the case of basic FGF, a 46% and 42.3% homology,
respectively, has been shown to exist with the predicted product of
Int-2 and the hst gene product (12,13). A lesser degree of structural

345

A. O. A. Miller (ed.), Advanced Research on Animal Cell Technology, 345–366.
© 1989 by Kluwer Academic Publishers.

homology exists between those two gene products and aFGF. While Int-2 has been implicated in the induction of virally induced mammary cancer (14), the hst gene was originally identified as a transforming gene in DNA's from human stomach cancer (15).

The FGF genes have been cloned and complementary DNA sequences of both bFGF and aFGF have been synthesized. The genomic organization of the genes encoding bFGF and aFGF has been described (16,17,18). The bFGF gene is localized on human chromosome 4, while that of aFGF is on chromosome 5 (3). This suggests that through a process of gene duplication and evolutionary divergence, bFGF and aFGF have become separate gene products. The basic FGF gene, with its size greater than 38 kbp, encodes two exons widely separated by two introns: the first one separates codons 60 and 61, and the second separates codons 94 and 95. The aFGF gene has a similar organization, with 2 large introns located in identical positions in the coding sequence once basic and aFGF are properly aligned. Southern blot analysis of human genomic DNA has shown that there is only one bFGF and one aFGF gene. Therefore, all of the characterized or uncharacterized heparin-binding endothelial cell mitogens related to bFGF or aFGF are the products of a single bFGF or aFGF gene (3,16). In various cultured cells and tissues, the bFGF gene gives rise to two polyadenylated mRNA's of approximately 3.7 and 7.0 kb (reviewed in ref.19). The aFGF gene appears to encode a single mRNA species of approximately 4.1 kb (20). The primary translation product for either bFGF or aFGF is composed of 155 amino acids. Proteolytic cleavage from the precursor molecule of the first 9 (bFGF) or 15 residues (aFGF) would result in the generation of the mature proteins which can then be cleaved further in homologous positions to give the NH_2-truncated form of bFGF (des.1-15) or aFGF (des.1-6) (reviewed in ref.3,19).

3. bFGF GENE EXPRESSION IN CELL TYPES WHICH DO NOT EXPRESS THE bFGF GENE, BUT RESPOND TO bFGF, RESULTS IN AUTONOMOUS CELL PROLIFERATION

The concept of autocrine stimulation of cell proliferation postulates that normal diploid cells can gain growth autonomy by acquiring the ability to produce, secrete, and respond to a given growth factor (21,22). Verification of the autocrine hypothesis, in the case of bFGF, requires demonstration that expression of an introduced FGF gene in non-tumorigenic cells results in or contributes to the malignant transformation of those cells.

The hypothesis that inappropriate expression of bFGF could lead to cell transformation, has been tested by introducing into BHK-21 cells a plasmid that directs the high-level expression of human bFGF. BHK-21 cells were chosen because they do not express the bFGF gene, and in previous studies, they have been shown to be totally dependent on exogenous bFGF in order to proliferate when maintained under serum-free conditions (23). Finally, exogenous bFGF induces anchorage-independent soft agar growth of BHK-21 cells. This effect, however, is transient, and cells revert to their normal phenotype once the mitogen is removed. High-level bFGF expression in BHK-21 cells might therefore be expected to lead to permanent anchorage-independent soft agar growth of BHK-21 cells.

Southern blot analysis demonstrates that in BHK-21 cells, following transfection, at least one complete bFGF gene copy per cell has been stably incorporated; a second, partial copy also appears to be present. That the integrated foreign bFGF genes are actively expressed

is indicated by the detection through Northern blot analysis of 0.6 and 4.3 kb RNA transcripts, which were not seen in the parental BHK-21 cell RNA, and which differed in size from those found in bovine, mouse, or human cells expressing the endogenous bFGF gene (3.7 and 7.0 kb, respectively). Additionally, transfected cells contained $6X10^6$ bFGF molecules per cell, while no detectable bioactive FGF was found in the parental BHK-21 cells (24).

The drastic effect of high-level expression of bFGF on the proliferation of cells which do not normally express the bFGF gene was evident from the cell behavior following transfection. BHK-21 cells, which, on solid substrate, have an absolute requirement for exogenous FGF in order to proliferate in serum-free medium, proliferated actively in the absence of exogenous bFGF once transfected with plasmids carrying the human bFGF coding sequence under the control of the SV40 enhancer and human metallothionein II$_A$ promoter. Likewise, in soft agar, transfected cells formed colonies, while parental cells required the addition of exogenous bFGF in order to do so. This therefore demonstrates that whether bFGF interacts with its receptor within the cells, or at a surface location, it is capable of triggering the autonomous growth of a single cell in which it is synthesized in high amounts (24).

4. EXPRESSION OF bFGF AND aFGF IN TISSUES AND CULTURED CELLS
4.1 Organs/Tissues

So far, basic FGF has been purified from a wide variety of mesoderm- and neuroectoderm-derived organs and tumors (Table 1)(reviewed in ref.1,2). Depending on the organ from which bFGF is isolated, the predominant forms are either the 146 or 131 amino acids-long form, with smaller amounts of the 155 amino acids-long form being present. It is not known whether these various forms coexist in the tissues, or if they are artefactually created by specific proteases during FGF extraction and isolation (25). The 140 and 136 amino acids-long truncated aFGF forms are the predominant molecular species, with a small amount of the 155 amino acids-long form present. So far, aFGF has only been found in brain, retina, bone matrix, osteosarcoma (1,2) and various gliomas. Since aFGF is 30- to 100-fold less potent than bFGF, it contributes to only 8% and 0.15%, respectively, of the total mitogenic activity present in crude brain or retinal extract, with the rest contributed by bFGF (2).

4.2 Cultured cells

All organs that contain bFGF are heavily vascularized. This suggests that cells of the vascular system might synthesize bFGF. In fact, vascular endothelial cells express the bFGF gene, and they synthesize bioactive bFGF (26). In contrast, these cells do not express the aFGF gene or bioactive aFGF. Thus, bFGF could act as an autocrine growth factor for vascular endothelial cells. The synthesis of bFGF in vascular endothelial cells also provides an explanation for the rather ubiquitous distribution of bFGF. Basic FGF is also expressed in a wide variety of other normal diploid cells, all of which are sensitive to bFGF in vitro (2,19) (Table 1). Various tumors derived from those cells also express bFGF (Table 1). This has led to the proposal that uncontrolled expression of bFGF could be involved in the development and in the progression of tumors (27; reviewed in ref.19).

Table 1 Normal and neoplastic tissues or normal and transformed cells containing bFGF[a]

Normal or Transformed Tissue	Cultured normal diploid cells	Cultured tumor cells
Brain	Corneal endothelial cells	Y-1 Adrenal cortex cells
Retina	Capillary endothelial cells	Osteosarcoma U2OS
Pituitary	Pituitary cells	Ewing's sarcoma
Kidney	Ovarian granulosa cells	Rhabdomyosarcoma
Placenta	Adrenal cortex cells	Melanoma
Corpus Luteum	Lens epithelial cells	Hepatoma (Sk HP-1)
Adrenal Glands	Uterine epithelial cells	Retinoblastoma
Immune System	Myoblasts	
(Macrophage-Monocyte)	Retinal pigmented epithelial cells	
Prostate	Vascular smooth muscle cells	
Bone	Astrocytes	
	Osteoblasts	
Cartilage		
Chondrosarcoma		
Melanoma		

[a] So far, aFGF has only been detected in brain, retina, bone matrix, osteosarcoma osteoblasts, astrocytes, and fetal vascular smooth muscle cells.

The demonstration of the presence of bFGF in various normal diploid or tumor cells has also provided clues for its possible function in the most unexpected locations. For example, bFGF is produced in cells derived from retinoblastoma (19), a tumor thought to be derived from photoreceptor cells. Thus, it is possible that photoreceptors are the source of bFGF within the normal retina. Although this localization of bFGF would be impossible to predict based on what was previously known of its biological activity, it has recently been demonstrated that bFGF is a component of the rod's outer segment, where it is strongly bound to rhodopsin (28). This suggests a possible role of bFGF in phototransduction. The localization of bFGF in cell types which have no function yet assigned to them provides new insights on their possible physiological functions. Pituitary glands are known to contain high concentrations of bFGF (29), but the cell types responsible for synthesis of pituitary bFGF was unknown. Recent studies have shown that the main cellular source of bFGF in pituitary are follicular cells which can contain as many as 5×10^5 bFGF molecules per cell (30). Previous studies (31) have suggested that follicular cells may play an important role in the restoration of degenerated pituitary glandular tissues during the early stage of transplantation. This is in agreement with the suggestion by Farquhar et al. (32), that follicular cells play the role of nurse cells in the pituitary gland, similar to that of the Sertoli cells in the testes. Follicular cells are also involved in extracellular matrix (ECM) synthesis (33), and there is a large amount of experimental evidence that indicates that ECM plays an important role in cell growth and differentiation (34). Previous studies have shown that FGF controls the production of ECM components and could become an integral part of such structures (see Section 7), thereby further supporting the growth and differentiation of cells becoming associated with newly produced ECM. The ability of follicular cells to support the restoration of pituitary granular cells could therefore reflect their

ability to synthesize and release FGF associated with ECM components, such as heparan sulfate proteoglycans. It has also been suggested that follicular cells, which are mainly concentrated in the pars tuberalis, would provide support for the portal vessels. The presence of an angiogenic factor such as bFGF in follicular cells, could therefore relate to the development and maintenance of the differentiated state of the pars tuberalis microvasculature.

5. THE FGF RECEPTORS

All cell types which respond to bFGF or aFGF bear specific FGF cell surface receptors. In the baby hamster kidney (BHK-21) cell line, the density of FGF cell surface receptors is 10- to 30-fold higher than that of normal diploid cells. In BHK-21 cells, bFGF binds to specific high-affinity cell surface receptors (k_d=0.27 nM; 1.2×10^5 binding sites per cell). This compares with a k_d for aFGF of 0.25 nM and 8.7×10^4 binding sites per BHK-21 cell. As expected from the high degree of structural homology between bFGF and aFGF, both mitogens bind to the same receptor. Basic FGF and aFGF do not bind to other growth factor receptors, nor do other growth factors bind to the FGF receptor (1,2). Cross-linking of bFGF or aFGF to the BHK-21 cell surface receptor indicates that qualitatively, both mitogens interact with the same two M_r 145 kD and 125 kD membrane components, which could differ by their degree of glycosylation. Quantitatively, bFGF appears to display a higher affinity than aFGF for the M_r=145 kD receptor species, while aFGF displays a higher affinity than bFGF for the M_r=125 kD receptor species (8). This could help in explaining the different biological potencies of bFGF versus aFGF (see below).

The FGF receptors of muscle cells, Swiss 3T3, lens epithelial cells, human umbilical vein endothelial cells, rhabdomyosarcoma, Ewing's sarcoma (SK-ES1), and PC-12 cells have also been characterized. They all share similar molecular weight ranges, but might differ by their degree of glycosylation. However, in all cases both bFGF and aFGF interacts with the same receptor, with k_d's ranging from 45 pM to 200 pM (reviewed in ref.19).

6. IN VITRO BIOLOGICAL EFFECT OF FGF

Most of the biological studies with FGF have been done with the basic form. Only recently has the biological activity of aFGF started to be investigated. As expected from its high degree of structural homology with bFGF, aFGF has a mitogenic effect identical to that of bFGF, although it is, depending on the cell type, 30- to 100-fold less potent (1,2). The lower potency of aFGF makes it a weak agonist of the basic form, and may reflect the qualitatively different interactions of aFGF versus bFGF with its cell surface receptor species (1,2). Basic FGF has both acute and long-term effects on the morphology and growth pattern of responsive cells. It increases their migratory activity (4), and it makes confluent cultures of BALB/c 3T3 cells look "transformed" in that it induces reduced cell-substratum adhesion, growth in crisscross pattern, and increased membrane ruffling (35). Basic FGF can also induce the growth in soft agar of non-transformed cells, and in that model it potentiates the effect of TGF_B (2).

Basic FGF is a potent mitogen for mesoderm-derived cells (19) (Table 2), triggering cell proliferation with half-maximal and maximal effects at 1.5 and 10 pM, respectively. Basic FGF is mitogenic both for cells seeded at clonal density and for low-density cultures (36),

and greatly reduces their average cellular doubling time. This is primarily due to a shortening of the G1 phase of the cell cycle (37).

Table 2. Cell Types for which FGF's are mitogenic or affect differentiation

Normal Diploid Cells	bFGF	aFGF		bFGF	aFGF
Glial and astroglial cells	+(D)	+	Granulosa cells	+	+
Oligodendrocytes	+(D)	+	Prostatic epithelial cells	+	+
Trabecular meshwork cells	+	?	Mesothelial cells	+	+
Endothelial cells from capillary, large vessel and endocardium	+(D)	+(D)	Neuronal cells	+	?
Corneal endothelial cells	+(D)	+(D)	**Established Cell Lines**		
Fibroblasts	+	+	Rat fibroblast-1	+	+
Myoblasts	+(D)	+(D)	Balb/c 3T3	+	+
Vascular smooth muscle	+	+	Swiss 3T3	+	+
Chondrocytes	+(D)	+(D)	BHK-21	+	+
Osteoblasts	+	+	A-204 Rhabdomyosarcoma	+	?
Blastema cells	+	?	PC-12	(D)	(D)
Adrenal cortex cells	+	+			

(D) = induces differentiation   + = positive effect on cell proliferation
 ? = effect on cell proliferation not determined

    Basic FGF stabilizes the phenotypic expression of cultured cells. (Table 2) (4). This property is particularly interesting, since it has made possible the long-term culturing of cell types that otherwise would lose their normal phenotype in culture when passaged repeatedly at low cell density (37). This biological effect of bFGF has been studied exhaustively in endothelial cells derived from large vascular vessels or cornea that were cloned and maintained in the presence of bFGF and then deprived of it for various time periods (reviewed in ref.4,36,38). This effect of bFGF on cell differentiation may be due to its ability to

control the synthesis and deposition of various ECM components that are known to affect cell surface polarity and gene expression. These include collagen, fibronectin, laminin, and proteoglycans (reviewed in ref.39,40). Basic FGF can also induce capillary endothelial cells to invade a three-dimensional collagen matrix and to organize themselves to form characteristic tubules that resemble blood capillaries. Concomitantly, bFGF stimulates endothelial cells to produce a urokinase-type plasminogen activator (PA), a protease that has been implicated in the neovascular response. Thus bFGF can stimulate processes that are characteristic of angiogenesis in vivo, including endothelial cell migration, invasion, and production of plasminogen activator (41).

When added to chondrocytes, bFGF can act as a mitogen as well as a differentiating agent (42). While costal chondrocytes grown in absence of bFGF soon assume a fibroblastic appearance and lose their ability to synthesize and release chondroitin sulfate, proteoglycans, and collagen type II, cells grown in the presence of bFGF retain these capabilities and at confluence become embedded in a thick ECM which has all of the characteristics of that produced in vivo (42). Interestingly enough, cells will express their correct phenotype only if exposed to bFGF when dividing actively; when added to confluent layers of dedifferentiated and resting chondrocytes, bFGF can no longer reverse their phenotype. bFGF also has a pronounced effect on astrocytes, stimulating both their proliferation and differentiation as reflected by its positive effect on the synthesis of glial fibrillary acidic protein (43,44).

Probably the most spectacular effect of bFGF on cell differentiation is observed with nerve cells. Togari and co-workers (45,46) first reported that bFGF acts as a differentiation factor in a rat pheochromocytoma (PC-12) cell line by inducing both neurite outgrowth and ornithine decarboxylase activity. Later it was shown that aFGF has similar properties (47,48), and that PC-12 cells express specific FGF receptor sites. Similar effects of bFGF on nerve cells have been reported by Walicke et al. (49), using highly purified populations of fetal rat hippocampal neurons. Under well-defined serum-free cell culture conditions, bFGF can increase both neuronal survival and neurite extensions. Furthermore, the addition of bFGF to rat cerebral cortical neurons markedly enhances their survival and the elaboration of neurites (50). These results suggest that bFGF may function as a neurotropic agent in the central nervous system.

Not all of the effects of bFGF on cell differentiation are positive effects. For example, bFGF can delay differentiation and fusion of myoblasts (51). In some established myoblast cell lines, bFGF and aFGF can induce a decrease in creatine phosphokinase expression (52). These inhibitory FGF effects on differentiation have been attributed to the ability of FGF to keep myoblast populations in an active proliferative stage, thereby decreasing the percentage of cells in slow growing populations which would enter during their extended G_0, G_1 phase into a stage of terminal differentiation (53).

The biological effects of FGF on various cell types can be positively or negatively modulated by various biological agents: these include TGF_B, tumor necrosis factor, Interferon, and heparin (reviewed in ref.19).

7. FGF AND THE EXTRACELLULAR MATRIX

In the early stage of embryonic development, different tissues composing an organ are formed as a result of strictly timed and

spatially interrelated proliferative and differentiative events. This involves the interaction of cells with newly formed ECM, a process that promotes their proliferation and which stabilizes their newly acquired phenotype (reviewed in ref.34). Studies have shown that ECM produced by vascular or corneal endothelial cells can mimic all of the effects of bFGF, including those on cell proliferation and/or differentiation (34,39; 54), suggesting that FGF associated with ECM components might be the active factor. Indeed, bFGF has a high affinity for heparin. This glycosaminoglycan is closely related to heparan sulfate, which is produced in large quantities by both corneal and vascular endothelial cells, and is a structural component of their ECM (reviewed in ref.19). The possibility therefore exists that FGF could be secreted by the cells in association with extracellular matrix components and become an integral part of the ECM. Indirect evidence for the integration of bFGF into an insoluble substrate such as the ECM can also be derived from the observation that neither media conditioned by capillary or corneal endothelial cells has a significant impact on their proliferation. In contrast, their own denuded ECM will induce them to rapidly proliferate and assume the proper phenotype once confluent. Thus bFGF, in contrast to other conventional growth factors such as TGF_B, EGF, and PDGF, may not be released in a soluble form (1,2). This is in agreement with the fact that neither bFGF nor aFGF are synthesized with a conventional signal peptide (16,17,20). Both growth factors, however, might be associated with ECM components, and as such, be transported to the cell exterior, where they could interact with specific FGF cell surface receptors to induce their biological effects (55), or be stored in the ECM and later released following hydrolysis of ECM components. In that context, it is interesting to note that during morphogenesis of lobular organs, the areas with the greatest mitotic activity are located in areas where hydrolysis of the ECM occurs. Similarly, in the kidney, angiogenesis correlates well with the hydrolysis of the kidney mesenchymal stroma. In the adult, heparan sulfate present in ECM could be degraded by heparitinase, an enzyme released either by platelets, when they attach to the subendothelium, or by macrophages once they are activated (reviewed in ref.2). This could ultimately lead to the solubilization of heparan sulfate/FGF complexes which would be biologically active, and could participate in various repair or developmental processes.

8. IN VIVO EFFECTS OF FGF

In early embryonic development, the basic body plan arises because cells in different regions of the egg become programmed to follow different pathways (56). During oogenesis, differences arise between the animal and vegetal halves of the eggs. Fertilization results in a subdivision of the vegetal half into a dorsal-vegetal and a ventro-vegetal region. Mesoderm is then induced from the animal hemisphere by signal(s) originating from the vegetal region of the egg (56,57). This induction is an instructive phenomenon that suppresses epidermal differentiation of cells from the animal pole, and directs them instead to differentiate into mesodermal cells. Signal(s) originating from the dorsal-vegetal region lead to the formation of dorsal-type mesoderm, mostly consisting of notochord and somites, while signal(s) originating from the ventro-vegetal region lead to the formation of ventral-type

mesoderm, consisting primarily of blood cells, mesenchyme, and mesothelium.

In recent studies Slack and his colleagues (5) have investigated the possibility of bFGF mimicking the effect of the ventro-vegetal signal(s) responsible for the formation of ventral-type mesoderm. Tissue explants isolated from animal pole of stage 8 Xenopus blastulae normally differentiate into epidermis or undifferentiated epidermal cells. When similar explants are exposed to bFGF, cells differentiated instead into mesodermal structures, forming concentric arrangements of loose mesenchyme, mesothelium, and blood cells within an epidermal jacket. Explants also contained significant amounts of muscle blocks (5). Therefore, in early embryo, bFGF can act as a primordial differentiation factor, inducing the ectoderm to become mesoderm. This is in close agreement with previous in vitro studies which have shown that FGF had a transforming activity and could act as a morphogen, as well as a mitogen, on practically all mesoderm-derived cells studied to date. In lower vertebrates (amphibians) bFGF can promote limb regeneration (58). This tends to support the concept that bFGF could be involved in the neurotropic control of this process (reviewed in ref.2). In addition to being a neurotrophic factor involved in limb regeneration, bFGF could also play a role in the early development of the nervous system. Basic FGF promotes both the survival and differentiation of nerve cells derived either from the hippocampal region or the cortex. Nerve cells have also been shown to contain bFGF (59), and preliminary studies have demonstrated that neuronal cell populations derived from early embryonic brain do proliferate in response to bFGF, and later express cholinergic differentiation. All these effects point towards bFGF's possible role in CNS development. FGF could also have pronounced effects on the proliferation and differentiation of other brain cell populations, such as astrocytes and oligodendrocytes, by influencing their glial properties during normal development or subsequent to a specific pathogenic event (19). Through its angiogenic properties bFGF could influence CNS development, since it has been reported that FGF could be responsible for capillary ingrowth into the brain (60).

Basic FGF, which has been detected in macrophages, could play a crucial role in the wound healing processes, following its release from the damaged cells. Interestingly, bFGF, unlike other growth factors such as EGF, PDGF or TGF$_B$, can stimulate both in vitro as well as in vivo the proliferation of all the cell types involved in wound healing (1,2). These include capillary endothelial cells, vascular smooth muscle, and fibroblasts, not withstanding other cell types which are involved in the wound healing of specialized territories, such as chondrocytes, myoblasts, etc. (4,36). Basic FGF also increases the formation of granulation tissue in vivo (reviewed in ref.19), and the synthetic function of fibroblasts and myoblasts. It also stimulates the rate of reepithelialization of the epidermis detached from dermis after blister induction. In other tissues such as cartilage, bFGF can promote chondrossification, and its presence in bone matrix could indicate that it could play an important role in the development and growth of osseous tissue (reviewed in ref.19).

Basic FGF acts as a potent angiogenic factor in vivo, as demonstrated by the the rabbit cornea, the chick chorioallantoic membrane (CAM) or the hamster cheek pouch assays. These observations are supported by the demonstration of bFGF as the major angiogenic agent in

healthy vascularized tissues such as corpus luteum, adrenal gland, kidney, and retina (reviewed in ref.12,19).

Recent studies have shown that bFGF is identical to the tumor angiogenic factor (61). Thus, bFGF might play an important role in tumor progression. By increasing capillary endothelial cell proliferation, and inducing the sprouting of new capillaries into solid tumors, bFGF might allow an increased supply of the tumors with O_2 and nutrients, and also the metastasis of the tumors. Basic FGF could also act at the level of the tumor cell itself. By increasing its PA level as well as by increasing the secreted levels of various proteases and collagenase, bFGF would facilitate the metastasic process and tumor invasion (reviewed in ref.19). It could also act as a mitogen for the tumor cells. The normal diploid cells listed in Table 2 are either uniquely dependent on bFGF or they require various growth factors including FGF, EGF, and PDGF in order to proliferate (reviewed in ref.37). Upon neoplastic transformation of cells that depend only on FGF, uncontrolled expression of bFGF in these cells could occur and make them divide in an uncontrolled manner. In the case of cells responding to multiple growth factors, uncontrolled expression of bFGF might make them independent of other exogenous growth factors during their growth phase.

9. CONCLUSION

The importance of FGF in the ontogeny of development and biopathology of mesenchymal tissues can no longer be ignored. The recent molecular characterization of FGF as well as the cloning and mapping of its genes, has lead to the general consensus that all the numerous heparin-binding growth factors were, in fact, represented by 2 single gene products: basic and acidic FGF. The further demonstration that both basic and acidic FGF have a high degree of structural homology and bind to the same receptor has lead to the conclusion that they had identical biological roles, although they differed in their specific activity.

Probably one of the most important questions to be resolved is the in vivo role of FGF. The high degree of structural conservation of bFGF through species as different as mammalian, avian and amphibian, as well as its presence in all vertebrates studied to date--including piscean--indicate that in vivo FGF could have a very primordial role. This is, in fact, what seems to be indicated by the studies of Slack et al. (5), which have shown that bFGF can act as a primordial morphogen at one of the earliest embryonic stages, inducing the transformation of cells destined to be ectodermal, into mesenchymal cells. This is consistent with the in vitro properties of bFGF, which has been shown to act as mitogen as well as morphogen for all mesenchymal cells studied to date. It is also in agreement with the ability of bFGF to support the regeneration process in lower vertebrates.

One of the most popularized aspects of the in vivo biology of FGF is based on its angiogenic activity. Basic FGF has been shown to induce in early embryo the appearance of blood islands and, its effects on capillary endothelial cells in vitro or in vivo, as well as its distribution in various organs or tumors known to have angiogenic potential provide a common denominator for its widespread angiogenic activity. However, narrowing the experimental focus to FGF's angiogenic properties would produce an extremely limited view of its potential targets in vivo. The cell types listed in Table II clearly indicate in which organ an effect of FGF, either in tissue formation or repair

process, should be investigated. In particular, the in vivo mitogenic and differentiation effects of bFGF on various cell types of the nervous system would be worth investigating, since it has been shown to control both the proliferation and differentiation of glial cells, oligodendrocytes, and astroglial cells, as well as acting as a triggering mechanism for the differentiation of nerve cells in vitro.

Although it is likely that bFGF could still control the proliferation and differentiation of mesenchymal cells later in the ontogeny of development, little is actually known about the implication of the activation of the FGF gene expression in neoplastic transformation. It has been speculated that activation of growth factor-controlling gene expressed early in embryogenesis and later repressed could lead to neoplastic transformation (11). Transfection of normal diploid cells such as vascular endothelial cells with plasmids carrying FGF-cDNA, resulting in the constitutive expression of bFGF, could be a useful approach to testing that hypothesis. Cells expressing high bFGF levels could also be used for studies dealing with the locus of action of FGF (either intra- or extracellular locations), the transcellular transport of FGF and pathways of FGF release from the cells. Of equal importance will be studies dealing with mechanisms and factors controlling the expression of the FGF receptor, since this process would ultimately determine the time, as well as the cell type, which would be FGF-responsive. Such studies cannot be initiated until we have a better understanding of the molecular properties of the FGF receptor, and in particular, until we have cloned FGF receptor cDNA, which can later be used to study its expression.

REFERENCES

1. Gospodarowicz D, Neufeld G, Schweigerer L: Fibroblast growth factor. Mol Cell Endocrin 46:187-206, 1986.
2. Gospodarowicz D, Neufeld G, Schweigerer L: Molecular and biological characterization of fibroblast growth factor: an angiogenic factor which also controls the proliferation and differentiation of mesoderm and neuroectoderm-derived cells. Cell Differ 19:1-17, 1986.
3. Abraham,A, Whang,L, Tumolo,A, Mergia,A, Fiddes,JC: Human basic fibroblast growth factor: nucleotide sequence genomic organization and expression in mammalian cells, in: Molecular Biology of Homo Sapiens, Cold Spring Harbor, New York, Vol.51, pp 657-668, 1987.
4. Gospodarowicz D: Biological activity in vivo and in vitro of pituitary and brain fibroblast growth factor. In: RJ Ford, AL Maizel (eds) Mediators in Cell Growth and Differentiation, Raven Press,New York,pp109-134, 1985.
5. Slack JM, Darlington B, Heath H, Godsave S: Heparin binding growth factors as agents of mesoderm induction in early Xenopus embryo. Nature 326:197-200, 1987.
6. Esch F, Baird A, Ling N, Ueno N, Hill F, Deneory R, Klepper R, Gospodarowicz D: Primary structure of bovine pituitary basic fibroblast growth factor (FGF) and comparison with the amino terminal sequence of bovine brain acidic FGF. Proc Natl Acad Sci USA 85:6507-6511, 1985.
7. Gospodarowicz D, Cheng J, Lui GM, Baird A, Esch F, Bühlen P: Corpus luteum angiogenic factor is related to fibroblast growth factor. Endocrinology 117:2283-2291, 1985.

356

8. Neufeld G, Gospodarowicz D: Basic and acidic fibroblast growth factor interact with the same cell surface receptor. J Biol Chem 261:5631-5637, 1987.
9. Esch F, Ueno N, Baird A, Hill F, Denoroy L, Ling N, Gospodarowicz D, Guillemin R: Primary structure of bovine brain acidic fibroblast growth factor (FGF). Biochem Biophys Res Commun, 133:554-562, 1985.
10. Gimenez-Gallego G, Conn G, Hatcher VB, Thomas KA: Human brain-derived acidic and basic fibroblast growth factors: amino terminal sequences and specific mitogenic activities. Biochem Biophys Res Commun 135:561-566, 1986.
11. Sporn MB, Roberts AB: Autocrine growth factor and cancer. Nature 313:745-747, 1985.
12. Dickson C, Peters G: Potential oncogene product related to growth factors. Nature 326:833, 1987.
13. Taira M, Yoshida T, Miyagawa K, Sakamoto H, Terada M, Sugimura T: cDNA sequence of human transforming gene hst and identification of the coding sequence required for transforming activities. Proc Natl Acad Sci USA 84:2980-2984, 1987.
14. Dickson C, Smith R, Brookes S, Peters G: Tumorogenesis by mouse mammary tumor virus: Proviral activation of a cellular gene in the common integration region Int-2. Cell 37:529-536, 1984.
15. Sakamoto H, Mori M, Taira M, Yoshida T, Matsukawa S, Shimizu K, Sekiguchi M, Terada M, Sugimura T: Transforming gene from human stomach cancers and a non-cancerous portion of stomach mucosa. Proc Natl Acad Sci USA 83:3997-4001, 1986.
16. Abraham,JA, Whang,JL, Tumolo,A, Mergia,A, Friedman,J, Gospodarowicz,D, Fiddes,JC: Human basic fibroblast growth factor: nucleotide sequence and genomic organization, EMBO J, 5:2523-2528, 1986.
17. Abraham,JA, Mergia,A, Whang,JL, Tuomolo,A, Friedman,J, Hjerrild,KA, Gospodarowicz,D, Fiddes,JC: Nucleotide sequence of a bovine clone encoding the angiogenic protein, basic fibroblast growth factor, Science, Washington D.C.,233:545-548, 1986.
18. Mergia A, Tumolo A, Haaparanta T, Whang JL, Gospodarowicz D, Abraham JA, Fiddes JC: Isolation and characterization of the human gene for acidic FGF. In preparation, 1987.
19. Gospodarowicz D, Ferrara N, Schweigerer L, Neufeld G: Structural characterization and biological functions of fibroblast growth factor. Endocrine Review 8:95-114, 1987.
20. Jaye M, Howk R, Burgess W, Ricca GA, Chiu IM, Ravera MW, O'Brien SJ, Modi WS, Maciag T, Drohan WN: Human endothelial cell growth factor: cloning, nucleotide sequence, and chromosome localization, Science. 233:541-544, 1986.
21. Todaro GJ, DeLarco E, Nissley SP, Rechler MM: MSA and EGF receptors on sarcoma virus-transformed cells and human fibrosarcoma cells in culture. Nature, London 267:526-528, 1977.
22. Sporn MB, Todaro G: Autocrine secretion and malignant transformation of cells. N Eng J Med 303:878-880, 1980.
23. Neufeld G, Massoglia S, Gospodarowicz D: Effect of lipoproteins and growth factors on the proliferation of BHK-21 cells in serum-free cultures. Regulatory Peptides 13:293-305, 1986.
24. Neufeld G, Ponte P, Mitchell R Gospodarowicz D: Expression of human basic fibroblast growth factor cDNA in baby hamster kidney-

derived cells results in autonomous cell growth. J Cell Biol (submitted), 1987.

25. Ueno K, Baird A, Esch F, Ling N, Guillemin R: Isolation of an amino acid terminal extended form of basic fibroblast growth factor. Biochem Biophys Res Commun 138:580-588, 1986.

26. Schweigerer L, Neufeld G, Friedman J, Abraham JA, Fiddes JC, Gospodarowicz D: Capillary endothelial cells express basic fibroblast growth factor, a mitogen that stimulates their own growth. Nature, London 325:257-259, 1987.

27. Schweigerer L, Neufeld G, Mergia A, Abraham JA, Fiddes JC, Gospodarowicz D: Basic fibroblast growth factor in human rhabdomyosarcoma cells: implications for the proliferation and neovascularization of myoblast-derived tumors. Proc Natl Acad Sci USA 84:842-846, 1987.

28. Plouet J, Mascarelli F, Lagente O, Dorey C, Lorans G, Faure J.-P, Courtois Y:) Eye derived growth factor: a component of rod outer sement implicated in phototransduction. In: (eds) Agardh E, Ehinger B. Retinal Signal Systems, Degenerations and Transplants. Elsevier Science Publishers BV, New York, pp 311-320, 1986.

29. Gospodarowicz D: Localization of a fibroblast growth factor and its effect alone and with hydrocortisone on 3T3 cell growth. Nature London 249:123-127, 1974.

30. Ferrara N, Schweigerer L, Neufeld G, Gospodarowicz D: A new function for pituitary follicular cells: The production of basic Fibroblast growth factor. Proc Natl Acad Sci, USA.In press., 1987

31. Gon G, Shirasawa N, Ishikawa H: Appearance of the cyst or ductile like structures and their role in the restoration of the rat pituitary autograft. Anat Rec 217:371-378, 1987.

32. Farquhar MG, Stutelsky EH, Hopkins CR: Structure and function of the anterior pituitary and dispersed pituitary cells. In Vitro Studies. In: A Tixer-Vidal, MG (eds) The Anterior Pituitary Gland. Farquhar, Academic Press, New York, pp 82-135, 1975.

33. Vila-Porcile E, Olivier L: The problem of the folliculo-stellate cells in the pituitary gland. In: (ed) PM Motta, Ultrastructure of Endocrine Cells and Tissues. Martinus Nijhoff Publishers, 6Boston, pp 64-76, 1984.

34. Gospodarowicz D, Cohen DC, Fujii DK: Regulation of cell growth by the basal lamina and plasma factors: relevance to embryonic control of cell proliferation. In: G Sato, A Pardee, and D Sirbasku, (eds) Cold Spring Harbor Conferences on Cell Proliferation Vol.9: Growth of cells in hormonally deficient media. Cold Spring Harbor, New York, pp95-124, 1982.

35. Gospodarowicz D, Moran J: Effect of a fibroblast growth factor, insulin, dexamethasone, and serum on the morphology of BALB/c 3T3 cells. Proc Natl Acad Sci USA, 71:4648-4652, 1974.

36. Gospodarowicz D: Fibroblast and epidermal growth factors: their uses in vivo and in vitro in studies on cell functions and cell transplantation. Mol Cell Biochem 25:79-110, 1979.

37. Gospodarowicz D, Greenburg G, Bialecki H: Factors involved in the modulation of cell proliferation in vivo and in vitro: the role of fibroblast and epidermal growth factors in the proliferative response of mammalian cells. In Vitro 14:85-118, 1978.

38. Gospodarowicz D: Purification of brain and pituitary FGF. In: D Barnes and D Sirbasku (eds) Methods in Enzymology: Peptide growth factors. Academic Press, Orlando, FL,147A:106-119, 1987.

358

39. Gospodarowicz D, Greenburg G: Growth control of mammalian cells. Growth factors and extracellular matrix. In: M Ritzen, A Aperia, K Hall, A Larsson, A Zetterberg, R Zetterstrom, (eds) The Biology of Normal Human Growth, Raven Press, New York, pp.1-21, 1981.

40. Gospodarowicz D: The control of mammalian cell proliferation by growth factors, extracellular matrix and lipoproteins. J Inv Derm 81: 41-50, 1983.

41. Montesano R, Vassali JD, Baird A, Guillemin R, Orci L: Basic fibroblast growth factor induces angiogenesis in vitro. Proc Natl Acad Sci USA 83:7297-7301, 1986.

42. Kato Y, Gospodarowicz D: Sulfated proteoglycan synthesis by rabbit costal chondrocytes grown in the presence and absence of fibroblast growth factor. J Cell Biol 100:477, 1985.

43. Pettman B, Weibel M, Sensenbrenner M, Labourdette G: Purification of two astroglial growth factors from bovine brain. FEBS Letters 189:102-108, 1985.

44. Morrison R, DeVeillis J, Lee YL, Bradshaw RA, Eng, LF: Hormones and growth factors induced the synthesis of glial fibrillary acidic protein in rat brain astrocytes. J Neurosci Res 14:167-172, 1985.

45. Togari A, Baker D, Dickens G, Guroff G: The neurite-promoting effect of fibroblast growth factor on PC-12 cells. Biochem Biophys Res Commun 144:1189-1196,1983.

46. Togari A, Dickens G, Huzuya J, Guroff G: The effect of fibroblast growth factor on PC-12 cells. J Neurosci 5:307-315, 1985.

47. Neufeld G, Gospodarowicz D, Dodge L, Fujii DK: Heparin modulation of the neurotropic effects of acidic and basic fibroblast growth factors and nerve growth factors on PC-12. J. Cell.Physiol., 131:131-140, 1986.

48. Wagner JA, D'Amore P: Neurite outgrowth induced by an endothelial cell mitogen isolated from retina. J Cell Biol 103:1363-1370, 1986.

49. Walicke P, Cowan M, Ueno K, Baird A, Guillemin R: Fibroblast growth factor pomotes survival of dissociated hippocampal neurons and enhances neurite extension. Proc Natl Acad Sci USA 83:3012, 1986.

50. Morrison RS, Sharma A, DeVeillis J, Bradshaw A: Basic fibroblast growth factor supports the survival of cerebral cortical neurons in primary culture. Proc Natl Acad Sci USA 83:7537-7541, 1986.

51. Gospodarowicz D, Weseman J, Moran J, Lindstrom J: Effect of fibroblast growth factor on the division and fusion of bovine myoblasts. J Cell Biol 70:395-405, 1976.

52. Lathrop B, Olson E, Glaser L: Control by fibroblast growth factor of differentiation in the BC3H1 muscle cell line. J Cell Biol 100:1540-1548, 1985.

53. Lathrop B, Olson E, Glaser L: Control of myogenic differentiation by fibroblast growth factor is mediated by position in the G1 phase of the cell cycle. J Cell Biol101:2194-2202, 1985.

54. Gospodarowicz D, and Tauber J-P: Growth factors and extracellular matrix. Endocrine Review 1:201-227, 1980.

55. Vlodavsky I, Folkman J, Sullivan R, Frieman R, Ishai R, Michaeli Sasse J, Klagsbrun M: Endothelial cell-derived basic fibroblast growth factor: Synthesis and deposition into subendothelial extracellular matrix. Proc Natl Acad Sci USA 84:2282-2296, 1987.

56. Slack JM: From egg to embryo: determinative events in early development. Cambridge University Press, Cambridge and London, 1983.
57. Nieuwkoop P: The formation of mesoderm in Urodelean amphibians. I.Induction by the endoderm. Wilhelm Roux' Arch Entw Mech Org 162:341-373, 1969.
58. Gospodarowicz, D, Rudland, P, Lindstrom, J, and Benirschke, K: Fibroblast growth factor: localization, purification, mode of action, and physiological significance. Nobel Symposium on Growth Factors. In Advances in Metabolic Disorders (H. Luft, ed.), Stockholm, v.8,302-341, 1975.
59. Pettman B, Labourdette G, Weibel M, Sensenbrenner M: The brain fibroblast growth factor (FGF) is localized in neurons. Neurosci. Letters 68:175-179, 1986.
60. Risau W: Developing brain produces an angiogenesis factor. Proc Natl Acad Sci USA 83:3855-3859, 1986.
61. Klagsbrun M, Sasse J, Sullivan R, Smith JA: Human tumor cells synthesize an endothelial cell growth factor that is structurally related to basic fibroblast growth factor. Proc Natl Acad Sci USA 83:2448-2452, 1986.

Gospodarowicz, D. : <u>Biochemistry and molecular biology of FGF</u>

Sherman :   I wonder if you could elaborate a little bit on the receptor situation. You mentioned that there are two receptors. I assume they can crossbind the FGF because the homology and possibly differing affinity. What is your view of the relationship of one receptor to the other and whether their localization on different cell types maybe the same or different ?

Gospodarowicz : Well, until now, there has been at least twelve different cell types which have been analyzed. I isolated several cell lines and normal diploid cells and different groups, including Schlessinger's obtained knowledge that basic and acidic FGF cross react with the same receptor. Now the form of the receptor varies with the cell type. In the BHK 21 cell, in the glioma cell and I think on the rhabdomyosarcoma cell, as well as the islet cell, there are two forms of the receptors which only differ by the degree of glycosylation. Whether they are being induced because we badly manipulated these cells or whether they really differ, we don't know since they are membrane preparations and there could be problems of degradation of the receptor. But it looks like if the interactions - based on the S-S crosslinking of basic and acidic GF's -, seem to differ.

Schlessinger agrees that the acidic form seems to have a higher affinity for the low molecular weight form of the receptor and the basic seems to have a higher affinity - apparent affinity I would say - for the high molecular weight form of that receptor.

Sherman :   Is the difference in affinity such that you might expect there to be some specificity involved ?

Gospodarowicz : It is difficult to say because these are apparent affinities. The way the experiment is being done is by the S-S linking and the chasing with increasing - you know before the cross linking you chase with increasing concentrations of gold which means that you get two little black bands which are radioactive and to quantify those kinds of data, is very difficult even to put an evaluation for the K_d of one receptor for basic versus acidic using that kind of technique.

Remacle : With the WI-38 fibroblasts, you could pass the Hayflick's limits of 50 doublings in presence of FGF.

Gospodarowicz : I think the limit of 50 population doublings is rather arbitrary. I also think even that HAYFLICK said it was arbitrary.

I think the main question in looking at these problems of senescence in vitro is first to keep in mind that the cells I showed you are bovine cells where chromosome rearrangements can take place easily. It does not mean they become polyploid. It is simply Robertsonian fusions I am speaking of. Under these conditions, FGF can indeed make them proliferate for extensive periods of time. Ordinarily it is rather the technician who quits the cells. That's the way we terminated the experiment anyway. That and contamination. But, in the case of HAYFLICK, he was working with human cells and human cell chromosomes are extremely stable and do not change that much, do not exchange much and therefore it is one of the first reason why human cells are so difficult to maintain in culture for long periods of time. It is rather due to chromosome stability of the two species.

Remacle : My question had also another implication. There is another way to increase the population doublings. It consists in using corticosteroids. Do you have any idea about the possible effects of corticosteroids upon FGF

expression or secretion because when CRISTOFALO did his work, there was the suggestion that the effect observed was an indirect one.

Gospodarowicz : That could be. That work was done in presence of serum anyway. But we see the same effect in serum-free medium also. But to answer your question as to glucocorticoid effect on cell senescence in the system I haven't looked at, I have'nt looked at it so I cannot even comment on it.

Schönherr : Is one of these effectors used in clinical trials and if yes, for what kind of purpose is it used ?

Gospodarowicz : There is actually an important use of FGF in humans. I mean the main use is evidently wound healing. That, every growth factor will accomplish anayway. However, due to the fact that it is an angiogenic factor, it is probably more useful and has more spectacular effects than other growth factors and these are undergoing clinical trials, for what I understand, from the CBI people, that means the group of John FIDDES.
So, one thing which is in my opinion more interesting was the recent clinical trial with FGF to look on receptor affinity showing the brain for those people having Parkinson disease and the third clinical trial - which is not going actually but many people are interested by it - evaluates in the central nervous system, the regeneration of peripheric spinal motor neurones. Well, FGF seems to have a rather drastic effect. But I am not involved in these things. That is commercial, so, CBI is doing these studies.

Spier : Can I ask another question, how available is FGF and what does it cost ?

Gospodarowicz : There is some from pituitary, and 127 mg from 5 g of bacteria.

Spier :   You mean from engineered bacteria ?

Gospodarowicz : It is true ! I mean when you have the cDNA vector... The main problem is that for scientists it will cost nothing because you can get the cDNA by writing to CBI and the understanding is that when you publish the cloning you are bound to give not the vector but at least the cDNA to scientists who required it. When you have the cDNA, you can put it in bacteria, it is not that difficult. It is evident that when you speak in terms of private industry, the situation is quite different because I don't know of any company who will be willing to sign an agreement with CBI as to the use of the cDNA. But they can make their own you know. You can make a synthetic gene also. That is the approach that Merck has chosen since they were unable to clone the acidic FGF. If you go to people like the group of AMGEN, I think now they are selling FGF for ... I think it was 10 $ per microgram which is outrageous evidently but pretty cheap compared to the others.

Cassiman :   You have two molecules, one acidic and the other basic which are clearly coded by two different genes and bind to the same receptor although with different affinities. Would you expect that maybe one of these two molecules would have a completely different function that you are not aware of at the present time ?

Gospodarowicz : No. Again I have to be cautious. I do not expect it for two reasons. First I have looked extensively in the body and there are very few tissues left to look at, frankly. The second reason is that there is a precedent. In the case of PDGF you have the A and the B chains and it is known that the A and the B chains are two gene products localized on different chromosomes. Now the PDGF can be

either an heterodimer or an homodimer. It has been reported rather recently in Stockholm that when you deal with PDGF naturally produced, which is the heterodimer composed of the A and B chains, it is not as potent as compared to basic FGF and it does not have transforming activity. But when you deal with the homodimer the BB chains, than you have transforming activity and it is extremely potent, almost at the level of basic FGF. So, growth factors can exist under two forms, have their genes localized on two different chromosomes and depending on the combination of AB or BB for PDGF, you can either have a physiological factor which is inocuous or a transforming factor. So, I think in the case of FGF, you have the same situation. The two genes are clearly derived from a common ancestral gene, are localized on different chromosomes. There are differences in primary structure but you know - not that much after all - ; they interact with the same receptor as the two forms of PDGF would, they react with the receptor quite differently but similarly.

Cassiman :   So, you are agreeing with what I am saying that there might be differences in the effects of the two factors.

Gospodarowicz : Yes.   There is only one thing : what I have not yet tested is whether or not acidic FGF has transforming activity.   This I do not know.   I know this with basic FGF because we were in a hurry to do the experiments and because we had it.   Now we have also acidic FGF.   It is at that level that you probably will see the difference if you are looking for one.   It will be probably like PDGF where the experiment has been done, where the heterodimer is not transforming when the homodimer is.

Sherman :   Another question aimed at the possible functions of one or both of these FGF's.   The data you presented from your experiments obviously suggest an interesting role in embryogenesis.   The question is what might be these FGF's

doing in the adult. You listed a series of activities that they can ostensibly have but these as you pointed out, to some degree, maybe unphysiological, the thing that interests me. It is my understanding that both FGF's do not have signal peptides. Is that correct ?

Gospodarowicz : It is too premature to say that yet, because as you may be aware since you seem to know the field, quite well the group of RIFKIN as well as another biotechnology company have reported that higher molecular weight forms of FGF could exist, as high as 25.000 molecular weight versus the 16.000 one that we have isolated. We don't know if this form has the signal peptide or not. Because of the commercial aspect, there are very few reports on the nucleotide sequence they have obtained. So, as RIFKIN says in his PNAS article, although the form we have isolated does not have the signal sequence, it is still premature to say whether FGF could be released or not and it could be that the higher molecular weight could have a signal peptide sequence.

Sherman : I was wondering whether the embryonic forms might have a signal sequence.

Gospodarowicz : Embryonic form does not have it. We have cloned - I mean not me but David KIMMELMAN in collaboration with CBI - the first and last exon - the middle one is missing - first in Xenopus laevis to follow the expression, to see if it was indeed the factor and from the structural data they particularly analyzed the first exon. There is no signal peptide. But you have to keep in mind again that the lack of a signal peptide here will fit very well with the role of FGF in the embryonic development because in early embryo, in order to get proper morphogenesis, the way the signal is being sent is always through the extracellular matrix according to the experiments of Merton BURNFIELD and others. It is a very short

signal, it goes one hundred angströms, in accordance with other experiments which showed these inductions to be extremely well limited. If FGF had a signal peptide, it would be relatively catastrophic because it could be a humoral factor being secreted and going everywhere in which case you can no longer get localization of the effect within a distance of angströms or microns when you really want the cell to be hit with extremely great selective ability.

Note from the Director. Due to bad tape recording, the comment by Dr M. Sherman and the final answer by Dr G. Gospodarowicz are missing from the discussion.

CULTURE OF HYBRIDOMAS - A SURVEY

O.-W. MERTEN

INSTITUT PASTEUR, Technologie Cellulaire,
25, Rue du Docteur Roux, F-75724 Paris Cedex 15

ABSTRACT
This survey describes today's hybridoma technology with special attention
to the mouse, rat, and human systems, and to possible bioreactor systems.
The different approaches to obtaining stable human hybridoma cell lines are
explained briefly, with special reference to product secretion and stability.
The implications of different cell-specific production kinetics on the
design of a fermentation process, in particular the process mode, are presented. The relationship between cell and bioreactor is discussed further
in the presentation of different bioreactor systems. Their advantages and
disadvantages are shown, with respect to achievable cell concentrations,
product titers, and bioreactor volume productivity.
Finally, some nutritional aspects are described with special attention to
the different bioreactor systems.

TABLE OF CONTENTS

1.INTRODUCTION
   In 1975 Köhler and Milstein (1) developed hybridoma technology which for
the first time allowed the production of monoclonal antibodies (mAb) recognizing specific antigens/epitopes of choice. The exquisite and nearly inva-
riable specificity of monoclonal antibodies, combined with the potential

A. O. A. Miller (ed.), Advanced Research on Animal Cell Technology, 367–400.
© 1989 by Kluwer Academic Publishers.

for producing them in unlimited quantities, made them very interesting for many aeras of biological, biochemical, and medical research.

Recently, mAbs have already found commercial applications in the field of diagnostic assays. In addition use of mAbs for medical purposes, mainly for in vivo imaging and immunotherapy (e.g.: Wistar symposium on immunodiagnosis and immunotherapy with CO17-1A in gastrointestinal cancer (139)), is increasing permanently. Their main biochemical/biomedical interest is their potential in affinity purification systems. Table 1 gives an idea of the amount of mAb that might be required for the different applications. These requirements range between 1g and 10kg of mAb per year with a purity of between 70% and more than 98%.

Table 1. Large-scale monoclonal antibody requirements (taken from 2)

Application	g/year	Quality/Purity
Diagnostic assay	1 - 200	70 - 95 %
In vivo imaging	1 - 100	more than 98 %, non toxic, free from pyrogen
Immunotherapy	$10^3 - 10^5$	more than 98 %, non toxic, free from pyrogen
Immunopurification	1 - 1000	85 - 98 %

When mAbs are used for in vivo imaging, immunotherapy, and to some degree immunoaffinity chromatography, high mAb-purities of more than 98% have to be provided by the purification procedure. In addition, these products have to be produced and purified according to the guidelines of good manufacturing practice (GMP). These requirements cannot be met by a classical ascites production method, because of the presence of high levels of mouse protein and non-specific mouse antibodies, high variability and inconsistency of the ascites system, and the high risk of contamination by adventitious agents, like viruses (e.g. 3)). Additional disadvantages of the ascites system are high costs, limits of the scale-up-ability, and problems when xenogeneic hybridoma cell lines have to be cultivated in vivo. For all these reasons, in vitro culture systems have become increasingly more interesting and their use is constantly extending.

In general, the in vitro culture system (= biorector system) has to full fill all nutritional and environmental requirements of the hybridoma cells. However, the following points, which will be discussed in detail later on, have to be considered when a culture system is used. Enough information about the cell line has to be avaiable in order to establish a successful in vitro culture. The specific characteristics of the cell line, like cell type, fusion parents, genetique stability of product generation, stability of the product in the culture supernatant, microheterogeneity of the product, and finally growth and production kinetics have to be known. A second group of points, concerning the relationship cell line - bioreactor system will be discussed, with special attention to nutrional requirements, oxygen transfer, mechanical fragility of the cells and the product, and the different operating modes in combination with cellular growth and production kinetics.

Finally, a comparison of different production systems with respect to mAb-production will be given, showing their main advantages and disadvantages.

2.CELL LINE SPECIFIC CHARACTERISTICS

2.1. Monoclonal antibodies producing cell lines

Besides the immunological characteristics of secreted mAbs, a monoclonal antibody producing cell line has to show certain additional characteristics, before it becomes interesting for in vitro production. The main features are the quantity and homogeneity of the secreted product and the stability of product secretion.

2.1.1. Rodent monoclonal antibodies.
Since the beginning of mAb technology in 1975 many attempts have been made to ensure the above criteria. The first was the mouse-mouse system, which was improved with the introduction of non-producing and non-secreting mouse myeloma cell lines (SP2/0-Ag14 (4), P3 X63-Ag8.653 (5), F0 (6)). Today, mouse-mouse hybridoma cell lines can provide product titers of 10-200 µg/ml (table 2). In the case of dead production (production during the dying phase of most of the hybridoma cells (see section 2.4.)) Ig-concentrations up to 600 µg/ml can be achieved (7). Production rates of up to 130 µg/10exp6 cellsxd were obtained permanently, whereas maximal rates of up to 1536 µg/10exp6 cellsxd could be obtained during the dying phase of batch culture (7). Similar mean values have been published for mouse-rat hybridomas (8). Pure rat-rat hybridoma cell lines showed lower product titers of 7 - 61 µg/ml, obtained during the dying phase of the culture (9). Generally, after cloning, the stability of pure mouse-mouse or rat-rat hybridoma cell lines is excellent, within the limitations described in section 2.2. No data are available, concerning mouse-rat heterohybridoma cell lines. Stability of production of the cell lines is reduced in comparison to mouse-mouse or rat-rat hybridoma cell lines (10). Inspite of the advantages of mouse and rat monoclonal antibody technology, some major disadvantages have to be kept in mind when the mAbs have to be administered to humans. First, multiple application of foreign (xenogeneic) proteins is always coupled with an immune reaction of the patient to the immunoglobulin. In the case of mAb based tumor therapy mAbs may become ineffective for further therapy because of the presence of human anti-mouse immunoglobulin antibodies (11). Second, human monoclonal antibodies are more suitable for inducing effector functions than mouse mAbs (e.g.: 41). Therefore, different approaches were used to produce human monoclonal antibodies of a certain affinity for a certain antigen.

2.1.2. Human monoclonal antibodies.
The first attempt was the combination of mouse myeloma and human B-lymphocytes (43). These heterohybridoma cell lines had quite good in vitro growth characteristics, and sometimes produced upto 100µg/ml. In general, however, product titers were 1 - 20 µg/ml and cell specific production rates ranged between 1 and 15 µg/ 10exp6 cells and day (12). Unfortunately, human hybridomas have one common disadvantage, which is the stability problem. Loss of production is mainly due to the overgrowth by non producer cells which have lost their genetic information for producing and secreting Ig. For instance, Cote et al. (14,15) found that about 47% of the heterohybridoma clones were lost after 7 months. In any case, repeated cloning is necessary to select stable clones.

Another attempt to produce human mAbs has been the EBV-approach, whereby human B-lymphocytes are transformed by Epstein Barr virus (EBV), produced by a marmoset cell line, B95-8 (e.g.: 44 - 47). The main disadvantage is the low productivity of EBV-transformed cell lines, which varies between 0 and 5 µg Ig per 10exp6 cells and day. Concentrations of 0.01 and 20 µg/ml are generally achieved, although maximal titers of 43 mg/l have been reported (18). Low cloning efficiency and limited stability are general problems. Non-producers overgrow producer cells very often and thereby may cause the loss of Ig-positive cultures. Crawford (18) reported that some

TABLE 2, Different types of hybridoma cell lines, their stability and their product secretion

Myeloma cell line[1]	B-lymphocyte[2]	Ig-production (µg/ml)	specific Ig-production (µg Ig/dx10exp6 cells)	Stability (months)	Comment	Reference
mouse (e.g.: 653 SP2/0 NS-1)	mouse spleno-cytes	10 - 200 - 600	- 172 - 1536	+	perma-nently deadpro-duction	(12),re-view(13) (7)
mouse (e.g.: 653)	rat splenocyte	100 - 250	?	?		(8)
rat (e.g.: IR983F)	rat splenocyte	3.7 - 20 [3] 7 - 61	?	+	perma-nently deadpro-duction	(9)
mouse (e.g.: 653 NS-1)	human PBL, human spleno-cytes	1 - 10 (70 - 75 % of the clones) 11 - 100 (25 - 30 % of the clones)	?	-/+, 53% stable after 7m	repeated cloning is neces-sary	(14,15)
	human PBL	1 - 100 (72h, 200000 cells/ml)	?	more than 18m		(16)
	human PBL	1.8 - ...	1 - 15			(12)
	human PBL	2	?	more than 24m		(17)
-	EBV-transformed human PBL	0.01 - 20 max. 43 (Airlift)	0 - 1 1 - 5	-/+, 3-6m up to 36m	a lot of problems	(12) (18)
mouse (e.g.: 653	EBV-transformed human PBL	25 - 50	25 - 50	more than 14m	see text	(19)
SP2/0 SBC-HLO)	EBV-transformed human PBL		10 10	more than 24m		(20)
human x mouse SPAZ 4	human PBL	25 - 50	?	-/+, some more than 22m		(21)
human x mouse HN 5	human PBL	1 - 10	0.6 - 2.4 µg/mlx dx10exp6 cells	more than 7m		(22)
human x mouse MHH 1	EBV-transformed human PBL	50 - 100 (in 48h)	?	2.5-6m		(23)
EBV-transformed hu-man PBL (LB4r) x mouse SHM-D33	human PBL		2 - 36 µg/mlx dx10exp6 cells	? more than 8m	IgM/IgG limitati-on of se-gregation of chro-mosomes	(24) (25) review: (26)
human plasmacytoma SKO-007	human spleen	3 - 11		+		(27)

Notes: [1] Only mouse, rat, or human myeloma, plasmacytoma or lymphoblastoid (LCL) cell lines are shown, EBV-transformed PBL = Epstein Barr Virus transformed peripheral blood lymphocyte.
[2] All sources of B-lymphoctes are included, in the case of the transfectomas the chimaeric mAb is shown.
[3] This was permanent production, every second day the cultures were refed.

Table 2. Continuation:

Myeloma cell line[1]	B-lymphocyte[2]	Ig-production (µg/ml)	specific Ig-production (µg Ig/ dx10exp6 cells)	Stability (months)	Comment	Reference
human LCL LICR-2	human PBL/ lymph nodes	1 - 10 (70 - 75 % of the clones) 11 - 100 (25 - 30 % of the clones)		52% stable after 7m		(14)
human LCL: generally	human PBL, tonsil, spleen,+/- in vitro immun.	0.1 - 20	(1-5)	+/-,4-77m	more IgM- than IgG- producers	review: (28,29)
human LCL GM1500	EBV-transformed human PBL	1 - 5		some m		(30)
human B-cell lymphoma RH-L4	human PBL/ human spleen	?	10 - 12.5, less than 1µg/ml of the RH-64 Ig	up to 24m		(31)
human LCL (GM1500) x human EBV-transf.PBL	EBV-transformed human PBL	0 - 44	?	up to 24m		Review: (32)
KR-4			1 - 10	more than 8m		(33)
human LCL (KR-4) x human plasmacytoma (RPMI-8226)	EBV-transformed human PBL	5	5 - 30	more than 10m		(34)
Bispecific monoclonal antibodies:						
mouse hybridoma	mouse spleen	?	?	+		(35)
mouse hybridoma NS-1	mouse spleen	?	?	+	50% of the mAb is bispecific	(36)
Self hybridisation:						
mouse hybridoma	mouse hybrid.	78 instead of 25 115 instead of 20	? ?	?	5% of the hybridomas up to 3x better production	(37)
Transfectomas:						
SP2/0	SP6-mouse Anti TNP IgM + human Cµ, k	5	?	?		(38)
J558L or P3	S107-mouse Anti phosphocholin IgG + human IgG1 or IgG2, k	?	?	?	ascites production	(39)
Ag8, SP2/0, J558L	B6.2-mouse Anti human tumor associated Antigen IgG1 + human Cγ, k	?	1µg/dx10exp6 cells	minimum 3-6 m	ascites: 0.3mg/ml	(40)
SP2/0	L6-mouse Anti carcinoma associated antigen IgG2A k + human IgG1, k	D7: k: 0.017,γ: 0.077 ? 3E3:k: 0.100,γ: 0.700 ?	?	? ?	ADCC: 100x more effective than the original mouse mAb	(41)
SP2/0	17-1A-mouse Anti carcinoma associated antigen IgG2A, k + human IgG3, k	20	?	?		(42)

of EBV-transformed cell lines showed natural senescence and were lost. Finally, the product itself - in many cases IgM - and the presence of EBNA (EBV-specific nuclear antigen) can pose problems. These reasons about a search for better producer cell lines and for other possibilities for producing human mAb.

To increase production, product titers, and fusion frequency (fusion frequency is about 25x better using EBV-transformed cell lines than normal PBLs (19,33)), EBV-transformed cell lines were fused with mouse myeloma cell lines. In some cases, productivities (25 - 50 µg/10exp6 cellsxd)(19) and stability (more than 24 months)(20) were higher than in pure EBV-transformed cells (productivities of 1 - 5 µg/10exp6 cellsxd, stability of 3 - 6 months, sometimes up to 36 months). The cloning efficiency was also improved (20).

Others have used mouse-human hybridoma cell lines as fusion partners for immortalization of human PBLs or spleen cells or improvement of human EBV-transformed PBLs with respect to stability and product titers. The human (PBL) x (mouse myeloma x human PBL) hybridoma cell lines sometimes showed an increased stability (more than 22 months)(21) combined with product titers of 25 to 50 µg/ml. In other cases productivity and stability (2.5 - 6 months) were too low for a technical production (26, unpublished results). Generally, achievable product titers range from 1 - 10 µg/ml for HN-5 based hybridomas (22) to 25 - 50 µg/ml for SPAZ-4 (21) and MHH 1 (23) based hybridomas.

An increased limitation of chromosome segregation in heterohybridomas was obtained using the cell line SHM-D33 as a malignant parent cell line (24-26). This cell line is itself a fusion product of an EBV-transformed B lymphocyte and a mouse myeloma cell line. Hybrid cell lines showed stable production, which ranged from 2 to 36 µgIgG/mlx10exp6 cellsxd over more than 8 months (24-26).

In 1980 Olsson and Kaplan (27) published the first human-human hybridomas, derived by fusing human spleen cells with a human plasmacytoma cell line (SKO-007). They showed good stability and an IgG-accumulation of 3-11 µg IgG/ml in the supernatant. Others have used EBV-transformed cell lines (LCL - lymphoblastoid cell lines) as a malignant fusion partner and obtained human hybridomas accumulating between 0.1 and 20 µg/ml, which is usual (for review: 28,29) or in some cases between 1 and 100 µg/ml (1 - 10 by 70 - 75% of the clones, 11 - 100 by 25 - 30% of the clones)(14). These hybridomas were relatively stable: between 4 and 77 months in culture (29).

The use of a human B-cell lymphoma cell line (RH-L4)(31) and hybrid cell lines which are fusion products between two EBV-transformed cell lines (32) or between one LCL and a plasmacytoma (34) and human PBLs (EBV-transformed or not) does not change the overall achieveable product concentrations nor long term culture stability of the hybridomas. IgG-concentrations ranged between 0 and 44 µg/ml and culture stabilities of up to 24 months were achieved. However, comparison of these hybridoma cell lines with mouse-mouse or rat-rat hybridomas shows that they are inferior with respect to product titers and stability.

   2.1.3.Other approaches: Finally, three other approaches should be mentioned briefly.

   The production of bispecific mAbs, which have two different antigen binding functions in one mAb molecule, should be very easy because of the relative stability of such cell lines. Up to 50% of the secreted product represents such bispecific mAb (36). In general, they are produced by fusing two different hybridoma cell lines producing two different mAb of interest. Their advantage is that the antigenic modulation can be reduced

during an immunotherapy (e.g.: tumor targeting), first, and second, that complex chemical modifications can be avoided because of the simple binding of, e.g., an enzyme for use in an ELISA (35) or a cytostaticum for use in tumor therapy to this bispecific mAb (e.g.: 36).

Self hybridization of hybridomas provides a 2- to 3-fold increased production in 5% of the derived clones. This might be an interesting method for a genetically based increase in the production kinetics of hybridoma cell lines of interest (37).

The last and the most-up-to date approach is the production of transfectomas (48). Transfectomas are lymphoid cells into which reconstituted immunoglobulin genes have been transfected. They provide the possibility of producing genetic manipulated monoclonal antibodies using transformed mouse myeloma cell lines, like SP2/0 (38, 40,41,42), Ag8 (40), or J558L (39,40). The advantages are for instance the production of antibodies consisting of a mouse variable region and a human constant region. This improves the biological effector function of the mAb because of the human part of the mAb when it is used in vivo in human beings, whereby the antigenic specificity, which often cannot be obtained with equal ease using human mAb-technology, is of mouse-origin. These chimeric mAbs are normally less antigenic in humans than classical mAbs are. Liu et al. (41) described a 100-fold more effective ADDC (antibody-dependent cellular cytotoxicity) than was obtained by the equivalent mouse mAb (17-1A). Their main disadvantages are still the low product titers - maximal: 20µg/ml - (38,40,41) and their sometimes unbalanced production of heavy and light chains (41). In one case the chimeric human-mouse mAb was 1/4 as effective in hemolysis of TNP-coupled sheep red blood cells than the homologous mouse IgM, probably due to steric hindrance (38).

Another attempt for producing mAbs in genetic engineered E.coli or yeast cells failed more or less (49,50, for review: 25) because glycosylation is not or only partially carried out by E.coli or by yeast, respectively, and in both cases the assembly of the chains is not efficient (less than 5%). In the case of E.coli the insoluble H and L chains have to be solubilized to obtain some antigen binding activity.

Although most of these attempts already provided useful human mAbs, there still exist large differences between rodent and human mAb-technology, mainly with respect to specific production rates and final product titers in the culture supernatant (Table 1). Stability is only one problem which is already solved in many human hybridoma systems. The number of the different attempts lets us know, however, that the ideal human mAb system was not found yet.

2.2. Stability and mutations of hybridoma cell lines

One of the most important points of large scale cell culture is stability of the cell lines used. In Fig. 1, the numbers of generations, a cell must necessarily undergo from a single cell clone to a

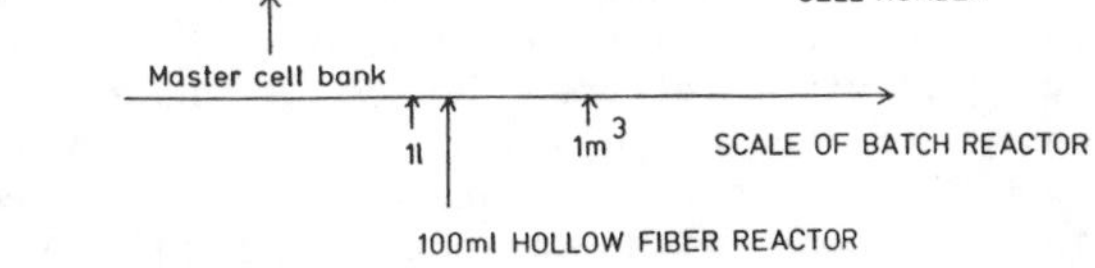

FIGURE 1. Importance of physiological stability of cells (taken from 12).

master cell bank, as well as to various production scales, are shown. It is evident that about 10exp7 cells are starting material of one vial for a fermentation process. This cell number increases during the propagation of cells for inoculating of the bioreactor. Using a batch process, e.g. 10001, this cell number has to increase upto 1-3 x 10exp12 to achieve the maximum cell number per fermentor unit. The use of continuous cultivation systems

TABLE 3. Mutations and alterations in cells lines producing monoclonal antibodies (taken from 51): Mouse hybridomas.

Selection	Mutation or alteration	Frequency/ cell x generation
Loss of isotype	Loss of H or L chain (mouse)	0.01
	Loss of H or L chain (rat)	0.0001
	Domain deletions	0.001
	Loss of lambda-chain secretion due to point mutation	0.0001
Change in IEF	Domain deletions, frameshift, point mutation	0.001
Cell sorter, positive selection for different isotype	Change of isotype, often associated with deletions	$10^{-5} - 10^{-7}$
Loss of idiotype, loss and gain of antigen binding	Possible gene conversion point mutation	0.01 - 0.001
Loss of lytic activity	Deletions, frameshifts, insertions	0.001 - 0.0001

increases the number of cell doublings significantly, depending on the duration of the process, the vessel volume, and the flow rate. Systems, like hollow fiber systems, which work with cell concentrations of 10exp8 cells/ml would reach a total cell number of 1-3 x 10exp12, when units of 101! (which do not exist) would be employed. Today's hollow fiber cartridges are used with an extracapillary space of 100ml, which corresponds to a total cell number of 10exp10.
Comparing these figures with aspects of stability like frequency of mutations and alterations in cell lines producing mAbs (Table 3)(51), some facts become evident. One has to count with a frequency of alterations ranging from 10exp-2 to 10exp-7 per cell and generation, depending on the type of mutation: loss of isotype, changes in the IEF-pattern (isoelectric focusing pattern) of the isotype (see 52), loss of lytic activity and idiotype, and loss or gain of antigen binding, i.e. changes in the affinity or avidity for an antigen of a mutated mAb compared to that of the unmutated variant (53-55). The alteration-frequency of 10exp-2 (loss of H and L chain in mouse plasmacytomas (140)) and 10exp-3 (domain deletions (141), loss of idiotype (53-55)) seems to be very high, with special respect to a large-scale in vitro culture. However, others reported losses of heavy and light chains in mouse myeloma cell lines which occured at a rate of about 0.001 to 0.0001 per cell and per generation (for references see (141)), which is in agreement with the observed frequency of loss of the production of heavy and light chains in rat-rat hybridoma cell lines (10). Generally, this frequency of chain loss is 10 to 100x reduced when compared with mouse-plasmacytoma cell line used by Coffino and Scharff (140). For more

details, the reader is referred to the Nobel lecture of G.Köhler (51).
No refernces about the frequency of changes concerning chimeric mAbs or
secondary hybrids are available, but Sahagan et al. (40) stated that transfectomas should be more stable than normal hybridomas because they are not
derived from cell fusions.
Two particular problems should be mentioned briefly:
The EBV-approach is accompanied by major drawbacks. That is low and unstable productivity, and intrinsic growth problems (56), like poor clonability,

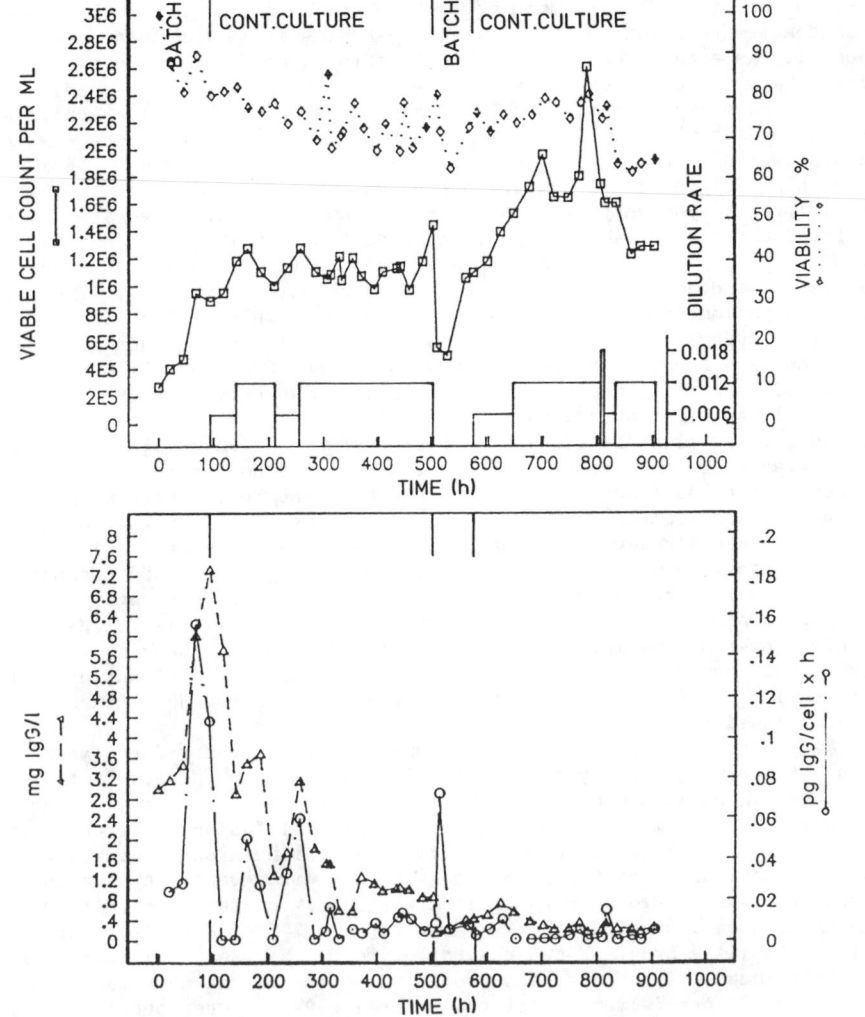

FIGURE 2. Bioreactor culture of an unstable human-mouse-human hybridoma
cell line: Bioreactor: Airlift-fermentor (10 l); medium: Dulbecco's Modified Eagle medium, supplemented with 5% fetal calf serum; pH: 7.0; pO2: 20-
50% air-saturated; gas-flow: 0.6 fermentor-volumes per hour; temperature:
37°C; operational modes: batch and chemostat (continuous culture). For
experimental details see Materials and Methods in (88).

(57) and sometimes naturally occuring senescence of the EBV-transformed cell lines (18). The last two points mean that production is limited. The low clonability often causes an overgrowth of non-producing, faster-growing, cell populations whereby the producer population is eventually lost (57). Improvements have been achieved by fusing EBV-transformed cells to human or mouse myeloma cell lines (section 2.1) or by in vitro selection techniques made before the transformation step (e.g.58,59).

A more serious problem is the instability of heterohybridoma cell lines (interspecies hybridomas). It is well documented that this type of mAb-producing cell line (e.g.: human-mouse hybridomas) looses human chromosomes (chromosome segregation) preferentially, and the mouse chromosomes are retained. It was established that the human chromosome 2, which codes for the k light chain, is preferentially expelled (60). Human chromosomes 14, which codes for the heavy chains, and 22, which codes for the λ light chain are retained preferentially or partially, respectively (61,60). One example illustrates the effect of the unstability of heterohybridomas (human x (mouse x human) on the fermentation process. Fig. 2 shows data obtained from a fermentation which was run in a continuous and in a repeated batch mode for 900hrs. After a relative good production (up to 7.7µg IgG/ml; 3.84 µg IgG/ d x 10exp6 cells) in the batch culture (0 - 100hrs) the process mode was changed to a continuous culture (chemostat culture). The IgG-concentration immediately decreased due a decrease in specific productivity. After 700hrs the fermentor population was selected for non-producers, as can be seen from the diagramme showing product concentration (Fig.2). Throughout, viability and the viable cell count were relatively constant, whereas cell specific productivity decreased rapidly.

Such cell lines might be stabilized by repeated subcloning (62) resulting in two events (19): First, human chromosomes coding for the human Ig are retained, which does not necessarily mean that the cell line continues to produce mAb. In some cases, human chromosomes were still detected in cell lines which had already ceased mAb-production. Raison et al. (17) stated that regulatory genes were lost resulting in cessation of production. Secondly, human chromosomes coding for the human Ig are not retained, but the genetic information coding for the human H and L chain is translocated to chromosomes of the malignant fusion partner (16,63). Levy et al. (64) stated that human x mouse hybridomas which ceased production of mouse light chains were more stable in the production of human Ig than those which continued the synthesis.

Other approaches are the use of only human fusion partners for immortalization of human B-lymphocytes (section 2.1) and/or the increase of human chromosomes in human x mouse heterohybridomas (section 2.1)(24), or the use of irradiated mouse myeloma cell lines for the cell fusion, thereby providing a predominant human chromosome pattern in the heterohybridoma (26).

These production stability problems might be circumvented by the use of culture systems providing more or less stationary cultures or cultures in which cells show a reduced mitosis rate (section 3.3). Finally, these production stability problems can be controlled by the use of immunofluorescent techniques (64) because mAb-producing cell lines have surface Ig in most cases. A pre-fermenter selection for a 100% producer population of a partially unstable cell line should be no problem when a flow cytometer is employed and might to some extent solve the problem of a production loss.

2.3.Microheterogeneity of monoclonal antibodies - product stability

The isoelectric point (pI) microheterogeneity of monoclonal antibodies is a well known fact and was already described for murine IgG2a secreted by a

murine plasma cell tumor (65). Hamilton et al. (66,67) studied this in combination with immunoreactivity and specificity of murine monoclonal anti-human IgG by using of isoelectric focussing (IEF)-affinity immuno-blotting. In the case of murine mAbs (IgG), the isoelectric points were located in the pH range of 5.5 to 8.0 with a 0.1 to 0.6 pH unit spread. Generally, 3 to 5 major dense bands flashed by 2 to 4 minor fainter bands were found (Fig. 3). These microheterogeneities were caused by posttranslational changes resulting from amidation and/or desamidation of amino acids of the two different chains (65,68,69) or differences in the glycosylation (68-72) of the H-chains.

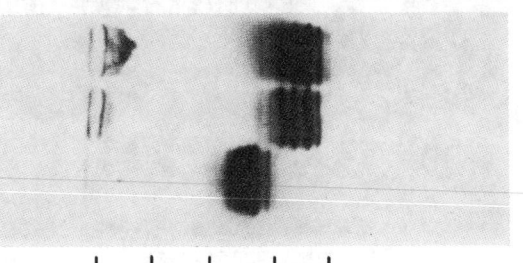

FIGURE 3. IEF-affinity immunoblot of mouse anti-human IgG mAbs for pI pattern (band, number) and specificity determination; two different lots of HP6002 ascites produced 3 years apart (a, b), and another ascites (HP6014) for comparison are shown (taken from 67).

These pI fingerprints are excellent tools for the control of the stability of mAb production by hybridomas (67)(compare with Table 3). Using the IEF-affinity immunoblot analysis, pI consistency, immunoreactivity, and specificity can be controlled over time. An example is given in Fig. 3 which shows the stability of a mAb-producing cell line over a period of 3 years (67). This method allows sensitive and excellent control of mAb-producing cell lines over years which is necessary when product is used for human therapy (66).

Although some people which not detect any proteolytic cleavage of mAb by cellular proteases (9), the presence of cell-released proteases should not be neglected, in particular, when serum-free media are used (section 3.2.2). Upto now only one study has been done with respect to this point (73). Schlaeger et al. (73) found that serum-free cell culture supernatant contains a lot of proteolytic activity at pH-values lower than 4.5, which may pose a problem when purification schemes employing low pH values are used. Monoclonal antibodies were cleaved giving F(ab') fragments. They are cellular proteases and they show some similarities with the lysosomal cathepsin D (see 73).

Recently, we were able to find some proteolytic activity in fibrin-bead encapsulated hybridoma cell cultures (Fig. 4a,b,c). It is evident that the same hybridoma culture contains different cell subclones, some of which are producing high proteolytic activities which can be detected by lysis of the fibrin surrounding the cell clone (Fig. 4a,b), others show nearly no proteolytic activity (Fig. 4c). In contrast to the data of Schlaeger et al. (73) these proteases were active at physiological pH values and may have some effect on the product itself.

These cellular proteases are of great importance when serum-free cultures or high density cultures with low viabilities are used, as is often the case with hollow fiber systems.

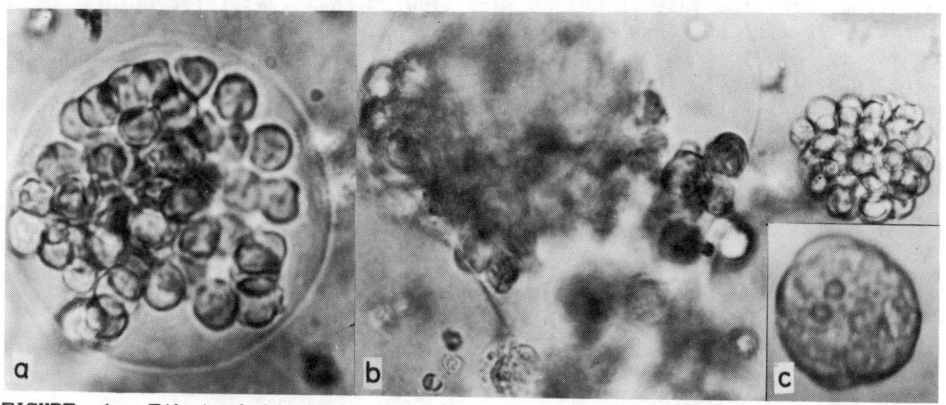

FIGURE 4. Fibrinolytic activity of mouse-mouse hybridomas (371) entrapped in fibrin beads after 8 days in culture: a), b) clones releasing proteases, c) another clone of the same culture showing no fibrinolytic activity (unpublished results).

2.4. Production kinetics of hybridomas and EBV-transformed cell lines in batch cultures

Product secretion characteristics are one important aspect in in vitro production of mAb. Upto now three different cell-specific production kinetic patterns of monoclonal antibody-producing hybridomas have been described (74)(Fig. 5a,b,c). Fig. 5a presents production pattern I which is characterized by a high cell-specific productivity at the beginning of a batch culture (during the lag phase and the onset of the log phase). This productivity decreases during continuation of the culture and no increase can be observed during the dying phase of the culture. This type of secretion-production kinetic has only been described for mouse x mouse hybridoma cell lines and may be found in 5 to 10% of the hybridoma cell lines (75-77,88). Cells belonging to the group II production pattern show a high specific productivity at the beginning of the batch culture, like group I. During maximum growth (highest growth rate) the specific productivity is significantly reduced. However, it increases again when the stationary and declining phases are reached, such that in some cases extraordinary high secretion rates are achieved during the death phase (see Table 2, Fig. 5b). Between 80 and 90% of the hybridoma cell lines show this production pattern II. Hybridoma cell lines of mouse x mouse- (78-88), rat x rat- (9), human PBL x mouse myeloma (89), and human x (mouse myeloma x human PBL)- (22) origin showing this production pattern, have been described. Cell lines of group III have a quite constant production of mAbs during their growth phase (Fig. 5c), independent of the growth rate. Their Ig-secretion is completely stopped after the beginning of the stationary phase. This type of production kinetics is relatively rare and has only been found for mouse x mouse- (77,88,90, Henno, pers commun.) and rat x mouse-hybridomas (90). Fig. 5d shows another type of production kinetics (IV) which was seen only with batch cultures of EBV-transformed PBLs (peripheral blood lymphocyte) (18), and fusion products of EBV-transformed PBLs and normal human lympho-

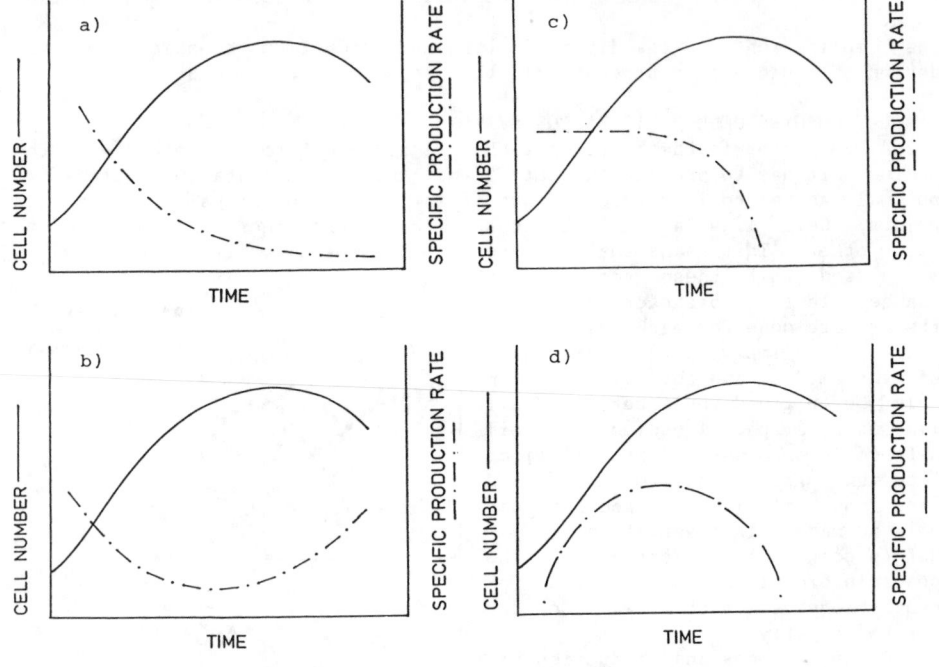

FIGURE 5. Different production kinetics of hybridomas and EBV-transformed cell lines in batch culture systems: a) Pattern I, b) Pattern II, c) Pattern III, d) Pattern IV.

cytes (28), or of EBV-transformed PBLs and human B-lymphoblastoid cell lines (30), and sometimes in cultures of human x (mouse myeloma x human PBL) hybridomas (22). In this case, the product secretion is more or less growth-associated. During the declining phase production is rarely detected. The reason for these different production kinetic patterns is not completely clear. Comparing the published data seems that the cultivation system has only a minor influence on the production kinetics. Wether medium composition and the environment itself have some influence is not yet clear. It is only known that the partial pressure of oxygen has some influence on deadproduction (=production during the declining phase) of type II cells (81). Finally, it seems that the fusion partner itself brings regulation of the product secretion pattern of the hybridoma cell line. One example might be the type IV production which was only seen in human intra- and interspecies hybridomas and EBV-transformed PBLs. That special pattern is associated with the human PBLs or EBV-transformed PBLs is only specula- tion. Ichimori et al (22) showed the growth and production curves of two human x (mouse myeloma x human PBL) hybridoma cell lines. The production kinetics of one have followed type IV (perhaps more influenced by the human side) and the production of the other cell line followed pattern II, which might be due to the mouse part of the hybridoma.

For a more detailed discussion about kinetic patterns I - III, the reader is referred to the paper of Merten (74).

In any case, it is too early to give a clear, detailed, and final picture of different batch culture production kinetics, of different factors affec-

380

ting these kinetics, and of the possibilties of influencing and changing them.
The implication of the different types of production kinetics for the design of a bioreactor process will be discussed in section 3.

3.RELATIONSHIPS CELL - BIOREACTOR SYSTEM

The bioreactor is the heart of technical cell culture. Together with the medium, it has to provide the nutritional and environmental parameters for optimal animal cell culture. Hence attention must be be paid to various points: Cell lines are unique and show some differnenes when compared to each other. This means that different cell lines have different requirements and that these requirements change with the bioreactor system and the process mode for each cell line. Both points imply that optimization of the medium and the process can only be obtained for a certain cell line using a special medium in a well defined bioprocess. Extrapolations from one process mode to another and from one cell line to another can only be made with reservations.

Before going into greater detail, the possible process modes and the different bioreactor systems will be described briefly:

3.1.Process modes and production kinetics

3.1.1.Process modes. For the selection of a relevant cultivation and production system it is necessary to be aware of the principle culture patterns of the systems in order to find the optimum of the respective process (91,92). Fig. 6 presents the most used process modes. A batch culture is characterized by decreasing substrate concentration. Cell growth is zero at the onset and passes through a maximum before reaching the highest cell concentration. It decreases at the end of the batch and may exhibit negative values, caused by cell death and desintegration (Fig. 6a).

A special case is the repeated batch, where, after a certain duration (normally the begin of a classical batch culture), a certain percentage (e.g.: 50%) of the supernatant including the cells is replaced by new medium. PH, pO2 and temperature are controlled. Another case is the fed-batch culture, which is characterized by constant (controlled) pH, pO₂, tempe-

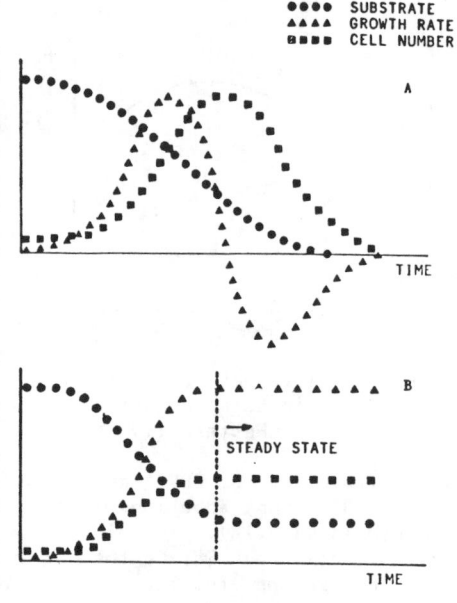

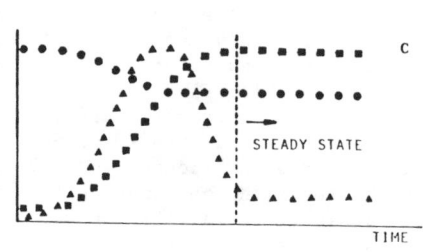

FIGURE 6. Principle culture characteristics of A) batch, B) continuous and C) perfusion (cell retention) culture with respect to substrate concentration, growth rate, and cell density (taken from 99).

rature, and nutrient-concentrations, and increasing waste-concentrations. These wastes may cause negative effects on cell proliferation and the production.

A continuous culture is characterized by steady states which are reached after the initial growth phase (Fig. 6b). The relationship of cell concentration, substrate concentration, and growth rate is given by the perfusion rate and by the operation as a chemostat or a turbidistat, respectively. The growth rate can be varied from the maximum in the turbidistat mode to about 25% of this in a strong chemostatic mode. However, the limiting nutrient for the chemostat mode is usually not defined, but it must not interfere with productivity.

A perfusion system with cell retention is also characterized by steady states which are reached after the initial growth phase (Fig. 6c). The cell concentration reaches a maximum, which is determined by space or other limitations, but usually not by a chemical one. The growth rate decreases to almost zero after confluency of the system. Some residual growth is always detected in practice because of the replacement of dead cells. The substrate concentration can be regulated by altering per fusion rate.

3.1.2.Process modes - production kinetics. From data published by Merten (74), it became evident that for the optimization of a bioreactor process, the proper choice of the process mode has to be made with respect to the specific production kinetics of the cell line.

The four different cell specific production kinetic patterns that were described above (section 2.4) have some implications on the conception of a production process. The comparison of the growth curve, the growth rate, and the specific production rate (Fig. 5a-d) already provides information for the design of a process, and for the physiological state of the cell optimal for a high cell specific production rate. Using low density culture systems, Merten et al. (88) showed that cell lines, showing production pattern I, had higher average cell specific productivities coupled with higher achievable product yields in a repeated batch mode than in a continuous culture (chemostat). All discontinuous process modes, like fed-batch or repeated batch should be superior to continuous modes, like chemostat or perfusion culture because in the former case the process mode supports the physiological states in which higher productivities are detectable. Depending on the slope of the curve representing the specific productivity (Fig. 5a), a continuous culture may be an acceptable process mode in some cases.

The production and growth of type II cells is more or less invert. This means that the culture system has to provide a situation in which the growth phase and the production phase are separated from each other. This can be achieved by the use of the following process modes: Batch, fed-batch, repeated batch with or without a second stage, or a continuous culture system with a second stage. The first stage provides the growth of the culture, the second one supports the production phase during the stationary and death phases of the cells. Merten et al. (88) showed that repeated batch was superior to one batch-culture which in turn was superior to a chemostat culture of the cells showing production pattern II. Reuveny et al. (86) presented data that proved the superiority of a two-stage, repeated batch process in comparison with a batch, fed-batch, and repeated batch process.

The production of cell lines showing production patterns III and IV is strongly growth associated. All process modes which support strong growth (chemostat or turbidistat) are superior to batch processes. Merten et al. (88) found that a continuous system was better than batch and repeated

batch processes for cell lines with production pattern type III.
Assuming that the differences between high and low density cell culture systems are not too large, the above mentioned implications of batch production kinetics might be extended to high cell density systems, too.
Reuveny et al. (86), Emery (93), Scheirer (pers. commun.), and van Wezel et al. (94) used homogeneously mixed cell culture systems (spinfilter: 86,94, Scheirer, pers. commun., homogeneous cell retention system using hollow fibers: 93) for the culture of type II hybridoma cell lines. In all cases, at least a two-fold increase of the cell specific productivity accompanied by increased product concentration could be detected because these systems hold cells in an almost stationary phase. Normally only the dying cells are replaced by new ones and a maximum cell concentration of 10 - 30 x 10exp6 cells per ml is possible over a long culture duration. The volume based productivity was 10 times higher than in low density culture systems (86). Using a homogeneous perfusion system Martin (95) achieved 27 times the antibody concentration obtained in a batch culture. Feder and Tolbert (96) used similar systems successfully, however nothing is known about the production kinetics in both cases.
Altshuler et al. (78) cultured cells showing batch production pattern II in hollow fiber units. The specific productivity was increased significantly in comparison with normal T-flask cultures.
Unfortunately, there are no results available concerning in particular cell lines showing batch production kinetics type I, III, or IV. Some problems might occur when these cells are cultured in high density systems, especially with types III and IV because production is more or less growth associated. However, the growth rate is reduced partially or completely in all high cell density systems.

3.2. Nutritional requirements

3.2.1. Oxygen demand and oxygen transfer problems. The oxygen demand of the culture, which increase directly with the cell density, may be critical because of the low solubility of oxygen in aqueous solutions (about 6 mg/l at 37°C)(97). This requirement ranges from 0.053 to 0.59 μmol O_2/mlxh (per 10exp6 cells)(98).
In homogeneous bioreactor systems, O_2-transfer (total oxygen transfer rate) is limited only by K_LA (mass transfer coefficient [mmol O_2/atmxlxh], where K_L is the rate of oxygen diffusion through the gas-liquid interphase, and A is the available aera of the gas-liquid interphase). The use of an aeration system via silicon tubing (99,100) and/or ceramic sinter pieces is recommended for spinfilter systems. However, O_2-transfer may be hindered in addition by O_2-diffusion through the silicon membrane, e.g. the K_LA for the O_2-transfer through a 1 inch (2.54 cm) silicon tubing is about 0.35 mmol O_2/atmxcm^2 x h which corresponds to an O_2-flux of 0.07 mmol O_2/hxcm^2, using 95% air and 5% CO_2. Nevertheless, the silicon tube aeration, the ceramic sinter pieces, and the increase of the vessel pressure (101) provide improved O_2-transfer.
The use of heterogeneous cell cultivation systems provides additional O_2-transfer problems. In all cases, the transfer of oxygen is hindered by the diffusion resistance (a) through the membranes (for all membrane systems without additional oxygenation system), (b) through the matrix of the microbeads, and (c) through the cell mass. Calculations of O_2-transfer, using the Thiele modulus and the effectiveness factor, were given by Glacken et al. (98). These transfer problems have to be considered for the nutrients, metabolites as well as the products.
3.2.2. Medium composition - some considerations. Generally, the medium

has to provide the cells with the necessary nutrients and has to take all released metabolites. The basic medium contains amino acids, organic acids, one or more carbon cources (glucose in most of the cases, glutamine to some extend), vitamines, inorganic ions, and generally serum, often fetal calf serum.

The fact that the cell culture media should imitate the in vivo environement of cells to some extent provides us with the problem that these media must be supplemented in most cases with serum, which functions as a source of hormones and growth factors in low density cell cultures (up to 3 x 10exp6 cells per ml), mainly. Not only because of relatively high cost of serum but mainly because serum is the major source of contaminants like viruses, mycoplasmas, and bacteria (97,102), and is not a chemically defined and therefore non standardizable reagent, a lot effort has been devoted, therefore to the replacement of the serum by defined compounds. In addition, it is difficult to obtain large quantities of serum of constant quality and composition, a necessity for industrial production and serum might be a source of inhibitory substances which can act unfavorably on the production of certain biologicals. The use of serum-free, protein supplemented or not, media is recommended and will be welcome by all people involved in purification. Table 4 presents the composition of one serum-free medium for lab-scale application which should provide cell proliferation of most of the established mouse x mouse and rat x mouse hybridoma

TABLE 4. Some formulations of serum-free media (taken from 103, changed)(a)

Substance	Basal medium			
	RPMI (103)	IMDM (136)	MEM (137)	F12:DMEM, 1:1 (138)
Hepes (mM)	25	25	–	15
Se (nM)	10	100	–	2.5
Pyr (mM)	1	1	–	0.5
Lipids (μg/ml)	SBL, 20 LDL, 2	SBL, 50-100 Chol., 0-8	–	Linoleate, 0.04
Protein (mg/ml)	BSA, 0.5	BSA, 0.5	–	–
Tf (μg/ml)	15	1	5	35
Ins (μg/ml)	–	–	5	5
2-AE	–	–	–	20
Other	B-ME, 50 μM GSH, 0.5mM	B-ME, 50 μM	Neaa, 1%	Put., 80 ng/ml
Cell types supported	PCTs, hybs, SP2/0 Ag8.653, P388, P388D1	LPS blasts, hyb (90)	Hybs from Ag8 only; cell-doubling time 3 days	Hybs, MPC-11 and SP2/0-derived
Cell types not supported			Ag8, Ag8.653	NS-1-derived hybs

(a) Abbreviations: Se: Na₂SeO₃; Pyr: pyruvate; Tf: transferrin; Ins: insulin; 2-AE: 2-aminoethanol; SBL: soybean lipids; LDL: low density lipoproteins; Chol.: cholesterol; BSA: bovine serum albumin; B-ME: B-mercaptoethanol; GSH: reduced glutathione; Neaa: non-essential amino acids; Put.: putrcine; PCTs: plasmacytomas; LPS: lipopolysaccharide; hybs: hybridomas.

cell lines (103). This is a particular characteristic, because most of earlier formulations supported only growth of hybridomas of certain myeloma parents (some examples are presented in table 4). Both the different composition of the supplements (some are essential, others only partially essential) and the different basal media greatly influence growth and production of hybridomas (104). Cole et al. (104) used a serum-free, protein-free medium for successful proliferating human x human hybridomas.

Using such media for homogeneous culture systems, where the cells grow in suspension, some substances have to be added in order to improve the protection of cells from shearing forces and to decrease the sedimentation rate of the cells in serum-free medium. Pluronic F-68 was used by many groups (86,105). Some used the addition of primatone RL, a protein hydrolysate for increasing the amino acid and peptide concentration (86,105,106). Both additives might not be necessary in heterogeneous systems because of the different physical environment of the cells.

Switching from a low to a high cell density culture system, the control of a probably (growth or production) limiting substance becomes necessary. The above mentioned medium compositions are sufficient in most cases of the low density culture systems, but may be inadequately composed for high cell density culturing systems. With development of new culture system configurations providing the possibility of high cell densities it is evident that cells in a high cell density environment have other nutritional demands and that their metabolism and physiology change. One example is given. Serum-free hybridoma cultures require the presence of transferrin in a low density culture system. Using high density culturing systems, like hollow-fibers, transferrin is an essential medium compound only at the start of a new culture, but it can be absent during the rest of the culture (93,107).

Some examples with respect to medium composition are given in the following: medium components, like the carbon source (mainly glucose) and some amino acids (like glutamine) are used in high concentrations and can cause some problems. Too high a glucose concentration causes the production of high amounts of lactic acid (Crabtree effect) which causes a strong acidification of the medium and an unnecessary waste of the energy source. Cells, cultivated in a medium with too high a concentration of glutamine (often main energy source (122) and a carbon source, too), produce high amouts of NH_4^+ which is toxic for cultured cells (108-110). A good feedback control system for these compounds would provide a means of adding both carbon sources to the culture according to the demand of the cells. Waste products would thus be minimized. Hu et al. (111) presented a feeding system for a forward regulation model system for hybridomas by keeping the concentrations of glucose and glutamine lower than usual. The prevention of production of lactic acid which is with the CO_2 the main cause of acidification of the supernatant in bioreactors is possible by replacing glucose by other sugars if the cell line of choice can be adapted to the new conditions. Replacing glucose by other sugars like galactose, maltose, or fructose would provide in many cases the same influence on the cell growth as glucose, but would be accompanied by a lower production of lactate (110,112). The Reduction of the sugar concentration in general is only possible when the glutamine concentration is elevated in parallel, which causes elevated NH_4^+-production.

The presence of amino acids is necessary but there is still question as to which amino acid must be present and at what concentrations. Upto now a lot of work was done in this field with the result that different cell lines have different demands and produce different amino acids. In addition, these demands depend on the cultivation system used and the physiological

state of the cells. Roberts et al. (113) presented some results for the MOPC-31C mouse plasmacytoma cell line in a stationary batch culture. It consumed glutamine, isoleucine, methionine, and valine, and some tyrosine and phenylalanine. Aspartic acid, glutamic acid, glycine, proline, and serine were produced. The other amino acids were not influenced.

Generally, hybridomas utilize glutamine and produce alanine (114, unpublished results). For instance, one mouse-mouse hybridoma consumed leucine, serine (totally), isoleucine, methionine, arginine (totally), glutamine (totally), and phenylalanine, tyrosine, and some valine in a static batch culture. In parallel, it produced alanine, asparagine, glycine (slightly), and glutamic acid (during the lag- and the beginning of the log-phase). Other amino acids were not affected. In comparison with these results, a human-mouse-human hybridoma consumed only glutamine and arginine and produced glutamic acid, alanine, and proline in a static batch culture. The concentration of the other amino acids decreased slightly during the log phase and increased at the end of the log phase and the beginning of the stationary phase (unpublished results).

These examples only concern static batch cultures of hybridoma cell lines. Unfortunately, no data about the consumption or production of amino acids in homogeneous or heterogeneous systems are available upto now. Certainly, the picture will change to some extend.

3.3.Bioreactor systems

3.3.1.Bioreactor systems, their advantages and disadvantages. In general the existing bioreactor systems can be classified into two main groups: the homogeneous, and the heterogeneous (or inhomogeneous) bioreactor systems (Table 5). The homogeneous systems are characterized by a more or less homogeneous mixing of cells (or microcarriers or beads) and medium. The following systems are used. The oldest type represents the agitated reactor (stirred tank) in which mixing is done by an agitator. Using this type, special care has to be paid to the geometry of the bioreactor and the shape and geometry of the stirrer (97). Generally, all process modes (section 3.1.1) can be employed using the stirred tank reactor. Katinger et al. (115) applied the airlift-type bioreactor for the first time for the culture of animal cells. The main advantages compared to the stirred tank reactor are the reduced power input ($10 - 15$ watt/m^3) effected by the injection of air, the almost total absence of shear forces, low turbulence, simple construction, and simple scale-up-ability (97,115). Unfortunately, there are also some disadvantages, namely, the invariability of the liquid level, which does not allow the use of the fed-batch mode (e.g.), and the tall conctruction, mainly, when used in an industrial scale. Both systems are limited by the relatively low cell concentration achievable ($1 - 4$ x 10exp6 cells per ml) and the dilution of the product.

One solution to this drawback (low cell number) is the use of spinfilter reactors (116), reactors equipped with a sedimentation tube (jar fermentor)(117), or the Diessel reactor, which is equipped with a hollow fiber membrane (118). In all cases, cells are retained and cell concentrations of upto 30 x 10exp6 cells/ml can be achieved. In addition, since the perfusion rate can be increased significantly, the volume based system productivity is therefore elevated.

The large group of heterogeneous bioreactor systems are characterized by a more or less complete separation of the cells from the culture medium, whereby the cells and, in the case of product retention, the product are protected from shearing forces which might be harmful to cells or product (119,120). Product retention is not recommended in cases where the integri-

TABLE 5. Comparison of low and high cell density cultivation systems for the production of monoclonal antibodies (taken from 121).

Parameters	Low density cultivation systems (homogeneous)	High cell density cultivation systems (cell retention)		
		For suspension cultures (homogeneous)	Immobilization systems (inhomogeneous)	
			With product retention	Without product retention
Reactor type	Stirred tank Airlift reactor	Spinfilter	Capsule systems Hollow fiber system Flat membrane system	Capsule systems Bead systems Opticell system Hollow fiber system Flat membrane system
Cell densities/ml	3×10^6	30×10^6	100×10^6	100×10^6
Product titer mg/l	2 - 600 (a)	2 - 600 (a)	900 - 10000 (b)	2 - 450 (a)
Product purity %	1 - 7	1 - 7	20 - 80	1 - 7
Reactor Productivity mg/l x d	0.25 - 62 (c)	19 - 660 (c)	21 - 1400 (c)	0.2 - 570 (c)
Product residence Time Th	Th=1/D (long)	Th=1/D (short)	The product is retained (d)	Th=1/D, depends also on the diffusion of the product through cell mass, membrane, or bead matrix
Shear protection of cells	No	No	Yes	Yes (e)
product	No	No	Yes	No
Cellular control	Direct	Direct	Indirect	Indirect
Cellular microenvironment	No	No	Yes	Yes (f)
Gradients (nutrients, metabolites, O2, product)	No	No	Yes	Yes
Oxygen transfer	$k_L A$	$k_L A$, and oxygen diffusion through the silicon aeration tubing (g)	$k_L A$, and oxygen diffusion through the membranes, and cell mass (g)	$k_L A$, and oxygen diffusion through the membranes, bead matrix, and cell mass (g)
Scale-up	already done (h)	already done	limited (i)	limited (i)

Notes: (a) depends on the dilution rate, the cell specific productivity, and the production kinetics (74,88). (b) depends on the cell specific productivity and the production kinetics (74,88). (c) depends mainly on the cell line used (see also (a) and (b)). (d) In the case of the hollow fiber and the flat membrane systems, there is the possibility of a product recovery independent of the cell recovery. In the case of the capsule systems, the process has to be stopped for the recovery of the product from the destroyed capsules! (e) only partially for fiber beads and Opticell. (f) except for Opticell and systems, where cellular products, like growth factors, etc., are diluted too fast (e.g.: fiber beads). (g) for details, see Ref. 98 and 99. (h) easy for the airlift system, can be difficult in agitated systems (97). (i) capsule systems: limited by the fragility of the capsules (maximal bioreactor volume: 40 - 100 l possible) and the encapsulation process. Opticell system: only linear. Hollow fiber systems: only linear. Flat membrane systems: easy, already done (MBR-Membroferm).

ty of the product cannot be guaranteed because of proteolytic activity released from the cells (see also section 2.3). Cell immobilization has the added advantage that the cells can build up a microenvironment which provides for more optimal cell proliferation and which reduces the requirement for serum compared to homogeneous systems. Generally, cell proliferation is reduced by confluency and the rate of mitosis decreased which are a great advantage for the culturing of partially unstable cell lines. It became evident that high density cell culture systems provide an increased cell-specific productivity (86,93, unpublished results). They are recommended for the culture of cell lines showing low specific productivity where homogeneous systems normally fail because of product dilution (see also section 3.1.2)(121).

The heterogeneous systems allow the retention of the cells with maximal achievable concentrations of 1 - 2 x 10exp8 cells per ml. This advantage is offset by difficulties in the re-use of cells from a heterogeneous system, the probable accumulation of toxic products, and the loss of the possibility of the direct cell control. Indirect methods have to be employed. The determination of the consumption of nutrients, the production of metabolites (e.g.: determination of the produced lactic acid (122)), and the calculation of the respiratory coefficient are estimations of the viable cell number. The release of lactate dehydrogenase is an estimation of the viability of the culture (123). In any case, these estimations may be erroneous and have to be considered with care. A more detailed view about the estimation of the cell number in reactor systems was given by Fleischaker et al. (124).

The following systems were developed: Cells were encapsulated (125-127) or entrapped in beads (128,129) and cultivated in a stirred tank reactor. The advantage is the retention of cells and, in the capsule systems with low molecular weight cut-off membranes, the retention of the product. However, the capsules providing this product retention have to be destroyed when the product is to be extracted. The process time is therefore limited (batch!).

Both systems have the main disadvantage of the establishment of gradients of nutrients, oxygen, and metabolites from the surface to the center of the capsules or beads, or vice versa.

Another group of heterogeneous systems is the group of membrane biorectors: the hollow fiber systems (HF), and the flat membrane systems. The principle of both types are quite similar. The medium is cycled from a conditioning vessel through the intracapillary space of the HF-membrane or the medium chamber of the flat membrane system, respectively, and back to the conditioning vessel (Fig. 7, taken from Ref. 78). Depending on the membranes used, the product is retained or not.

In 1972, Knazek et al. (130) described for the first time the use of HF-modules for the culture of animal cells, in which

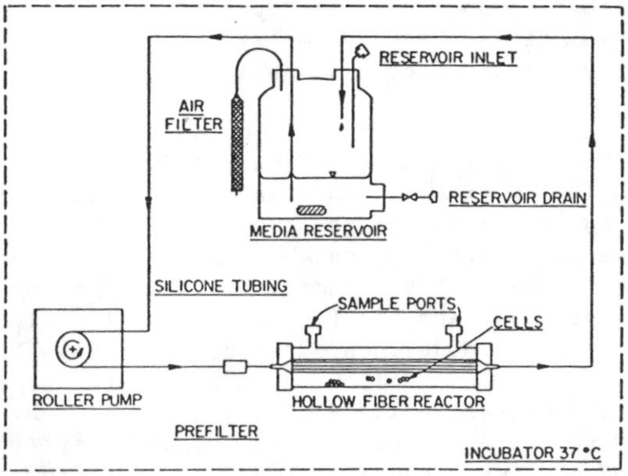

FIGURE 7. Flow diagramme of a hollow fiber reactor system placed inside an incubator at 37°C (taken from 78).

388

TABLE 6. Operation modes of hollow fiber systems (see also Fig. 8)

Mode	Open shell ultra-filtration (a)	Closed shell fil-tration (b)	Crossflow ultra-filtration (c)
Problems	Membrane clogging, Unequal pressure distribution, Shear forces (?) Cell wash out	Unequal distribu-tion (pressure, flow) Gradients	Membrane clogging, Shear forces (?)
Advantages	Better distribution of pressure and flow than closed shell	Lower membrane clogging than other two modes	Equal distribution of pressure and flow Better transmembrane flow than other two modes

they obtained tissue like cell densities. Nowadays, this type of bioreactor is widely used for small and medium scale animal cell cultures (121). The main disadvantages are limitations in the scale-up-ability, gradient crea-tion as with encapsulation systems, different flow rates in different fi-bers, and the possibility of membrane clogging. These disad-vantages can be overcome to some extend when the closed shell filtration mode is replaced by an open shell ultrafiltraion or cross flow ultrafiltration mode (131,132)(Table 6, Fig. 8), or when both modes are used in an alternating manner done by Endo-tronics. Both concepts operate 3 - 4 times more optimally than the classical closed shell fil-tration mode (131).

Flat membrane systems, using a two chamber configuration, show only few differences when compa-red to the HF-modules. The ad-vantages are the scale-up-abili-ty (133), free choice of the membrane, and the flexibilty of operation modes. A pure diffu-sion mode, as well as real cross flow ultrafiltration mode can be used more optimally than using HF-modules because of the better geometry (12,99). The MBR-Mem-broferm offers a special confi-guration: the three chamber sys-tem (Fig. 9). Here, the product

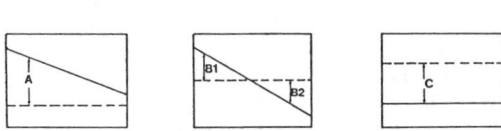

(a) Open shell ultra-filtration. (b) Closed shell ultra-filtration. (c) Crossflow ultra-filtration.

FIGURE 8. Schematic representation of the three modes of operation of hollow fiber systems and their pressure distribution. The open arrows indicate feed inlet and the closed arrows indicate effluent stre-ams. The tube side pressure is represen-ted by a solid line and shell side pres-sure by a dashed line: a) open shell ul-trafiltration and transmembrane pressure difference, A; b) closed shell ultrafil-tration and transmembrane pressure diffe-rences, B1 and B2; c) crossflow ultrafil-tration and transmembrane pressure diffe-rence, C (taken from 131).

chamber is separated from the cell chamber by a microfiltration membrane, which allows product recovery without loosing cells. Cell chamber and medium chamber are separated by an ultrafiltration membrane for preventing the loss of product or contamination of the product by serum-derived proteins.

The last bioreactor type, the Opticell system, will be mentioned briefly. It is based on a ceramic cartridge as support for the adherence of mammalian cells. Medium is cycled from a conditioning vessel through the channels of this cartridge and back to the vessel. It is a small to medium-scale system, with the disadvantages of product dilution, partial cell retention, no protection of cells and/or product from shearing, and a linear scale-upability.

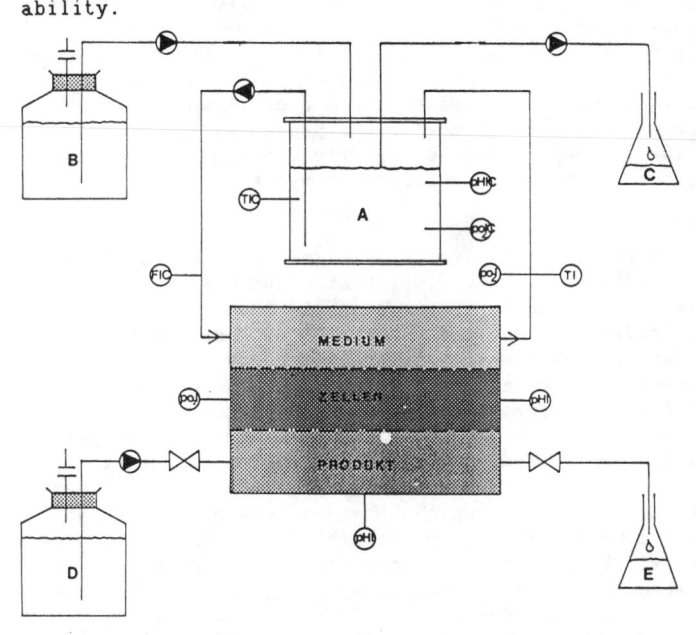

FIGURE 9. Pheripherial installation of a Membroferm with conditioning vessel (A), reservoir of growth medium (B), spent medium (C), product collecting medium (D), and product collection (E). Probes are indicated as follows: TI temperature indicator, FI flow rate indicator, pO2I oxygen tension indicator, pHI pH indicator, (C) shows an automated control circuit (taken from 99).

These heterogeneous systems are mainly used for continuous perfusion cultures. A more detailed view was published recently (121).

3.3.2. Comparison of the different bioreactor systems for the production of monoclonal antibodies. Table 5 gives a comparison of the different bioreactor systems with respect to the production of mAbs. As already mentioned (section 3.2.1), cell densities up to 10exp8 cells per ml can be achieved in immobilization systems which provides for the build up of a microenvironment (except for the Opticell) and higher productivity per volume (21 - 1400 mg Ig/l x d in product retention systems; 0.2 - 570 mg Ig/l x d systems without product retention) than in homogeneous systems (0.25 - 62 mg Ig/l x d and 19 - 660 mg Ig/l x d for homogeneous low and high density systems, respectively). This may decrease volume costs of the reactor system. The large ranges are explicable by the great differences of product secretion by the different hybridoma cell lines, as it was shown in table 2.

It should also be mentioned that these values were obtained mostly from mouse - mouse hybridoma cultures largely (13). Therefore, care has to be paid when human hybridomas are cultured in the different production systems, because of their general lower productivity (e.g.: 900 µg human Ig/ml

(134) in comparison to 1500 - 10000 µg mouse Ig/ml (135), both in Damon capsules).
Maximal product concentrations range from 2 to 600 mg/l for the homogeneous systems and from 900 and 10000 mg/l (ascites like concentrations) to 2 and 450 mg/l for heterogeneous systems with or without product retention, respectively. The differences in product titers obtained are due to the cell specific productivity (Table 2, section 3.1), production kinetics (sections 2.4, 3.1), and the process mode used (section 3.1). In those cases where the product is not retained, product accumulation depends on the dilution/perfusion rate (D). The product residence time (Th) can be easily calculated (Th = 1/D). The product residence time is influenced, in addition, by diffusion of the product through the cell mass, membrane, or bead matrix in immobilizations systems without product retention.
Generally, product purity ranges between 1 and 7% for all systems, except for reactor types with product retention (purities of 20 - 80% are achievable). These low values may be increased by the use of serum-free medium (section 3.2), but with the disadvantage of the possible appearance of shearing problems and increased proteolytic acitvity (section 2.3) normally inhibited by serum inhibitors in serum supplemented media.

4. FINAL REMARKS

This survey presented the state of art of hybridoma technology to some extend, with special attention to the mouse-, rat-, and human- inter- and intraspecies hybridomas in small scale cultures and to the critical points which have to be considered when one starts hybridoma cultures in bioreactors. The different approaches in establishing new human hybridoma cell lines using different partners and new fusion partner combinations, as well as in the waste field of bioreactor systems demonstrate that many problems still have to be solved. This is also valid for the different approaches in the field of serum-free media and the medium composition in general.
Today, it is not yet possible to say which hybridoma system is the best a classical one or perhaps a new one using the possiblity of genetic engineering. Certainly, there will be some optimal ones for the different applications. This can be extended to the bioreactor systems. As already said recently (121), there are some systems which will be generally accepted in the future because of their large application possibilities, whereas others will disappear because their applicability is too limited and therefore uneconomical.
In any case, the future will show, which are the best or optimal systems (type of cell line as well as bioreactor system as well as medium composition for a particular bioreactor).

Acknowledgments: The author has to thank Mrs. Michelson for critical reading of the manuscript and Mrs. Yvert for helping preparing the manuscript.

5. REFERENCES
1. Köhler G, Milstein C: Continuous cultures of fused cells secreting antibody of predefined specificity. Nature 256 (1975): 495-497.
2. Birch JR, Edwards GO: Production industrielle d'anticorps monoclonaux. Biofutur Mars (1986): 29-34.
3. Carthew P: Is rodent virus contamination of monoclonal antibody preparaions for use in human therapy a hazard? J gen Virol 67 (1986): 963-974.
4. Shulman M, Wilde CD, Köhler G: A better cell line for making hybridomas

secreting specific antibodies. Nature 276 (1978): 269-270.

5. Kearney JF, Radbruch A, Liesegang B, Rajewsky K: A new mouse myeloma cell line that has lost immunoglobulin expression but permits the construction of antibody-secreting hybrid cell lines. J Immunol 123 (1979): 1548-1550.

6. Fazekas de St. Groth S, Scheidegger D: Production of monoclonal antibodies: Strategy and tactics. J Immuol Method 35 (1980): 1-21.

7. Merten O-W, Palfi GE, Klement G, Steindl F: Specific kinetic patterns of production of monoclonal antibodies in batch cultures and consequences on fermentation processes. Presented at the 8th ESACT - 32nd OHOLO - Meeting on Modern Approaches to Animan Cell Technology, 6th - 10th April 1987, Tiberias Il.

8. Ware CF, Donato NJ, Dorshkind K: Human, rat or mouse hybridomas secrete high levels of monclonal antibodies following transplantation into mice with severe combined immunodeficiency disease (SCID). J Immunol Method 85 (1985): 353-361.

9. Bodeus M, Burtonboy G, Bazin H: Rat monoclonal antibodies.IV. Easy method for in vitro production. J Immunol Method 79 (1985): 1-6.

10. Galfré G, Butcher GW, Howard JC, Wilde CD, Milstein C: Clonal competition and stability of hybrid myelomas of mouse and rat origin. Transplant Proc 12 (1980): 371-375.

11. Shawler DL, Bartholomew RM, Smith LM, Dillman, RO.: Human immune response to multiple injections of murine monoclonal IgG. J Immunol 135 (1985): 1530-1535.

12. Katinger H: Animal cell culture: Biological and technological aspects. Presented at the 4th European Congress on Biotechnology, 14th-19th June 1987, Amsterdam Nl.

13. Merten O-W: The use of different types of fermentor for the production of monoclonal antibodies. Presented at the 2nd European Symposium on Protein Purification Technologies, 29th September - 2nd October 1986, Nancy F.

14. Cote RJ, Morrissey DM, Houghton AN, Beattie EJ Jr, Oettgen HF, Old LJ: Generation of human monoclonal antibodies reactive with cellular antigens. Proc Natl Acad Sci USA 80 (1983): 2026-2030.

15. Cote RJ, Morrissey DM, Oettgen HF, Old LJ: Analysis of human monoclonal antibodies derived from lymphocytes of patients with cancer. Fed Proc 43 (1984): 2465-2469.

16. Yoshikawa K, Ueda, R, Obata, Y, Utsumi KR, Notake K, Takahashi T: Human monoclonal antibody reactive to stomach cancer produced by mouse-human hybridoma technique. Jpn J Cancer Res (GANN) 77 (1986): 1122-1133.

17. Raison RL, Walker KZ, Halnan CRE, Briscoe D, Basten A: Loss of secretion in mouse-human hybrids need not to be due to the loss of a structural gene. J Exp Med 156 (1982): 1380-1389.

18. Crawford DH: Production of human monoclonal antibodies using Epstein-Barr Virus. In: Human hybridomas and monoclonal antibodies (eds. Engleman EG, Foung SKH, Larrick J, Raubitschek A), (1985): pp. 37-53, Plenum Press.

19. Thompson KM, Melamed MD, Eagle K, Gorick BD, Gibson T, Holburn AM, Hughes-Jones NC: Production of human monoclonal IgG and IgM antibodies with anti-D (rhesus) specificity using heterohybridomas. Immunology 58 (1986): 157-160.

20. Foung SKH, Perkins S, Arvin A, Lifson J, Mohagheghpour N, Fishwild D, Grumet FC, Engleman EG: Production of human monoclonal antibodies using a human-mouse fusion partner. In: Human hybridomas and monoclonal antibodies (eds. Engleman EG, Foung SKH, Larrick J, Raubitschek A), (1985): pp. 135-148, Plenum Press.

21.Östberg L, Pursch E: Human x (mouse x human) hybridomas stably producing human antibodies. Hybridoma 2 (1983): 361-367;

22.Ichimori Y, Sasano K, Itoh H, Hitotsumachi S, Kimura Y, Kaneko K, Kida M, Tsukamoto K: Establishment of hybridomas secreting human monoclonal antibodies against tetanus toxin and hepatitis B virus surface antigen. Biochem Biophys Res Commun 129 (1985): 26-33.

23.Van Meel FCM, Steenbakkers PGA, Oomen JCH: Human and chimpanzee monoclonal antibodies. J Immunol Method 80 (1985): 267-276.

24.Teng NNH, Lam KS, Riera FC, Kaplan HS: Construction and testing of mouse-human heteromyelomas for human monoclonal antibody production. Proc Natl Acad Sci USA 80 (1983): 7308-7312.

25.Bron D, Feinberg MB, Teng NNH, Kaplan HS: Production of human monoclonal IgG antibodies against rhesus (D) antigen. Proc Natl Acad Sci USA 81 (1984): 3214-3217.

26.Teng NNH, Reyes GR, Bieber M, Fry, KE, Lam KS, Hebert JM: Strategies for stable human monoclonal antibody production: Construction of heteromyelomas, in vitro sensitization, and molecular cloning of human immunoglobulin genes. In: Human hybridomas and monoclonal antibodies (eds. Engleman EG, Foung SKH, Larrick J, Raubitschek A), (1985): pp. 71-91, Plenum Press.

27.Olsson L, Kaplan HS: Human-human hybridomas producing monoclonal antibodies of predefined antigenic specificity. Proc Natl Acad Sci USA 77 (1980): 5429-5431.

28.Buck DW, Larrick JW, Raubitschek A, Truitt KE, Senyk G, Wang J, Dyer B: Production of human monoclonal antibodies. In: Monoclonal antibodies and functional cell lines. Progress and applications (eds. Kennett RH, Bechtol KB, McKearn TJ), (1984): pp. 275-309, Plenum Press.

29.Kozbor D, Croce CM: Fusion partners for production of human monoclonal antibodies. In: Human hybridomas and monoclonal antibodies (eds. Engleman EG, Foung SKH, Larrick J, Raubitschek A), (1985): pp. 21-36, Plenum Press.

30.Hilfenhaus J, Kanzy E-J, Köhler R, Willems WR: Generation of human-anti-rubella monoclonal antibodies from human hybridomas constructed with antigen-specific Epstein-Barr virus transformed cell lines. Behring Inst Mitt 80 (1986): 31-41.

31.Andreasen RB, Olsson L: Antibody-producing human-human hybridomas.III. Derivation and characterization of two antibodies with specificity for human myeloid cells. J Immunol 137 (1986): 1083-1090.

32.Roder JC, Cole SPC, Atlaw T, Campling BC, McGarry RC, Kozbor D: The Epstein-Barr virus-hybridoma technology. In: Human hybridomas and monoclonal antibodies (eds. Engleman EG, Foung SKH, Larrick J, Raubitschek A), (1985): 55-70, Plenum Press.

33.Kozbor D, Lagarde AE, Roder JC: Human hybridomas constructed with antigen-specific Epstein-Barr virus-transformed cell lines. Proc Natl Acad Sci USA 79 (1982): 6651-6655.

34.Kozbor D, Triputti P, Roder JC, Croce CM: A human hybrid myeloma is an efficient fusion partner that enhances monoclonal antibody production. J Immunol 133 (1984): 3001-3005.

35.Milstein C, Cuello AC: Hybrid hybridomas and the production of bi-specific monoclonal antibodies. Immunol Today 5 (1984): 299- 304.

36.Corvalan JRF, Smith W: Construction and characterisation of a hybrid-hybrid monoclonal antibody recognizing both carcinoembryonic antigen (CEA) and vinca alkaloids. Cancer Immunol Immunother 24 (1987): 127-132.

37.Rüker F, Reiter S, Jungbauer A, Liegl W, Himmler G, Steinkellner H,

Wenisch E, Steindl F, Wagner K, Katinger H: Self-hybridization of hybridomas leads to stabilization of clones and increased yield of monoclonal antibodies. Dev biol Standard 66, 71-74.

38. Boulianne GL, Hozumi N, Shulman MJ: Production of functional chimaeric mouse/human antibody. Nature 312 (1984): 643-646.

39. Morrison SL, Johnson MJ, Herzenberg LA, Oi VT: Chimeric human antibody molecules: Mouse antigen-binding domains with human constant region domains. Proc Natl Acad Sci USA 81 (1984): 6851-6855.

40. Sahagan BG, Dorai H, Saltzgaber-Muller J, Toneguzzo F, Guindon CA, Lilly SP, McDonald KW, Morrissey DV, Stone BA, Davis GL, McIntosh PK, Moore GP: A genetically engineered murine/human chimeric antibody retains specificity for human tumor-associated antigen. J Immunol 137 (1986): 1066-1074.

41. Liu AY, Robinson RR, Hellström KE, Murray ED Jr, Chang CP, Hellström I: Chimeric mouse-human IgG1 antibody that can mediate lysis of cancer cells. Proc Natl Acad Sci USA 84 (1987): 3439-3443.

42. Sun LK, Curtis P, Rakowicz-Szulczynska E, Ghrayeb J, Chang N, Morrison SL, Koprowski H: Chimeric antibody with human constant regions and mouse variable regions directed against carcinoma-associated antigen 17-1A. Proc Natl Acad Sci USA 84 (1987): 214-218.

43. Schwaber J, Cohen EP: Pattern of immunoglobulin synthesis and assembly in a mouse-human somatic cell hybrid clone. Proc Natl Acad Sci USA 71 (1974): 2203-2207.

44. Steinitz M, Klein G, Koskimies S, Makel O: EB virus-induced B lymphocyte cell lines producing specific antibody. Nature 269 (1977): 420-422.

45. Kozbor D, Steinitz M, Klein G, Koskimies S, Mäkelä O: Establishment of anti-TNP antibody producing human lymphoid lines by preselection for hapten binding followed by EBV-transformation. Scand J Immunol 10 (1977): 181-194.

46. Steinitz M, Seppälä I, Eichmann K, Klein G: Establishment of a human lymphoblastoid cell line with specific antibody production against group A streptococcal carbohydrate. Immunobiol 156 (1979): 41-47.

47. Kozbor D, Roder JC: Requirements for the establishment of high-titered human monoclonal antibodies against tetanus toxoid using the Epstein-Barr virus technique. J Immunol 127 (1981): 1275-1280.

48. Morrison SL: Transfectomas provide novel chimeric antibodies. Science 229 (1985): 1202-1207.

49. Boss MA, Kenten JH, Wood CR, Emtage JS: Assembly of functional antibodies from immunoglobulin heavy and light chains synthesised in E.coli. Nucl Acid Res 12 (1984): 3791-3806.

50. Wood CR, Boss MA, Kenten JH, Calvert JE, Roberts NA, Emtage JS: The synthesis and in vivo assembly of functional antibodies in yeast. Nature 314 (1985): 446-449.

51. Köhler G: Derivation and diversification of monoclonal antibodies. Science 233 (1986): 1281-1286.

52. Neuberger MS, Rajewsky K: Swith from hapten-specific immunoglobulin M to immunoglobulin D secretion in a hybrid mouse cell line. Proc Natl Acad Sci USA 78 (1981): 1138-1142.

53. Krawinkel U, Zoebelein G, Brüggemann M, Radbruch A, Rajewsky K: Recombination between antibody heavy chain variable region genes: Evidence for gene conversion. Proc Natl Acad Sci USA 81 (1983): 4997-5001.

54. Cook WD, Scharff MD: Antigen-binding mutants of mouse myeloma cells. Proc Natl Acad Sci USA 74 (1977): 5687-5691.

55. Rudikoff S, Giusti AM, Cook WD, Scharff MD: Single amino acid substitution altering antigen-binding specificity. Proc Natl Acad Sci USA 79 (1982): 1979-1983.

56.Viallat JR, Kourilsky FM: Induction of antibody-producing cell lines by Epstein-Barr virus. Path Biol 30 (1982): 232-242.

57.Zurawski VR Jr, Haber E, Black PH: Production of antibody to tetanus toxoid by continuous human lymphoblastoid cell lines. Science 199 (1978): 1439-1441.

58.Steinitz M, Klein E: Human monoclonal antibodies produced by immortalization with Epstein-Barr virus. Immunol Today 2 (1981): 38-39.

59.Winger L, Winger C, Shastry P, Russell A, Longenecker M: Efficient generation in vitro, from human peripheral blood cells, of monoclonal Epstein-Barr virus transformants producing specific antibody to a variety of antigens without prior deliberate immunization. Proc Natl Acad Sci USA 80 (1983): 4484-4488.

60.Erikson J, Martinis J, Croce CM: Assignment of the genes for human l immunoglobulin chains to chromosomes 22. Nature 294 (1981): 173-178.

61.Croce CM, Shander M, Martinis J, Cicurel L, D'Ancona GG, Koprowski H: Preferential retention of human chromosome 14 in mouse x human B cell hybrids. Eur J Immunol 10 (1980): 486-488.

62.Olsson L, Andreasen RB, Ost A, Christensen B, Biberfeld P: Antibody producing human-human hybridomas. II. Derivation and characterization of an antibody specific for human leukemia cells. J Exp Med 159 (1984): 537-550.

63.Yarmush ML, Gates III FT, Weisfogel DR, Kindt TJ: Identification and characterization of rabbit-mouse hybridomas secreting rabbit immunoglobulin chains. Proc Natl Acad Sci USA 77 (1980): 2899-2903.

64.Gardner JS, Chiu ALH, Maki NE, Harris JF: A quantitative stability analysis of human monoclonal antibody production by heteromyeloma hybridomas, using an immunofluorescent technique. J Immunol Method 85 (1985): 335-346.

65.Awdeh ZL, Williamson R, Askonas BA: One cell-one immunoglobulin: Origin of limited heterogeneity of myeloma proteins. Biochem J 116 (1970): 241-248.

66.Hamilton RG, Roebber M, Reimer CB, Rodkey LS: Quality control of murine monoclonal antibodies using isoelectric focusing affinity immunoblot analysis. Hybridoma 6 (1987): 205-217.

67.Hamilton RG, Roebber M, Reimer CB, Rodkey LS: Isoelectric focusing-affinity immunoblot analysis of mouse monoclonal antibodies to the four human IgG subclasses. Electrophoresis 8 (1987): 127-134.

68.Pearson TW, Anderson NL: Use of high resolution two-dimensional gel electrophoresis for analysis of monoclonal antibodies and their specific antigens. Methods Enzymol 92 (1983): 196-220.

69.Tracy RP, Katzmann JA, Kimlinger TK, Hurst GA, Young DS: Development of monoclonal antibodies to proteins separated by two-dimensional gel electrophoresis. J Immunol Method 65 (1983): 97-107.

70.Tracy RP, Currie RM, Kyle RA, Young DS: Two dimensional gel electrophoresis of serum specimens from patients with monoclonal gammopathies. Clin Chem 28 (1982): 900-907.

71.Reisfeld RA: Heterogeneity of rabbit light-polypeptide chains. Cold Spring Harbor Symposium quant Biol 32 (1967): 291-298.

72.Elkon KB: Isoelectric focusing of human IgA and secretory proteins using thin layer agarose gels and nitrocellulose capillary blotting. J Immunol Method 66 (1984): 313-321.

73.Schlaeger EJ, Eggimann B, Gast A: Proteolytic activity in the culture supernatants of mouse hybridoma cells. Dev biol Standard 66 (1987): 403-408.

74.Merten O-W: Batch production and growth kinetics of hybridomas: Cyto-

technology 1 (1987): in press.

75. Williams JA: Effects of medium concentration on antibody production. J Tissue Culture Method 8 (1984): 115-118.

76. Lavery M, Kearns MJ, Price DG, Emery AN, Jefferis R, Nienow AW: Physical conditions during batch culture of hybridomas in laboratory scale stirred tank reactors. Dev biol Standard 60 (1985): 199-206.

77. Merten O-W, Reiter S, Himmler G, Scheirer W, Katinger H: Production kinetics of monoclonal antibodies. Dev biol Standard 60 (1985): 219-227.

78. Altshuler GL, Dziewulski DM, Sowek JA, Belfort G: Continuous hybridoma growth and monoclonal antibody production in hollow fiber reactors-separators. Biotechnol Bioeng 28 (1986): 646-658.

79. Birch JR, Thompson PW, Boraston R, Oliver S, Lambert K: The large-scale production of monoclonal antibodies in airlift fermenters. In: Plant and animal cells - process possibilities (eds. Webb C, Mavituna F), (1987): pp. 162-171, Ellis Horwood Ltd.

80. Emery AN, Lavery M, Williams B, Handa A: Large-scale hybridoma culture; In: Plant and animal cells - process possibilities (eds. Webb C, Mavituna F), (1987), pp. 137-146, Ellis Horwood Ltd.

81. Reuveny S, Velez D, Macmillan JD, Miller L: Factors affecting cell growth and monoclonal antibody production in stirred reactors. J Immunol Method 86 (1986): 53-59.

82. Velez D, Reuveny S, Miller L, Macmillan JD: Kinetics of monoclonal antibody production in low serum growth medium. J Immunol Method 86 (1986): 45-52.

83. Birch JR, Boraston R, Wood L: Bulk production of monoclonal antibodies in fermenters. Trends Biotechnol 3 (1985): 162-166.

84. Reuveny S, Velez D, Riske F, Macmillan JD, Miller L: Production of monoclonal antibodies in culture: Dev biol Standard 60 (1985): 185-197.

85. Boraston R, Thompson PW, Garland S, Birch JR: Growth and oxygen requirements of antibody producing mouse hybridoma cells in suspension culture. Dev biol Standard 55 (1984): 103-111.

86. Reuveny S, Velez D, Miller L, Macmillan JD: Comparison of cell propagation methods for their effect on monoclonal antibodies yield in fermenters. J Immunol Method 86 (1986): 61-69.

87. Low K, Harbour C: Growth kinetics of hybridoma cells: (1) The effects of varying foetal calf serum levels. Dev biol Standard 60 (1985): 17-24.

88. Merten O-W, Palfi GE, Klement G, Steindl F: Specific kinetic patterns of production of monoclonal antibodies in batch cultures and consequences on fermentation processes. Presented at: 8th ESACT - 32nd OHOLO - Meeting on Modern Approaches to Animal Cell Technology, 6th - 10th April 1987, Tiberias Il.

89. Merten O-W, Reiter S, Katinger H: Stabilizing effect of reduced cultivation temperature on human-mouse hybridomas. Dev biol Standard 60 (1985): 509-512.

90. Fazekas de St. Groth S: Automated production of monoclonal antibodies in a cytostat. J Immunol Method 57 (1983): 121-136.

91. Glacken MW, Fleischaker RJ, Sinskey AJ: Large-scale production of mammalian cells and their products. Engineering principles and barriers to scale-up. Ann N Y Acad Sci 413 (1983): 355-372.

92. Katinger H, Scheirer W: Mass cultivation and production of animal cells. In: Animal cell biotechnology, Vol.1 (eds. Spier RE, Griffiths JB), (1985): pp. 167-194, Academic Press.

93. Emery AN: Growth of hybridomas and secretion of monoclonal antibodies in vitro. Presented at the Society of Chemical Industry-Symposium on Lerga-scale production of monoclonal antibodies, 9th December 1986, London GB.

94.Van Wezel AL, van der Velden-de Groot CAM, de Haan HH, van den Heuvel N, Schasfort R: Large scale animal cell cultivation for production of cellular biologicals. Dev biol Standard 60 (1985): 229-236.
95.Martin N: High productivity in mammalian cell culture. Bio/Technology 5 (1987): 838-840.
96.Feder J, Tolbert WR: Mass culture of mammalian cells in perfusion systems. Am Biotechnol Lab 3 (1985): 24-36.
97.Katinger HWD, Scheirer W: Status and developments of animal cell technology using suspension culture techniques. Acta Biotechnol 2 (1982): 3-41.
98.Glacken MW, Fleischaker RJ, Sinskey AJ: Mammalian cell culture: Engineering principles and scale-up. Trends Biotechnol 1 (1983): 102-108.
99.Scheirer W: High density growth of animal cells within cell retention fermentors equipped with membranes. In: Animal Cell Biotechnology, Vol. III (eds. Spier RE, Griffiths JB), (1987): in press, Academic Press.
100.Fleischaker RJ Jr, Sinskey AJ: Oxygen demand and supply in cell culture. European J Appl Microbiol Biotechnol 12 (1981): 193-197.
101.Onken U, Kiese S, Jostmann Th: An airlift fermenter for continuous culture at elevated pressures. Biotechnol Lett 6 (1984): 283-288.
102.Maldonado RL, Fulbright JG: Processed serum: A consistent growth support for hybridomas. Int Biotechnol Lab 2 (1984): 34-36.
103.Shacter E: Serum-free medium for growth factor-dependent and -independent plasmacytomas and hybridomas. J Immunol Method 99 (1987): 259-270.
104.Cole SPC, Vreeken EH, Mirski SEL, Campling BG: Growth of human x human hybridomas in protein-free medium supplemented with ethanolamine. J Immunol Method 97 (1987): 29-35.
105.Reuveny S, Bino T, Rosenberg H, Traub A, Mizrahi A: Pilot plant scale production of human lymphoblastoid interferon. Dev biol Standard 46 (1980): 281-288.
106.Mizrahi A: Primatone RL in mammalian cell culture media. Biotechnol Bioeng 19 (1977): 1557-1561.
107.Schönherr OT, van Gelder PTJA, van Hees PJ, van Os AMJM, Roelofs HWM: A hollow fiber dialysis system for the in vitro production of monoclonal antibodies replacing in vivo production in mice. Dev biol Standard 66 (1987): 211-220.
108.Butler M, Spier RE: The effects of glutamine utilisation and ammonia production on the growth of BHK cells in microcarrier cultures. J Biotechnol 1 (1984): 187-196.
109.Butler M: Growth limitations in high density microcarries cultures. Dev biol Standard 60 (1985): 269-280.
110.Butler M: Nutrition of hybridoma cells. Presented at the Society of Chemical Industry-symposium on Large-scale production of monoclonal antibodies, 9th December 1986, London GB.
111.Hu WS, Dodge TC, Frame KK, Himes VB: Effect of glucose on the cultivation of mammalian cells. Dev biol Standard 66 (1987): 279-290.
112.Eagle H, Barban S, Levy M, Schulze HO: The utilization of carbohydrates by human cell cultures. J biol Chem 233 (1958): 551-558.
113.Roberts RS, Hsu HW, Lin KD, Yang TJ: Amino acid metabolism of myeloma cells in culture. J Cell Sci 21 (1976): 609-615.
114.Seaver SS, Rudolph JL, Gabriels JE Jr: A rapid HPLC technique for monitoring amino acid utilization in cell culture. Bio Techniques 2 (1984): 254-260.
115.Katinger HWD, Scheirer W, Krömer E: Bubble column reactor for mass propagation of animal cells in suspension culture. Ger Chem Eng 2 (1979): 31-38.
116.Himmelfarb P, Thayer PS, Martin HE: Spin filter culture: The propagati-

on of mammalian cells in suspension. Science 164 (1969): 555-557.

117. Kitano K, Shintani Y, Ichimori Y, Tsukamoto K, Sasai S, Kida M: Production of human monoclonal antibodies bu heterohybridomas. Appl Microbiol Biotechnol 24 (1986): 282-286.

118. Lehmann J, Piehl GW, Schulz R: Bubble free cell culture aeration with porous moving membranes. Dev biol Standard 66 (1987): 227-240.

119. Stathopoulos NA, Hellums JD: Shear stress effects on human embryonic kidney cells in vitro. Biotechnol Bioeng 27 (1985): 1021-1026.

120. Tramper J, Vlak JM: Some engineering and economic aspects of continuous cultivation of insect cells for the production of Baculoviruses. Ann N Y Acad Sci 469 (1986): 279-288.

121. Merten O-W: Concentrating mammalian cells I. Large-scale animal cell culture. Trends Biotechnol 5 (1987): 230-237.

122. Reitzer LJ, Wice BM, Kennell D: Evidence that glutamine, not sugar, is the major energy source for cultured HeLa cells. J biol Chem 254 (1979): 2669-2676.

123. Arathoon WR, Birch JR: Large-scale cell culture in biotechnology. Science 232 (1986): 1390-1395.

124. Fleischaker RJ, Weaver JC, Sinskey AJ: Instrumentation for process control in cell culture. Adv Appl Microbiol 27 (1981): 137-167.

125. Jarvis AP Jr, Grdina TA: Production of biologicals from microencapsulated living cells. Bio Techniques 1 (1983): 22-27.

126. Gharapetian H, Davies NA, Sun AM: Encapsulation of viable cells within polyacrylate membranes. Biotechnol Bioeng 28 (1986): 1595-1600.

127. Pommerening K, Ristau O, Dautzenberg H, Loth F: Immobilisierung von Proteinen und Zellfragmenten durch ein neues Verfahren der Mikrokapsulierung. Biomed Biochim Acta 42 (1983): 813-823.

128. Nilsson K, Scheirer W, Merten O-W, Östberg L, Liehl E, Katinger HWD, Mosbach K: Entrapment of animal cells for production of monoclonal antibodies and other biomolecules. Nature 302 (1983): 629-630.

129. Scheirer W, Nilsson K, Merten O-W, Katinger HWD, Mosbach K: Entrapment of animal cells for the production of biomolecules such as monoclonal antibodies. Dev biol Standard 55 (1984): 155-161.

130. Knazek RA, Gullino PM, Kohler PO, Dedrick RL: Cell culture on artificial capillaries: An approach to tissue growth in vitro. Science 178 (1972): 65-67.

131. Tharakan JP, Chau PC: Operation and pressure distribution of immobilized cell hollow fiber bioreactors. Biotechnol Bioeng 28 (1986): 1064-1071).

132. Tharakan JP, Chau PC: A radial flow hollow fiber bioreactor for the large-scale culture of mammalian cells. Biotechnol Bioeng 28 (1986): 329-342.

133. Klement G, Scheirer W, Katinger HWD: Construction of a large scale membrane reactor system with different compartments for cells, medium and product. Dev biol Standard 66 (1987): 221-226.

134. Grdina TA, Jarvis AP Jr: Microencapsulation of human x human hybridoma cells: cell growth & monoclonal antibody production. Presented at BIOTECH '84 USA, (1984), pp. A235-A246, Online Publications.

135. Rupp RG: Use of cellular microencapsulation in large-scale production of monoclonal antibodies. In: Large-scale mammalian cell culture (eds. Feder J, Tolbert WR), (1985): pp. 19-38, Academic Press.

136. Iscove NN, Melchers F: Complete replacement of serum by albumin, transferrin, and soybean lipid in cultures of lipopolysaccharide-reactive B-lymphocytes. J Exp Med 147 (1978): 923-933.

137. Chang TH, Steplewski Z, Koprowski H: Production of monoclonal antibo-

dies in serum free medium. J Immunol Method 39 (1980): 369-375.

138. Murakami H, Masui H, Sato GH, Sueoka N, Chow TP, Kano-Sueoka T: Growth of hybridoma cells in serum-free medium: Ethanolamine is an essential component. Proc Natl Acad Sci USA 79 (1982): 1158-1162.

139. Wistar symposium on immunodiagnosis and immunotherapy with C017-1A mab in gastrointestinal cancer, April 2-3, 1986, in: Hybridoma 5 (1986): S1-S185.

140. Coffino P, Scharff MD: Rate of somatic mutation in immunoglobulin production by mouse myeloma cells. Proc Natl Acad Sci USA 68 (1971): 219-223.

141. Morrison SL: Sequentially derived mutants of the constant region of the heavy chain of murine immunoglobulins. J Immunol 123 (1979): 793-800.

Merten, O. : Culture of hybridomas. A survey

Hope : I'd like to underline the point that you can make twenty milligrams per ml of monoclonal antibody but they will be of no use at all if you have a 0.1 % recovery and so, the bottom line really is from the practical point of view, how much stuff you can actually put in a vial that comes out of the system and it is the combination of downstream processing but also the state of the material when it comes out of a bioreactor.

Spier : It is the quality of the material times the quantity of the material both in terms of the biological activity and the specific activity.

Merten : Yes, I said this. Perfusion is not a continuous process. Continuous process is a chemostat where you have no retention of cells, where you have normally no dead cells in it.

Spier : That's a question of definition...

Merten : Yes, but in real perfusion system where you have cell retention, then you always stay in the stationary phase more or less.

Van der Velden-De Groot : It does not go far from continuous perfusion system. You always have some cell growth and some cell loss.

Merten : Yes of course but the growth rate is different. In the chemostat you have a really high dilution rate and a high growth rate but in a perfusion system not. In a perfusion

system you have a high dilution rate but a low growth rate. So you are working in another physiological state of the cell.

Spier :   This is an important point.

Merten :   Yes.

APPLICATIONS OF CHROMATOGRAPHY IN DOWNSTREAM PROCESSING.

E. BOSCHETTI

IBF-Biotechnics
35 avenue Jean Jaurès, F-92390 Villeneuve la Garenne, France.
Tel. (1) 47 98 83 53 - Telex 610719 Rhone

1. INTRODUCTION

The eukaryotic cells have the natural property of synthetizing or secreting proteins into the medium and by that way, they represent an excellent tool for producing biomolecules. This principle is being widely applied in the preparation of proteins of therapeutic or diagnostic interest (lymphokines, interferons, tissue plasminogen activator, monoclonal antibodies, ...).

However, the protein production from animal cell culture is conditioned by the nature of the culture medium which contains serum proteins. Actually, the proteins brought in the medium may contaminate significantly the cell-secreted biological molecules and under this condition, purification could be a difficult task even when performant purification techniques are used. In this respect, the low concentration of serum or the use of serum substitutes have been suggested (1) (2). Whether they are animal sera or substitution products, the supplements necessary to basal medium represent obstacles in the isolation of secreted proteins, particularly in case of low protein concentration. As the use of medium constitutes a compulsory passage in the processus of protein biosynthesis by cell culture, it is necessary to devise technologies adapted to the definition of culture medium as well as to the purification of secreted proteins (3).

2. STEPS IN DOWSTREAM PROCESSING

The objective of downstream processing is to reach a highly pure protein starting from a crude material, by assembling different available separation technologies in a best configuration ; that means overcoming the problems connected with the nature of the starting raw material, the characteristics of the protein to be isolated, the suited degree of purification and the cost of entire process.

A purification process can be divided into three main steps as illustrated on Fig. 1 :

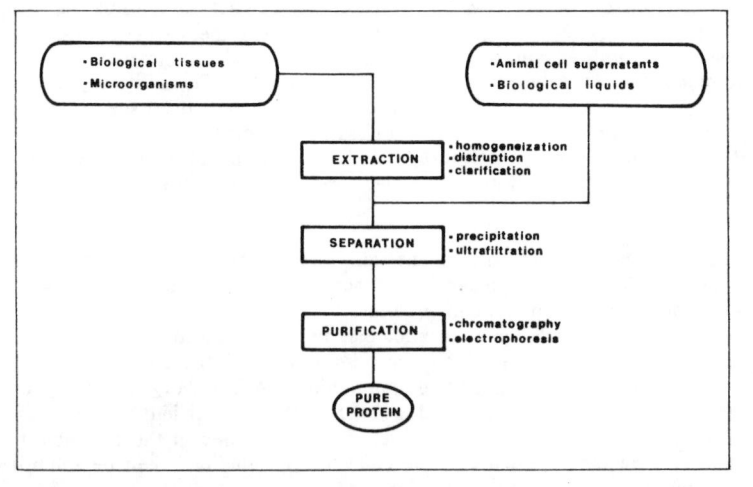

Fig. 1   SCHEMATIC REPRESENTATION OF THE THREE MAIN STEPS IN PROTEIN DOWNSTREAM PROCESSING.

401

A. O. A. Miller (ed.), Advanced Research on Animal Cell Technology, 401–418.
© 1989 by Kluwer Academic Publishers.

extraction, separation and purification.

The common feature to different purification steps is playing upon the intrinsic properties of proteins. This leads, on the one hand, to the recognition and separation between proteins and salts, and on the other hand, to the separation between proteins themselves.

Among the primary factors affecting the fractionation of a mixture, there are in fact the molecule size and shape, differential solubility in aqueous buffers, diffusivity, ionic charge and biological activity.

Fig. 2 represents schematically the possible useful technologies to be adopted according to the type of molecule and the exploited intrinsic property.

PRIMARY EXPLOITED FACTOR	TECHNOLOGIES COURRENTLY APPLIED
ionic charge	├──── electrophoresis ────┤ ├──── ion exchange chromatography ────┤
hydrophobicity	├─ hydrophobic chromat. ─┤ ├──── precipitation ────┤
solubility	├──── precipitation ────┤ ├──── microfiltration ────┤
molec. size	├──── gel filtration ────┤ ├──── ultrafiltration ────┤
diffusivity	├──── dialysis ────┤ ├─reverse osmosis ──┤
MOL. WEIGHT	10^3 10^4 10^5 10^6 10^7

Fig. 2   SEPARATION AND PURIFICATION TECHNIQUES AS A FUNCTION OF THE PROTEIN PROPERTIES

2.1. Extraction :

This step is of course restricted to the insoluble starting material such as cells, cell debris, animal or vegetal tissues. Extraction involves the use of crushing machines homogenisers, stirrers and then filters or centrifuges in order to separate clear water-soluble material from the solid pellet. As it is a very classical operation, this step needs not really sophisticated material , the most important imperatives being temperature control, choice of machines compatible with the stability of investigated protein and the right choice of aqueous solution (composition, ionic strength, pH, additives) so as to reach a high extraction yield. Sometimes, the destruction of cell wall or tissues is optimized by alternating steps of freezing and thawing or ultrasonic treatments.

The characteristic of an aqueous crude extract is to be diluted in term of protein content, to be rich in salts, small organic molecules and often in pigments.

2.2. Separation :

It can be summarized as an enrichment step where proteins are separated from other small undesirable molecules and yielded in a concentrated clear aqueous solution. This step can also remove a group of contaminant proteins.

The most known and performant technologies used in separation are precipitation and ultrafiltration. The former is based on the insolubility of proteins when the aqueous solution contains some specific salts (*e.g.* ammonium sulfate), organic compounds (*e.g.* polyethylene glycol, ethanol, acetone) and when the pH is modified by acidic or alkaline solutions. Since precipitation is an easy way to recover proteins in the insoluble phase, all other molecules are in the supernatant. The collected proteins can then be solubilized in a minimum volume of buffer.

Fractionated precipitation can also be carried out when separating two groups of proteins. This is particularly done for instance, by increasing progressively the quantities of ammonium sulphate.

In downstream processing, ultrafiltration involves the use of microporous flat membranes or microporous hollow fibers that separate in the aqueous phase, the large molecules (proteins) from the small ones (*e.g.* salts). This method is advantageous as it acts like a real molecular sieving and at the same time, excess water is eliminated from the diluted protein solutions. Ultrafiltration can be used not only as an alternative precipitation (separation) method, but also as a complement to precipitation (elimination of remaining salts or solvents).

Ultrafiltration, diafiltration and microfiltration are three different aspects of this technology and are applied at different levels of downstream processing as indicated on Fig. 3.

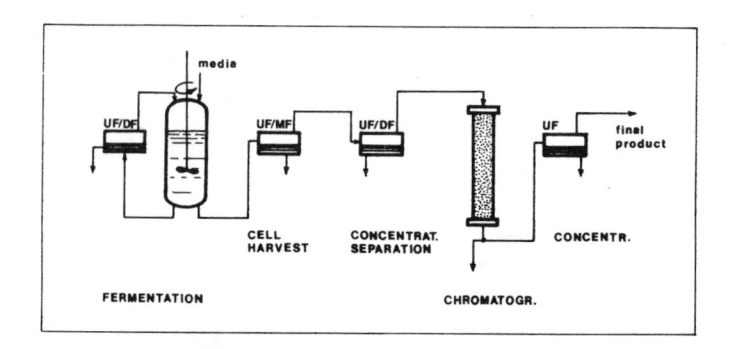

Fig. 3   POSITIONING OF ULTRAFILTRATION (UF), MICROFILTRATION (MF) AND DIAFILTRATION (DF) UNITS IN A GENERAL BIOTECHNOLOGICAL PROCESS.

2.3. Purification :

For the purification step in downstream processing, the most suitable techniques for proteins are liquid chromatography methods. Except gel filtration, liquid chromatography is based on the differential interaction between proteins solubilized in the buffer and the column solid phase. When the protein solution runs through the column, the proteins interact differently with the solid phase according to their composition and can thus be separated.

Among different types of chromatography, the most important ones are ion exchange chromatography, adsorption chromatography, affinity chromatography and hydrophobic chromatography.

Ion exchange chromatography is based on the ionic interaction between a support electrically charged and the net charge of a protein which may be modified by the pH and ionic strength of the aqueous solution. The isoelectric point and the charge density command the type of ion exchanger to be used.

Affinity chromatography is based on the highly specific interaction between the molecule to be purified and a ligand chemically immobilized on the solid support.

Hydrophobic chromatography exploits its ability of creating hydrophobic complexes between the proteins possessing long-chain aliphatic amino acids and the hydrophobic ligands immobilized on the solid phase. This association occurs in particular when the proteins medium has high ionic strength.

3. APPLICATION STRATEGY FOR PURIFICATION METHODS IN DOWNSTREAM PROCESSING

In the purification of a cell-produced protein, the first operation to be effected is separating the cells from the medium. This is easily done by simple centrifugation or filtration.

A culture surpernatant which contains between 5 and 20 % serum, may be submitted to an ammonium sulfate precipitation. This approach permits to leave in the supernatant many small molecules while concentrating the proteins. The protein precipitate is redissolved in a minimum buffer volume and is therefore in a highly saline environment which is a suitable condition for hydrophobic chromatography. The involved protein is via this step separated from contaminating proteins and its further purification may be carried out by other chromatographic steps.

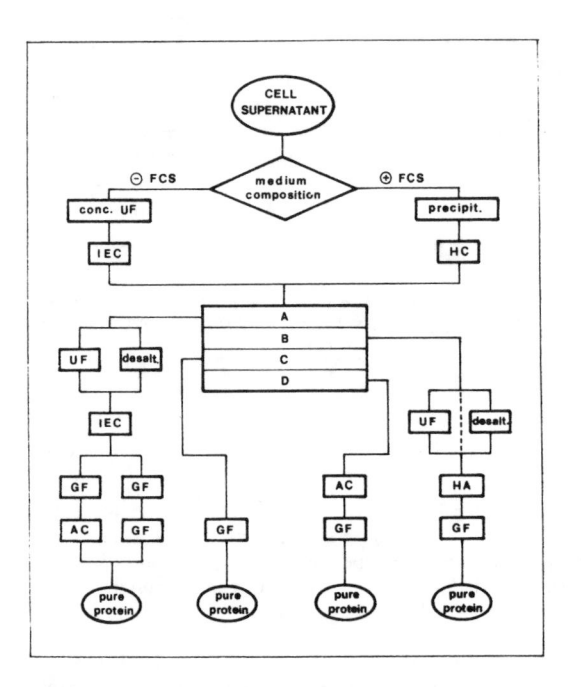

Fig. 4 *PURIFICATION POSSIBILITIES OF A PROTEIN ISSUED FROM IN VITRO CELL CULTURE.*
FCS : fetal calf serum ; IEC : ion exchange chromatography ; HC : hydrophobic chromatography ; UF : ultrafiltration ; GF : gel filtration ; AF : affinity chromatography HA : adsorption chromatography.
A = *the protein fraction is quite complex.* **B** = *the protein to be purified is quite unstable.* **C** = *the protein mixture is simple and of different molecule size.* **D** = *specific ligand easy to define and available.*

When the culture supernatant is poor in proteins (medium supplemented with a serum substitute), it is advisable to first proceed to a concentration by ultrafiltration which will remove small molecules (according to the membrane porosity) and besides, equilibrate the proteins in the buffer chosen for the following chromatography step. At this stage, the most appropriate chromatography is ion exchange and more particularly, anion exchange due to its good selectivity, its high capacity and a rapid setup.

In either situation (hydrophobic chromatography or ion exchange chromatography), the purification is followed by other chromatographic techniques with different specificities. The choice is made according to different criteria as indicated on Fig. 4. A protein fraction which is too complex to be submitted to a gel filtration, will be separated by ion exchange chromatography. Afterwards, gel filtration and/or affinity chromatography will follow in an order to be defined. A protein fraction which contains a limited number of different-molecular weight proteins, may be directly purified by gel filtration. The most appropriate gel and the column geometry are easily chosen for each case. No prior treatment (desalting) of the fraction to be purified is necessary. A protein fraction which contains a particularly labile protein, may be advantageously separated by hydroxyapatite chromatography. Indeed, the conditions of separation are quasi physiologic (phosphate buffer, neutral pH, weak ionic strength). The final step of purification is effected on a gel filtration column.

Beside the possibilities cited above, affinity chromatography may represent an interesting alternative when it associates selectivity and easy setup. Generally, such a technique does not always require sample equilibration before injection into the column. It is often followed by gel filtration when groups of proteins are to be separated or when the purification factor achieved is not sufficient .

In these sequences of techniques, the compatibility of different buffers from column to column must be born in mind. Effectively, it is not always possible to pass directly from a chromatography to another, due to the nature of the buffers used and the protein concentration.

TABLE I . EXAMPLE OF PROTEINS PRODUCED FROM CULTURE OF ANIMAL CELLS
AND PURIFIED BY CHROMATOGRAPHY

PROTEIN	PRODUCER CELL	CHROMATOGRAPHIC TECHNIQUE
Interleukin 1	Macrophages stimulated by immuno-modulateurs	GF , IEC , HC
Interleukin 2	Spleen cells stimulated by mitogens	GF , IAC
Plasminogen activator	Epithelium cells or melanome	GF , IEC , AC , IAC , MCAC
Colony-stimulating factor	Lung cells	GF , HC
Interferon	Leucocytes, Fibroblasts, Lymphocytes	GF , IEC , HC , AC , IAC , MCAC
Tumor necrosis Factor	Lymphoblstoid lines	GF , IEC , AC
Monoclonal antibodies	Hybridoms	IEC , HAC , AC , GF

GF :	Gel filtration	IAC :	Immunoaffinity chromatography
IEC :	Ion exchange chromatography	HAC :	Hydroxyapatite chromatography
HC :	Hydrophobic chromatography	MCAC :	Metal chelating chromatography
AC :	Affinity chromatography		

The preliminary operations such as extraction and separation are aimed at achieving a solution enriched with the investigated protein. However, its concentration with regard to the remaining impurities frequently very low ; in this situation, the first chromatographic step is chosen in such a way that the investigated protein is adsorbed on the column. The adsorption conditions are defined so as to obtain most undesirable proteins in the flowthrough.

The protein is then desorbed by modifying the ionic strength and/or pH, and the obtained fraction injected into a second column. The latter will be chosen in view of adsorbing only impurities. Effectively, as the impurities present at this stage are quantitatively minor, the adequate column can be of small size.

4. PURIFICATION OF MONOCLONAL ANTIBODIES

The aspect of monoclonal antibodies purification is particularly interesting in dowstream processing. A possible purification scheme is represented on Fig. 5.
Though the extraction step is practically inexisting since it is limited to the separation of cells, the two other steps are well developed.

The separation and concentration on hollow fibers or tangential flow allow the isolation of small molecules which may considerably diminish the chromatographic column performances. Gel filtration can also be used for desalting a concentrated protein solution (ascite fluid). Among the purification steps , many possibilities are available by chromatography (6). Immobilized antigens can be used when available in large quantities. Alternatively, specific anti-immunoglobulin antibodies or bacterial specific proteins (Protein A, Protein G) can also be used after immobilisation on solid support (4,5).

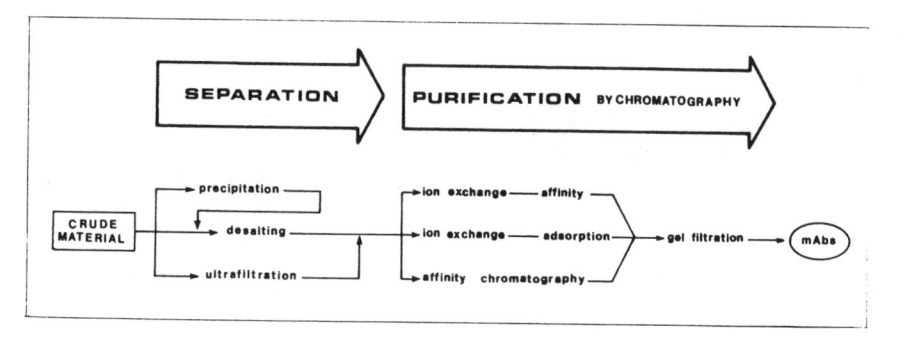

Fig. 5   SCHEMATIC PURIFICATION PROCESS FOR MONOCLONAL ANTIBODIES (mAbs)

These techniques that involve protein-protein interaction can however associate certain disadvantages : a too strong complex to be dissociated, unstability of the ligand in long term, possibilities of disruption of the protein-support bond, or partial protein hydrolysis with consequent leakage of undesirable products.
In this situation where the costs may also be high, it is preferable to use ion exchange columns eventually in association with hydroxyapatite (7). These systems present the advantage of being easily handled by the labs with low experience in chromatography. Moreover, the columns are extremely stable and not expensive. An example of this association is illustrated on Fig. 6 . In ion exchange, the polyanionic supports are

preferable to polycationic ones. Effectively, the conditions of pH and ionic strength may be adapted so as to have few contaminants adsorbed on the column ; this is made possible by the notable difference in pI between IgG , albumin and transferrin which represent the major contaminants.

Sometimes, extremely pure monoclonal antibodies are required ; in this case, gel filtration may be used in the last step of purification. This technique permits the separation of smaller molecules (albumin traces for instance) but also possible traces of other small molecular weight products of protein origin.

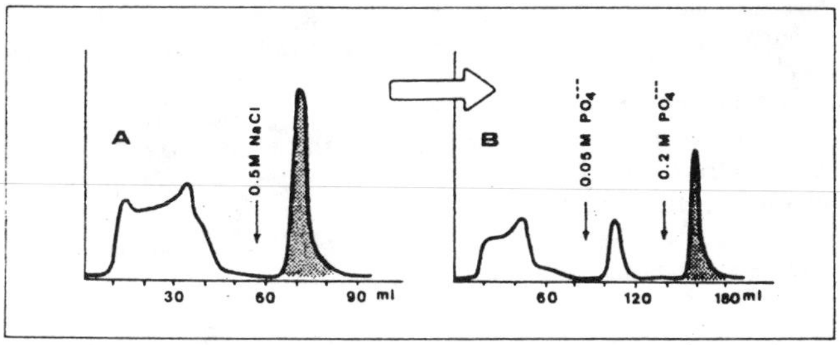

Fig. 6   *EXAMPLE OF APPLICATION OF AN IgG$_2$ MOUSE MONOCLONAL ANTIBODY USING A CHROMATOGRAPHIC TANDEM SYSTEM COMPOSED OF A CATION EXCHANGE COLUMN (A) (SP-TRISACRYL) AND OF HYDROXYAPATITE ADSORPTION COLUMN (B) (HA-ULTROGEL)*
   A) *Column : 1.14 x 10 cm ; adsorption buffer : 0.05 M acetate pH 5.5 ; elution buffer : 0.05 M acetate, 0.5 M sodium chloride pH 5.5. ; flow rate : 30 ml/hr ; sample : 2.5 ml ascite fluid diluted 20-times with the adsorption buffer.*
   B) *Column : 2.5 x 7.2 cm ; adsorption buffer : 0.01 M phosphate pH 6.8 ; elution buffer : 0.2 phosphate pH 6.8 ; flow rate : 30 ml/hr ; sample : 25 ml SP-TRISACRYL fraction equilibrated in the adsorption buffer.*
The purity of final mAb was superior to 90 % (SDS-PAGE).

5.   HYGIENIC AND REGULATORY ASPECTS IN DOWNSTREAM PROCESSING.

The qualitative and quantitative aspects in the purification of a cell-secreted protein depend totally on the intended application.

The production of proteins for in vivo use requires very stringent conditions because the cultured cells are often of cancer origin or considered as products from the DNA recombinant technology, and on the other hand, because the culture implied the use of animal proteins-based media.

A purification process of therapeutic proteins must be capable of eliminating any component which may generate undesirable reactions of either protein or virus or nucleic acid origin. In the precise case of monoclonal antibodies, the risk of carrying contaminant proteins and viruses along with nucleic acids must be considered when producer cells are of malignant origin.

Production of biologicals from cell culture involves in fact possible protein contaminants from cell metabolism , media constituents or affinity columns. These proteins represent a potential risk because they may be recognised as antigens stimulating immunoresponses. The technical problem of protein contaminants is related with the difficulty in measuring them qualitatively and quantitatively. In fact, the clinical

significance of protein contaminants may be unknown if each of them is not identified and clinically tested.

The presence of viruses is another major safety issue ; the origin of such a contamination can be the animal cells themselves and/or the serum added to the basal growth medium. The main concern is contamination of the final product by retroviruses which are capable of activating oncogens from cells.

As regard with the risk of presence of residual cellular DNA (hybridomas loose sometimes chromosomes), what is important to consider is :
- the amount of genetic material present after the purification process,
- the ability of residual DNA to induce malignant transformation related to the quantity.
The importance of bacterial toxines presence (pyrogens) must not be neglected either.

All these substances must imperatively be removed and all the purification steps be effected under aseptic conditions. This implies systematic in-process controls and consequently, setting validation methods and regulatory standards. It is not sufficient to demonstrate that the final product corresponds to the norms fixed, but it must be proved that in-process controls permit to obtain in all cases, a safe pure protein, specific, consistent and therapeutically efficient for the purposed application (8). To this end, the contamination due to possible leakage of ligands in affinity chromatography (immobilized proteins) should be systematically considered.

REFERENCES

1. Sato G.H., Pardee A.B., Sirbasku D.A. : Growth of cells in hormonally defined media . Cold Spring Arbor Conference on Cell Proliferation , 1982, p. 9.
2. Barnes D.W., Sirbasku D.A., Sato G. H., : Cell culture methods for molecular and cell biology , N.Y. Alan R. Liss (ed), 1984, Vol. 1, 2, 3 and 4.
3. Sené C., Boschetti E., : The place of chromatography in the separation and purification of proteins produced from cultured cells , Advan. Biotech. Processes, Alan R. Liss, Inc. (ed), in press.
4. Manil L., Motte P., Pernas P., Troalen F. et al., J. Immunol. Meth., 90, 1986, 25.
5. Stephenson J.R., Lee J.M., Wilton-Smith P.D., Anal. Biochem., 142, 1984, 189.
6. Boschetti E., Egly J.M., Monsigny M., Trends in Anal. Chem., 5, 1986, 4.
7. Pouradier Duteil X., Sené C., Boschetti E., in preparation.
8. Edmond S.K., Grady L.T., Oiutschoorn A.S., RHODES C.T., Drug Dev. Ind. Pharm., 12, 1986, 107.

Boschetti, E. : Chromatography : the downstream processing

Van Meel : You again accentuated the problems you might have with using monoclonal antibodies in clinical situations but you have also solutions for these problems. Have you suggestions on how to remove DNA from such products and can you maybe say something more about the stability of monoclonal antibodies, the quality of the purified products and in storage conditions, the final purification level ? You mentioned 95 % but I think this value is unacceptably low.

Boschetti : Of course. Unfortunately, I do not have all the solutions to propose for the final purification of proteins. The elimination of DNA for example, is not solved. Some standards accept very low quantities of DNA : its removal cannot be realized by chromatography. I don't think so. There are maybe other possibilities to apply at the beginning of the treatment and there are also the same problems during purification of viruses. Personally, I think - and we are working in this direction now - that it is necessary to eliminate during the first steps by chromatography, the very difficult impurities. I don't know at all whether this methodology will meet with success or not. Unfortunately.

The second part concerns the purification of monoclonal antibodies. In eighty percent of the cases, a purity of 95 % is enough because the monoclonal antibodies are used today in diagnostic. Of course it won't be enough for therapeutic applications. I would say that for therepeutical approaches you need 101 % purity.

Kedinger : You get some general rules to follow during purification and your advice would be to try to precipitate at the beginning of the procedure and to gel filtrate at the end.

You commented about why you should refine at the end using gel filtration, but what are the reasons of precipitating at the beginning besides the fact that there are more proteins around, acting as carrier ?

Boschetti : It is not systematically possible to precipitate in the first step. In fact, for example, purification of mono- clonals of the IgM class is not possible because they aggregate and then it is very difficult to separate them. To me, this technique is applicable only to biological liquids containing a high proportion of proteins, for which - because of its viscosity - ultrafiltration is difficult to apply. In certain cases however, precipita- tion can be replaced by ultrafiltration. Precipitation is more easy and less expensive than ultrafiltration and when it is possible to perform this operation, the best is to precipitate with ammonium sulfate.

Kedinger : But why at the beginning ?

Boschetti : First of all because you reduce the volume of your sample and second because afterwards, if you systematically repeat chromatographic separations using a wholy automated procedure, you can't put in between two successive columns your ethanol- or ammonium sulfate precipitation step.

Spier : That is automated physically then ?

Boschetti : Yes.

Miller : I have lots of comments. First there is a rather recent paper in Bio/Technology where it is shown that analysis of the purification steps used to purify several proteins showed indeed a tendency towards happening in a certain order. I think for people who are not specialized in

protein purification, it is really a hell of a business to try protein purification because for each protein, like TPA for instance, there might be as many as five different purification steps. It is really difficult for the layman to find his way into these procedures. There is one way I think though, which - thanks to new chromatographic supports - will supersede all these different purification steps and that is High Pressure Liquid Affinity Chromatography (HPLAC). The drawbacks of affinity chromatography are of course preparation of the ligand that you have to purify and its covalent immobilization. There is certainly some leakage but in view of the advantage afforded by the purification which is now made in one step - a 1.500^{th} - 2.000^{th} - fold purification, this approach is worth being considered. Another advantage is that with HPLAC, the risk of denaturing the protein is diminished since elution from the affinity support occurs now within 5-8 minutes, a time much shorter than the usual 30 min of conventional affinity chromatography which left plenty of time to destroy the protein.

The last advantage is that automation of the whole process is now made possible. Precipitation with ammonium sulfate is no longer required, avoiding therefore subtle modifications being brought in the structure of the protein.

It has been shown in certain cases, that if the protein afterwards must be crystallized for crystallographic purposes, you run into many problems because this ammonium sulfate precipitation somewhat modifies the protein. I think that HPLAC will really constitute a major evolution in the forthcoming years. I won't say it will replace completely all the procedures that you mentioned but it certainly will modify significantly today's strategy. What do you think of that ?

Boschetti : I did not mention at all the difference between HPLC and classical chromatography. In fact, in this case, it is

possible to do hydroxyapatite high performance liquid chromatography and gel filtration high performance liquid chromatography. Personally I do not see any basic difference between the separation mechanisms of HPLC and normal liquid chromatography. The separation principles are the same. The time taken by the separation is clearly not the same. The problem however is different from laboratory scale to large scale. It is different because unfortunately supports for large scale HPLAC are not yet available or when they are, they are terribly expensive making it impossible to use them in an economical way. Anyway, I could call this kind of chromatography High Speed Chromatography instead of High Pressure.

Boschetti : I think we have to choose a compromise between the costs and the performance of the chromatography. The good point, the interesting point of the high speed or high performance chromatography is that the time of contact between the protein and the solvent is limited. This can result in very big advantages because sometimes, interaction of the protein with the solvent - even when it is an ion exchanger -, can modify the structure of the protein or can activate some protease present or can adsorb some proteases and the column then acts as an enzymatic reactor.

Chowdhury : I do not think that your optimism about being able to eliminate the precipitation step completely, is borne out by the present experience. Affinity chromatography whether performed under gravity or high pressure, often requires exposure to low salt concentrations otherways the protein doesn't attach. As a result, high performance affinity chromatography, although it is very quick and all that, leads to more loss of material and because of the fast speed does not allow completely biological attachment, specially of enzymes. So, I don't like precipitation steps but I am not sure that affinity chromatography will

obviate all that. What concerns the short contact time with the HPLC column, that is certainly an advantage because it not only cuts down the time for breakdown, it also decreases the diffusion time. So, it is a competition between the chromatographic forces and diffusion. As compared to conventional columns, many of the new high pressure chromatographic columns are causing additional problems such as the accentuated loss of material. I speak from personal experience. So, separation depends not only on the speed and pressure, flow and all that, but also of the material with which the columns are made. Therefore generalization that HPLC will be systematically superior to conventional matrices is not valid.

Spier :   How does FPLC perform ?

Boschetti : I think FPLC is another version of HPLC or - in between the classical chromatography and HPLC. I think it is a very interesting technique. There exists today several different columns mainly ion-exchanger- and gel filtration columns. I do not know really the performances of the affinity chromatographic supports or FPLC's. Some of them are available from different american companies but I don't know their capacity, efficiency, degree of purification you obtain after having performed one affinity chromatographic step, etc...

Chowdhury : We have gained some experience with FPLC columns although not on an industrial scale and their advantage is their broad range of pH. They tolerate pH varying between 1 and 12 allowing therefore cleaning and sterilization, elimination of RNA or DNA using very harsh methods. Affinity chromatography would be more feasible with FPLC than with HPLAC because one does'nt need the very high pressure - resisting material. For instance, you can use the ordinary cyanogen bromide activated Sepharose to prepare one's own affinity column and have then broader columns to

further purify because as was already mentioned, it is the flow and the speed which are important, not the pressure.

Merten : Normally, the affinity chromatography is not only dependent on the speed but also on the incubation time. You need some sort of equilibration between the ligand and the protein you want to purify.

If you have an HPLC system, it is too fast perhaps, then it doesn't work anymore.

Like Alain said already, I think the precipitation system are not so interesting but I think one should try to develop concentrating systems like the ultra filtration. Not only to remedy to ammonium sulfate precipitation but also other precipitation techniques.

Drillien : What do you think about the contribution of genetic manipulation techniques, gene fusion, making fusion proteins where you associate a ligand, allowing you therefore to perform affinity chromatography. At the end, it is only necessary to clip the ligand. Is there any industrial application for that that you see coming ahead ? Certainly in the laboratory it is very useful.

Boschetti : Unfortunately, I think this system is not yet developed on a large scale. The more interesting developments along this approach is the combination of the protein A - like fragment in order to obtain affinity towards IgG. This would be a very good example of a final purification by gel filtration because one needs to separate the two fragments after having cut the linkage between the biological activities and the ligand.

I think it is still expensive at two levels. First of all, in the upstream process, the genetic manipulations involved. I do not know the difficulties associated with the genetic incorporation of a large ligand fragment. Second, you have always to perform an affinity chromatography except when you have a polyarginine in which case

sulfopropyl (SP) resins can be used for example. Finally, a supplementary cleavage reaction ought to be performed and you must be sure that cleavage occurs at the right position, is 100 % effective and that the final separation is also 100 % effective. But I think this is a very interesting technique indeed.

Spier :   Our biological colleagues have to realize that every step you put in a process can go wrong and usually results in a yield effect which is generally negative on total overall recovery in spite of everything else.

Miller :   I have one comment and one question. My comment is : more than often samples are heavily contaminated with cell DNA, broken pieces of DNA which are really difficult to get rid of. The reason for this is that people use sonication and harsh procedures which tend to break the high molecular weight DNA. I remember that people isolating polysomes, used cavitation. This is a very gentle procedure which does not break DNA and which may contribute to a more easy purification procedure afterwards.

Horaud :   What is cavitation ?

Miller :   You load the cells with gas and make them decompress suddenly afterwards. Bursting bubbles then rupture the cell membrane.
My question now is as follows : I noticed that you made a tandem purification on DEAE-Trisacryl/Hydroxyapatite Ultrogel. The first peak to be eluted at 0.1 M NaCl was your monoclonal antibody and then you elute the impurities in two steps. I wonder why you don't elute them in one step ?

Boschetti : Yes, this was one experience only. Because we did not know whether 0.2 M NaCl was good enough to eliminate all the impurities. Normally, 1 M NaCl is strong enough to

eliminate all impurities. Your first comment is indeed a good suggestion.

Merten :   Coming back to the genetically engineered proteins, it is a very interesting approach indeed because the transfectomas that I showed you, constitute another attempt along these lines. These experiments were done I think by MORRISON's group. They coupled an enzyme to the constant region of immunoglobulins and then incorporated the whole construction within a plasmid which ultimately was used to transfect a myeloma cell line. Monoclonal antibody was secreted which contained the enzyme bound in the F_C region. So, I think it won't be too difficult to tag the protein you want with other fragments, not an enzyme but a ligand perhaps which would stick with high specificity to an affinity system you don't find normally in your supernatant. That would afford an easy purification indeed.

Spier :   Yes, but I think we have to abide to the question which was raised I think by Dr Boschetti that you get another step in the process : separating the unwanted bit of the antibody from the rest.

Merten :   Yes, but perhaps it is easier. You have one additional step but you can get rid of say two other steps.

Spier :   If you can get rid of two steps, that's great. Generally you add on steps. Heavy sophistication costs you money. It is very difficult to get things onto the stop floor.

Merten :   ... and then you have to be aware of the control steps because if you cut, you have to be sure that...

Spier :   ... You have to control, you have to analyze a beginning, an end and our official of the GMP and all the rest, it doubles the assays, it doubles the costs...

Edy :   It's a long time since I did any protein purification but
            one thing I remember being told and reading in all the
            books was that you will get far better resolution if you
            use gradient elution in all your techniques whenever
            possible. In your processes that I saw on the screen, you
            are using stepwise elution. Why is that ?

Boschetti : You are right. These diagrams were obtained after optimi-
            zation of ion-exchange chromatography.   In fact, for
            monoclonal antibodies for example, the isoelectric point
            (pH) dictates the ionic strength. It varies between 5.2
            and 8.5 for antibodies in general. This is a difference
            large enough to try and make different chromatographic
            experiments at different concentrations of salts with
            different salts, not only sodium chloride, and also using
            several gradient slopes. After such optimization you can
            use a stepwise elution and increase the size of the
            column. I did not mention the scaling up and the calcula-
            tions related to it but this is another important issue
            for economizing money and time.

Chowdhury : I just wanted to ask you something since you are so
            interested about step cutting and cost cutting, do you -
            or anybody else - know or have experience with the new
            ampholite-free industrial   scale isoelectric   focusing
            system ?

Boschetti : No, there is no practical example in large scale.

Chowdhury : Because there is such an instrument now available in the
            United States... well, it is difficult to explain it in a
            short time but it does not require ampholites and whole
            protein stays on the membrane and gets focussed.   It
            concentrates and purifies in the same step.

Boschetti : Yes, it is a quite new technology and it is very promising
            but I don't know at all any large scale application yet.

Merten :   Coming back to gradient elution. Three to four years ago,
we purified hepatitis B antigen using monoclonal antibody
affinity chromatography. Production of the antibody was
superior with stepwise elution than using a gradient.

GENERAL DISCUSSION - CONCLUDING REMARKS

Miller :   I would like first to stress the fact that biotechnology
            is really a joint venture, you can't do that on your own,
            you need many talented people. We are here to bear
            testimony on that !
            Because so many talents are needed I would ask Florian to
            talk about the possibility in the future to establish some
            type of european network to analyze, validate the products
            of biotechnology.

Horaud :   I am unable to tell you what the future will be. I don't
            know how such an european network will develop. I can
            tell you what happens here in Brussels about the products
            made by biotechnology, what is the european procedure,
            etc... It is very complicated. I am a member of the
            Committee of Biotechnology of Pharmacy. Pharmacy belongs
            to the Division of Pharmaceutical Products.   This
            Committee has nothing in common with the Biotechnology
            Division which is research-oriented.
            This Committee is a part of the european system to licence
            products issued by biotechnology. So far, this Committee
            elaborated three documents which - if I remember correctly
            - will be published as a supplement in Trends in Bio/
            Technology. These documents concern first "Note   to
            Applicants for the production of monoclonal antibodies for
            human use", second   "The products   obtained by   DNA
            recombinant technology" and third "Preclinical testing of
            the products obtained by biotechnology".
            The philosophy behind these three documents is the same
            as the one behind the requirements we made for WHO. There
            is one document published in 1983 dealing with continuous
            cell cultures. John Petricciani knows more about that.
            The situation in Europe is a highly complicated one. If I
            understand correctly, the problem is that one product has
            to be registered and licensed in twelve countries. Some
            of these have big institutions for the control of biolo-

gical products, some other have smaller institutions if any. The issue at stake is whether the products will be registered in all 12 countries, controlled by the same twelve countries or not. In the first instance, the price of each individual biological will be crazy. So, the balance between the right of a country to judge and to licence a product and the european planning is being slowly established. It is not yet established so far. Since 1st July 1987 the new products issued by biotechnology and registered in any one of the european countries are discussed in Brussels.

The Brussels Committee is giving its opinion about the product and each country can take this opinion in consideration. The suggestion has been made to create before 1992 a central european organization which would have authority on the evaluation of the products issued by biotechnology. This organism has not yet been created.

Spier : But presumably the three documents that you are talking about will only relate to the kind of attitude the european bureaucracy will take when the national registrar or licencing agency has actually enabled one particular country to manufacture a given product.

Horaud : There were guidelines, requirements, recommendations, notes to the producer, I don't know. It establishes and stipulates what it is possible to have and to licence made by the technology of monoclonal antibodies. What are the principles of the quality control and of the characterization of the use of such products. If such documents are not drafted, you run the risk of contest, people saying substances purified with monoclonal antibodies will induce cancer and so on...

Authorities in different european countries do not always meet the same scientific standards. So, the EEC recommends the creation in each country of a multidisciplinary committee made of molecular biologists, chemists, etc...

able to evaluate and take a decision concerning a series
of products issued by biotechnology. Such a Committee
exists already in France.
The responsability is national. That's the problem. The
responsability is not european. If something is wrong
with a product, today, it is not Brussel's responsability.
It is the responsability of the authorities in Paris,
etc... This situation creates a very complex problem. In
the United States on the other hand, although everything
is decentralized at the level of the States, drug regulation is federal.

Miller : If I understand you properly, in 1992 we have a chance to
see order emerge from today's chaos.

Horaud : I hope, I am not taking the decisions. There are big
tensions between industry, the national authorities, it is
really a mess.

Spier : There should be one european authority which combines the
best practices of the nation states. What we have to do
is to have a centralized system and let's get on with it
and do it.